COMPUTER SCIENCE, TECHNOLOGY AND APPLICATIONS

RECONFIGURABLE MULTIPIPE
COMPUTING STRUCTU

IGOR A.

ILY

EVGEN

COMPUTER SCIENCE, TECHNOLOGY AND APPLICATIONS

RECONFIGURABLE MULTIPIPELINE COMPUTING STRUCTURES

COMPUTER SCIENCE, TECHNOLOGY AND APPLICATIONS

Additional books in this series can be found on Nova's website under the Series tab.

Additional E-books in this series can be found on Nova's website under the E-book tab.

For permission to use material from this book please contact us:
Telephone 631-231-7269; Fax 631-231-8175
Web Site: http://www.novapublishers.com

Additional color graphics may be available in the e-book version of this book.

Library of Congress Cataloging-in-Publication Data

Rekonfiguriruemye multikonveiernye vychislitel nye struktury. English.
Reconfigurable multipipeline computing structures / [edited by] Igor A. Kalyaev ... [et al.].
p. cm.
Includes bibliographical references and index.
ISBN 978-1-61942-854-6 (hardcover)
1. Multiprocessors. 2. System analysis--Data processing. I. Kaliaev, I. A. (Igor Anatol evich) II. Title.
QA76.5.R4513 2011
004'.35--dc23

2012011769

Published by Nova Science Publishers, In

CONTENTS

PREFACE

The book covers theoretical aspects of design of reconfigurable multipipeline computing structures of various types and their practical implementation, based on large-scale integration FPGAs.

The book consists of the introduction and six chapters. The first chapter covers general design principles of reconfigurable multipipeline computing structure organization. The authors show that, depending on complexity of operations, carried out by pipeline stages, computing structures may be classified as multimicropipeline and multimacropipeline.

The second chapter covers organization principles of multimicropipeline computing structures within homogeneous environments. The authors view the architecture of homogeneous computing environments, and the structure of a cell. They give examples of solving tasks of digital signal and image processing within homogeneous environments.

The third chapter covers structural-procedural methods of organization of computations in multimacropipeline computing structures. The authors suggest a method of transformation of various application problems into a cadr form, and a method of evaluation of effectiveness of structural-procedural organization of computations.

The fourth chapter is devoted to problems of practical implementation of reconfigurable multipipeline computing structures, based on large-scale integration FPGAs. Owing to the FPGA-technology, it is possible to create large computation fields, whose architecture can be adjusted to the task structure. The authors consider principles of design and give examples of implementation of modular-scalable reconfigurable computer systems on basis of FPGA computation fields.

The fifth chapter covers design problems of reconfigurable multipipeline computing structures, based on the macroobject architecture, which provides considerable increasing of effectiveness of computational resource use. The authors consider the generalized structure of the macroobject and give examples of implemented macroobjects for such problem domains, as mathematical physics, digital signal processing, linear algebra. Besides, the structure of the general-purpose macroobject is also given.

In the sixth chapter the authors give the description of the system software suit for reconfigurable multipipeline computing structure programming. They describe the general structure of system software for reconfigurable multipipeline structures, and an assembler language for programming of structural-procedural computations. Much attention is paid to a high level language COLAMO with implicit description of parallelism. There is given the description of parallel program development environment for reconfigurable multipipeline computing structures.

INTRODUCTION

During the last decades, organization of scientific process has been considerably changed. The domain of computer modeling and experiment was remarkably extended owing to wide use of computer equipment. Progress of supercomputer technologies and their spreading into new spheres of human activity prove, that supercomputers are a powerful tool, that can speed up development of science and technology in numerous domains.

Experience of advancement of computer engineering shows, that the gap between the performance of desktop personal computers and the performance of supercomputers is about 13 years, i.e. the performance of up-to-date professional personal computers corresponds to the performance of supercomputers of 13 years' prescription. It may seem, that necessity of supercomputers will be reduced with increasing of the performance of desktop computers. However, it has become evident, that new applications, continuously appearing and requiring computer equipment with high and ultrahigh performance for solving tasks during the acceptable time, are a stable tendency of scientific and technological progress.

Traditional spheres of supercomputer application have always been scientific research works in such domains as plasma physics, condensed-matter physics, atomic physics, theory of elementary particles, astrophysics, etc. Supercomputers are used for solving such technical problems as problems of aerospace and motor-car industry, and problems of nuclear power engineering. Supercomputers are necessary for forecasting in such domains as finances and economics, oil and gas exploring, weather, global climate fluctuations, optimization of transport flows in megalopolises, management of large enterprises. Advanced research in many other areas such as pharmacology, biology, genetics, etc., are also impossible without supercomputers. Supercomputers also penetrate into commercial fields. Here are some examples of applications: graphic applications for cinema and television, applications for processing of super-large databases, tools for visualization of processing results, etc. Besides, supercomputers are widely used in chemistry, including calculations of electronic structures for design of new materials, for example, catalysts and superconductors.

A list of five hundred the most highly-performance computers, so-called TOP-500, which is being published twice a year, illustrates progress of supercomputer design. The leading experts from National Energy Research Scientific Computing Center's (NERSC's), and from University of Manheim and Tennessee have been making this rating list since 1993. The lines in the TOP-500 list are placed according to the performance, that was provided by supercomputer systems during execution of the Linpack test. According to this rating, the performance of the supercomputers, included into the TOP-500 list, rapidly grows. It seems, that quite recently, in 1997, the list of the most highly-performance computers in the world was headed by the Intel Asci Red with the performance of several teraflops (1 teraflops is a billion or 10^{12} floating-point operations per second), but in 2005 all systems of the TOP-500 list had the performance, measured in teraflops.

First supercomputers with the performance over 1 petaflops (a quadrillion or 10^{15} floating-point operations per second) appeared in 2008. It is expected, that in 2015 the TOP-500 list will contain only systems with the performance of several petaflops, and that 2019 will be the year of systems with the exaflops performance (10^{18} flop/sec). Specialists assume, that systems with the performance of 1 sextillion operations per second (10^{21} flop/sec) will be created in 2029.

The TOP-500 list illustrates not only the current situation in the field of supercomputer technology in the world, but also its development trends. Analysis shows, that the majority of supercomputer systems, included into the TOP-500 list, has cluster architecture and, as a rule, consists of a set of traditional processing nodes, united with the help of standard network technologies. The number of cluster supercomputers of the TOP-500 list is 80% from the total number of supercomputers, included in the last release of the list. This is the result of relatively simple organization and low price of such systems. Besides, software, developed and debugged on personal computers, can be easily adapted to cluster systems. Of course, it provides wide opportunities of use of cluster supercomputers for solving such problems as modeling, design, computing experiment, etc.

However, cluster supercomputers have considerable disadvantages, caused by relatively low speed of data exchange between their processors, limited capacity of data transfer networks, necessity of synchronization of many interrelated processes, each of which is performed by a separate processor, etc. As a result, cluster supercomputers generally provide high real performance only for loosely-coupled tasks, that need no intensive data exchange between processor nodes. But if the task is tightly-coupled, then real performance of cluster systems does not exceed 5-10% from the declared peak one, and even drops down, because non-productive costs, that are necessary for organization of parallel computational process, sharply grow.

Reconfigurable multiprocessor computer systems are devoid of these shortcomings. Unlike multiprocessor computer systems with hard architecture, particularly cluster supercomputers, the architecture of reconfigurable systems can be changed during operation. So, the application programmer can adapt the computer system architecture to the solving task structure, or in other words, the application programmer can create a special-purpose multiprocessor computing structure within a general-purpose computing environment. Research shows [11], that implementation of the concept provides the high real performance of multiprocessor computer system, close to the peak one, for various problems; it also provides practically linear growth of its performance with increasing of its number of processors.

The concept of reconfigurable multiprocessor computer systems has been developing in two general directions. The first direction is homogeneous computing environments and structures [2, 4, 12, 7]. The second direction is multiprocessor computer systems with programmable architecture [1, 13, 14]. A homogeneous computing environment (HCE) consists of similar and similarly connected general-purpose computing elements (cells), each of which is programmed for execution of some logic function and some function of communication with adjacent cells. In addition, it was usually assumed that the general-purpose processing element is a primitive single-bit processor, that sequentially processed digits of input data, and that is connected with four adjacent quite similar processors.

In contrast to homogeneous computing environments, the authors of the concept of multiprocessor computer systems with programmable architecture (MCS PA) suggested to organize the processor node of a reconfigurable system as a set of primitive multi-digit processors, united by a full-accessible communication structure [15]. In other words, the processor node of such computer system, called a macroprocessor, was a low-level reconfigurable structure. The macroprocessors were connected into an upper-level reconfigurable computing structure by a system switch. As for the HCE, so for the MCS with programmable architecture, the process of task implementation consisted in organization of multipipeline chains, on which the information graph of a solving task was mapped [11, 16, 17]. This is the reason, why computer systems of such type may be considered as reconfigurable multipipeline computing structures.

Implementation and wide practical use of the concept of reconfigurable multiprocessor computer systems was restrained by lack of circuit components, satisfying this concept. Owing to programmable gate arrays (FPGAs) of very large-scale integration, that appeared in the 1990-s of the 20th century and became the main components for creation of large computation fields, it became possible to entirely implement all advantages of reconfigurable multipipeline computing structures. Research results show [18, 19] that, using FPGAs, it is possible to design, with substantially low expense, reconfigurable multipipeline computational structures, that considerably exceed traditional MCSs with hard architecture by the following characteristics: "real performance/peak performance" (effectiveness), "real performance/volume" (compactness) and "real performance/power consumption" (power efficiency).

It should be noted, that many leading vendors [20, 21] follow this direction. Cray Inc. suggests to supply Dual Core Opteron processors with an auxiliary FPGA-module, which will be used for solution of the tasks that are too laborious for a general-purpose processor. Reconfigurable devices, implemented as co-processor boards for digital signal processing and modeling of special-purpose devices, are designed by Alpha Data Systems, Annapolis Micro System, Aptix, Nallatech, and Virtual Computer Corporation. Stretch Inc. has introduced its soft-configurable S5000 processor, which combines a well-known RISC-architecture and an FPGA. The compiler, developed by Stretch Inc., automatically defines the fragments of program code, that require intensive calculations, and implements them in the FPGA. Starbridge Systems Inc. was one of the first companies, that introduced a supercomputer, that was based on reconfigurable technology, and that was built on Xilinx FPGAs. Scientists of Berkeley Wireless Research Center designed the FPGA-based High-End Reconfigurable Computing System. In opinion of developers, use of FPGA chips provide increasing of performance more than in 10 times in comparison with systems, built on general-purpose microprocessors, though their power consumption and price remain unchanged.

However, western vendors traditionally use FPGAs only as co-processors to standard computing nodes, implemented by means of general-purpose microprocessors. The concept of reconfigurable multipipeline computing structures widen opportunities of use of FPGAs as components for creation of large computation fields, within which the application programmer can create special-purpose computer systems, adapted to the structure of the solving task. This is the method, which will provide the most possible real performance of the system for computational problems.

The book covers theoretical aspects of design of reconfigurable multipipeline computational structures of various types, and their practical implementation, based on large-scale integration FPGAs.

The book consists of the introduction and six chapters. The first chapter covers general design principles of reconfigurable multipipeline computational structure organization. The authors show the advantage of reconfigurable computer systems in comparison with MCSs with hard architecture, and draw a conclusion about the maximum effectiveness of systems of this class for dataflow tasks, i.e. the tasks of data flow processing according to the one and the same algorithm. The authors analyze various methods of data flow problem solving, and prove that the multipipeline method of computations provides the minimum time of the data stream processing. They consider various methods of organization of reconfigurable computing structures, that perform the multipipeline computational process, and show that, depending on complexity of operations, carried out by pipeline stages, computing structures may be classified as multimicropipeline and multimacropipeline. The chapter also covers principles of organization of structural-procedural multipipeline computations, when hardware resources of some multipipeline structure are insufficient for mapping of the whole information graph of the solving task.

The second chapter covers organization principles of multimicropipeline computing structures within homogeneous environments. The authors view the architecture of homogeneous computing environments, and the structure of a cell, and describe organization of the computational process in the homogeneous computing environment and its programming, based on microprogram modules of special-purpose processors of various operations. The chapter contains various examples of solving tasks of digital signal and image processing within homogeneous environments.

The third chapter covers structural-procedural methods of organization of computations in multimacropipeline computing structures. According to the structural-procedural method, the informational graph of a task is partitioned into a sequence of isomorphic information-independent or directly dependent subgraphs. The subgraph is selected according to the following condition: system hardware resources must be sufficient for structural implementation of the subgraph. The authors suggest a method of transformation of the isomorphic subgraph sequence into a cadr form, which can be naturally mapped on the architecture of macropipeline computing structures. Such organization of parallel computations provides determinism, i.e. the sequence of cadr execution and the processing order of cadr operands remain unchanged for different starts of the parallel program. There are shown methods of transformation of various information graphs into the cadr form. In addition, there are given examples of application transformation and evaluation of effectiveness of structural-procedural organization of computations.

In the fourth chapter, the authors analyze modern components, used for design of multiprocessor systems, and make conclusion that field programmable gates arrays (FPGA) of large-scale integration perfectly suit the concept of reconfigurable multipipeline computing structure design. Owing to the FPGA-technology, it is possible to create large computation fields, whose architecture can be adjusted to the task structure. As it is shown, several design and technological problems of MPCS design can be solved by modular-scalable organization of reconfigurable computing structures. The authors consider design principles of modular-scalable multipipeline computing structures, based on general-purpose basic modules, and suggest a generalized structure of FPGA-based basic modules. The chapter contains examples of implementation of basic modules of various generations, and examples of highly-performance reconfigurable computing system families.

Owing to implementation of FPGA-based basic modules, it was possible to pass from a macroprocessor architecture of reconfigurable multipipeline computing structures to a macroobject one, that can provide considerable increasing of effectiveness of computational resource use.

The fifth chapter covers the design problems of reconfigurable multipipeline computing structures, based on the macroobject architecture. The authors give a definition of a macroobject according to its structural and program implementation, and consider the generalized structure of the macroobject. The chapter contains examples of implemented macroobjects for such problem domains as mathematical physics, digital signal processing, linear algebra. Besides, the structure of the general-purpose macroobject is also given.

Circuit programming of FPGA-based reconfigurable multipipeline structures is a sufficiently complicated process, that requires special knowledge and skills. That is why the main purpose of development of the system software suit for reconfigurable multipipeline structures, described in the sixth chapter, was to simplify development of applications for multipipeline structures, and to approach it to programming of cluster multiprocessor systems. There is described a general structure of system software for reconfigurable multipipeline structures. Much attention is paid to a high level language COLAMO with implicit description of parallelism. COLAMO-applications can be parallelized according to the type of access to data arrays, and by indexing of their elements. The authors suggest principles of translation from the COLAMO-language, that provide synthesis of structural, stream, procedural and control components within the single language for applications, implemented within reconfigurable multipipeline computing structures. The chapter contains the description of parallel program development environment for reconfigurable multipipeline computing structures.

Chapter 1

PRINCIPLES OF ORGANISATION OF RECONFIGURABLE MULTIPIPELINE COMPUTING STRUCTURES

1.1. THE PROBLEM OF MULTIPROCESSOR SYSTEM PERFORMANCE

Multiprocessor computer systems (MCS) are intended for solving complicated highly intensive computing problems that require processing of huge volumes of information and extremely high performance. Such problems cannot be solved on systems with low performance, because their solution takes extremely huge time. As a result, the obtained solution becomes practically useless. However, there are several ways to achieve high performance of multiprocessor computer systems.

At present multiprocessor computer systems has been successfully developing, for the most part owing to technological progress, and particularly owing to reduction of topological size of silicon chips. This leads to increasing of a number of gates on a chip, and to growth of processor speed. In spite of onrush of new computer technologies of the 21st century, including optical and quantum technologies, theoretical and experimental results are far from manufacture [23].

In addition to technological methods, computer system performance may be increased by means of algorithmic, software and architectural methods. The algorithmic methods consist in development of more effective mathematical methods of problem solving. The software methods are based on development of software tools that provide effective use of computer system components, including memory, communication channels, etc.

The architectural methods are the most important direction of computer performance increasing. During the last half of the century specialists of world benchmark companies implemented pipeline, vector, vector-pipeline, matrix, toroidal, hypercube, hierarchical, cluster, and other architectures of computer systems [6,9,24-28]. Theorists made several attempts to classify a huge variety of computer systems. The classification that became the most well-known, was suggested by American specialist Michael J. Flynn in the middle 1960s of the 20th century. According to Flynn's classification, all computer systems can be placed into 4 groups: SISD (single instruction, single data stream), SIMD (single instruction, multiple data streams), MISD (multiple instructions, single data stream), MIMD (multiple instructions, multiple data streams). In spite of simplicity, Flynn's classification was of little practical use, because real computer systems often could not be related to the restricted Flynn's groups. Majority of existing architectures of multiprocessor computer systems has an important shortcoming. For certain classes of problems system performance may be close to the peak one, but it may drop decreasing in tens times, when solving problems of other classes (see Fig. 1.1).

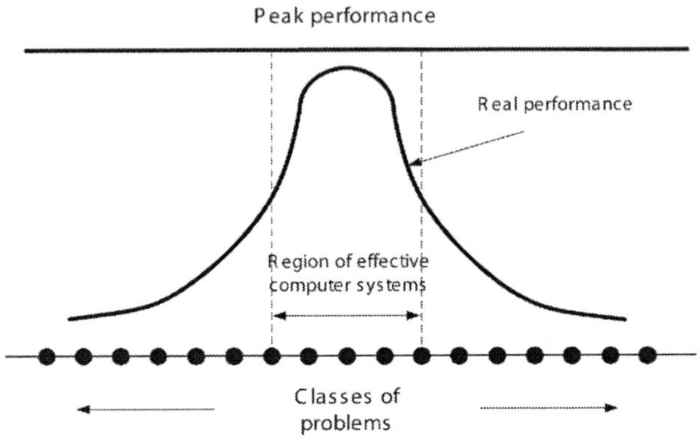

Figure 1.1. The dependence of the system performance on the class of problems.

This is consequence of dissimilarity between the architecture of the multiprocessor system and the internal structure of the solving problem. In general, any problem and any process that occurs in a real system S, may be described by a set of mathematical and logical dependences, or in the form of a certain graph $G(Q,X)$ (see Fig. 1.2). The graph $G(Q,X)$ contains a set of vertices $q_i \in Q$, each of which corresponds to a certain operation O_i, included in a set of feasible operations O. Edges $x(q_i,q_{i+1}) \in X$ define the sequence of operations, which correspond to the vertices of $G(Q,X)$. If two vertices q_i and q_{i+1} are connected by the edge $x(q_i,q_{i+1})$, then it means that the result of the operation O_i is the input data of the operation O_{i+1}. The graph $G(Q,X)$ also contains a set of input edges $x(q_{0i},q_i)$, which define the source of input data, and a set of output edges $x(q_i,q_k)$, which define the target of operation results. Henceforth, $G(Q,X)$ will be called *an information graph* or *a graph of a problem algorithm*.

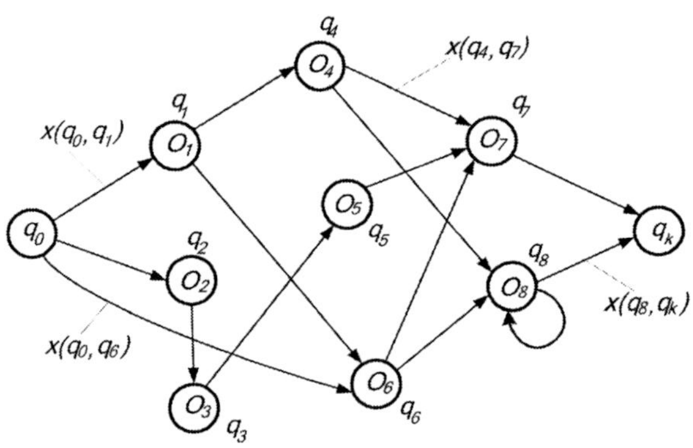

Figure 1.2. An information graph of a problem.

In a parallel multiprocessor computer system, which is used for modeling of processes that occur in the system S, a correspondent computational process is organized. In most cases, the computational process is described by a quite different graph $G^*(Q^*, X^*)$ (see Fig. 1.3).

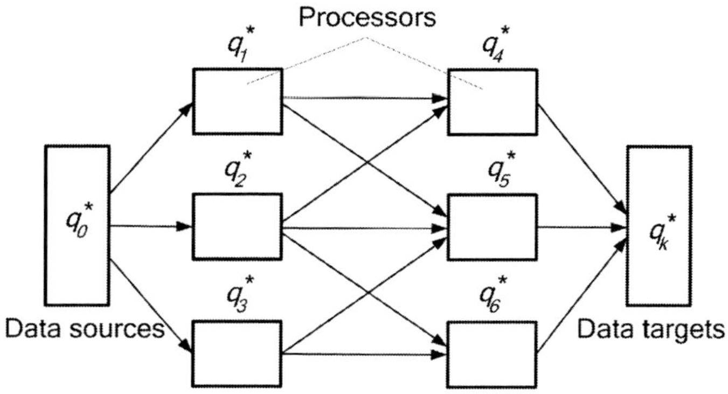

Figure 1.3. A graph of some computational process in a system.

Here the set of vertices Q^* is defined by the set of processing elements of the computer system. The set of edges X^* is the set of communication channels between the processing elements. The input and output edges of $G^*(Q^*, X^*)$ are defined by the communication channels, that connect the data sources with the output data targets (system memory). The operations O_i, contained in the set O, are implemented in the processing elements of the system as successive procedures of data processing.

As a rule, if the multiprocessor computer system has a "hard" architecture, then the graphs G and G^* considerably differ from each other. The difference between G and G^* may be explained by a considerable dissimilarity (in the majority of cases) between the "hard" architecture of the parallel multiprocessor computer system and the structure of the model-based system S. As a result, to model the system S in the multiprocessor computer with the "hard" architecture, a multiprocedural computational process is to be organized. The computational process, simulating the system S, has structural dissimilarities with the system and is described by G^*, which also considerably differs from the graph G of the real simulated system S.

In the majority of cases, the graph G^* of the multiprocessor system with the "hard" architecture is significantly more complex than the initial graph G of the simulated system or the solving problem. In addition, its system of the internal communication channels X^* is more complex than the system of the information edges X of G. This leads to essential "non-productive" time loss, because MCS implementation of the application requires execution of complicated procedures such as task distribution between the processing elements, solving the problem of queues and conflicts of data exchange or access of the processing elements to memory, etc. [29]. As a result, in comparison with the peak performance, the real performance of such computer systems is considerably lower, and hence, according to opinions of users of the most widely used cluster supercomputers, the real performance does not exceed 10% from the peak performance of the system rather often [9].

Furthermore, the real performance of such systems does not grow at increasing of the number of processors, but drops quite often (see Fig. 1.4). It is explained by a sharp increase of "non-productive" computations, that are oriented not on data processing, but on organization of the computational process in the MCS and take the most part of the total time of problem implementation, when the number of processing elements is increasing.

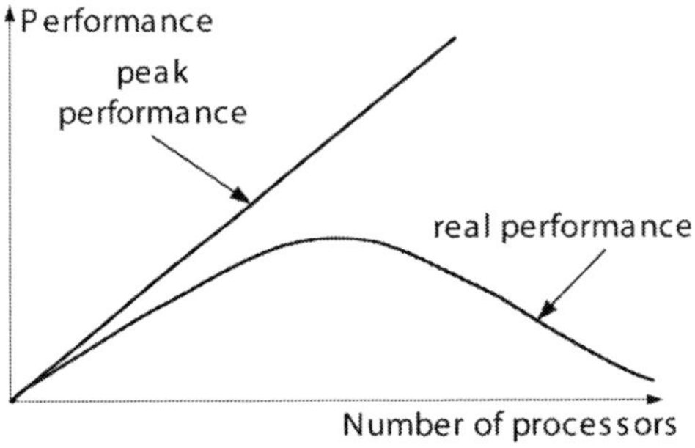

Figure 1.4. The dependence of the performance of the system with the "hard" architecture on the number of processors.

The abovementioned shortcoming of computer systems with "hard" architectures can be avoided, if there is provided reconfigurability of the MCS architecture and the most possible similarity between the graph G^* of the computational process and the graph G of the modeling process (or problem solving). In this case, the high real performance of the MCS is achievable owing to optimization of its computing structure according to the solving problem. In addition, the "non-productive" time, taken for organization of the computational process, may be reduced owing to minimization of data exchange. In other words, the concept of the MCS architecture reconfigurability allows variation of the peak of system performance according to the class of the solving problem (see Fig. 1.5) and practically linearly growth of the MCS real performance at increasing of the number of processing elements of the system (see Fig. 1.6).

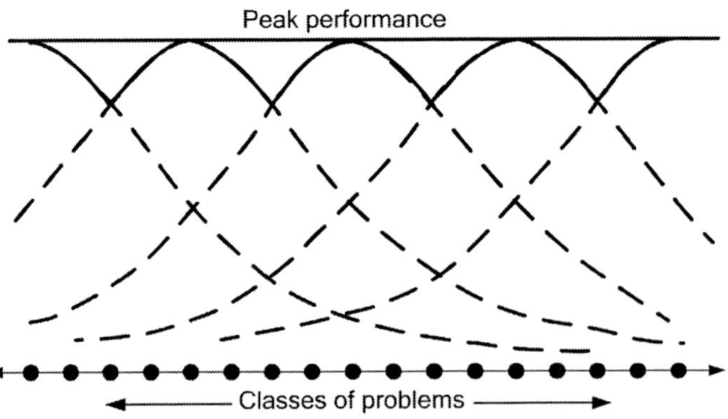

Figure 1.5. Approaching of the real performance to the peak performance.

This idea is the basis of reconfigurable MCSs of all types in spite of their apparent differences. Nevertheless, let us consider various methods of organization of reconfigurable MCSs in greater details.

Analogue computers and so-called digital differential analyzers may be considered as the first real reconfigurable computers [30]. Architecture reconfiguration in these devices was made by the programming specialist, who used a plug-in patch board and switched connections between the processing elements of the system according to the structure of the solving problem.

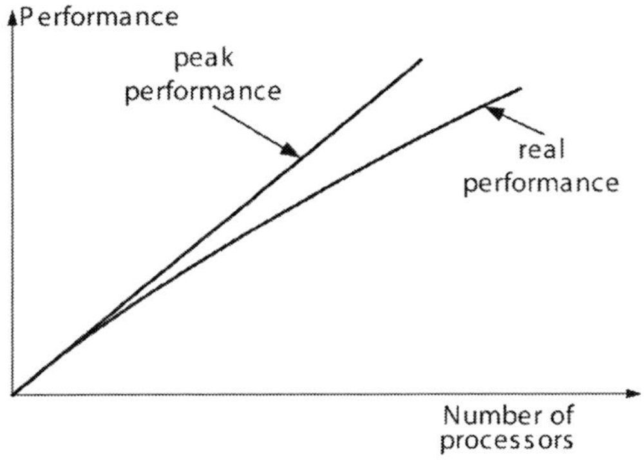

Figure 1.6. The dependence between the performance and the number of processing elements of a computer system with a reconfigurable architecture.

In 1966 in Novosibirsk (Russia) E.V. Evreinov and Y.G. Kosarev published a book "Homogeneous general-purpose systems of high performance" [12]. In the book the authors suggested to create a computer system as a "homogeneous general-purpose computer system, containing similar and similarly connected general-purpose computers, with software reconfiguration of its structure". They also considered *homogeneous computing environments (HCEs)*, that contained similar and similarly connected elementary universal elements that could be programmed to perform any operation from a set of logical and memory operations or could be connected with any adjacent element.

The concept of design of homogeneous computer systems and environments was thoroughly scrutinized at Moscow Trapeznikov Institute of Control Sciences of RAS (ICS RAS) under the leadership of I.V. Prangishvili. In 1967 I.V. Prangishvili with a group of colleagues published an interesting monograph "Microelectronic and homogeneous structures for design of logical and computer devices" [3], in which practical implementation of homogeneous computing structures was directly connected with microelectronic, an incipient area of science in those years. In 1972 E.V. Evreinov and I.V. Prangishvili published a book "Digital automatons with adjustable architecture" [10]. Generalizing research results, the authors formulate the main principles of homogeneous computing structures:

- concurrent execution of a large number of operations;
- pipeline data processing;
- variability of logical structure;
- homogeneity of elements and connections between them.

The concept of homogeneous computing environments and structures was the base of several practical projects in the end of 1970s. For several years Elva Enterprise (Tbilisi, Georgia, former Soviet Union) manufactured the computer PS-300, that was designed at

Moscow Trapeznikov Institute of Control Sciences of RAS under the leadership of I.V. Prangishvili. The processor node of the PS-300 was designed on the base of a homogeneous computing environment that might be adjusted to a large number of operations [31]. It should be noted, that the cells of the homogeneous environment were assembled from various chips (K-155 series), and hence, effectiveness of the entire environment was rather low. On-board computer devices, manufactured by Astrofizika Scientific Production Association, were designed on the base of serial special-purpose chips of the homogeneous computing environment, manufactured by Alpha Enterprise (Riga, Latvia, former Soviet Union). M.P. Bogachev was chief designer of the HCE computer devices.

Since the middle of 1980s Lvov (Ukraine, former Soviet Union) had become the research center in the field of HCEs. Large projects, concerning HCE software and prospective architectures, were accomplished [27, 32-35] in Lvov institutes. In Scientific Research Institute of Precise Technology (Zelenograd, Russia) a fault-tolerant HCE chip, which consisted of 130 processing elements, had been manufactured.

In 1980s American scientists published papers, concerning so-called systolic structures [36]. Systolic processing was represented as an original method of organization of computations. However, systolic processing had been naturally implemented on homogeneous environments [38] long before those publications [37]. According to flexibility and effectiveness the considered systolic structures were much worse and could be viewed as merely a subkind of homogeneous environments.

A concept of multiprocessor computer systems with programmable architecture became another direction of research of reconfigurable computer systems in Russia. For the first time academician A.V. Kalyaev had formulated the concept in 1970s of the XX century and became the pioneer of this research [1, 30]. In contrast to the HCE developers, who implemented processor nodes of environments as elementary single-digit processors connected with adjacent processors by direct links, A.V. Kalyaev suggested to create a processor node of a computer system as an elementary structure with reconfigurable architecture. Such processor nodes were called *macroprocessors*. Each macroprocessor contained a set of processing elements, united by a full-accessible communication system [39]. Multi-core processors that appeared in the beginning of the XXI century and were represented as a completely modern concept, *per se* are quite similar to the macroprocessors suggested by A.V. Kalyaev thirty years ago. By means of an upper level communication system, the macroprocessors are configured into a multipipeline structure, similar to the graph of the solving problem. Such systems have two levels of pipelining: the level of macroprocessors and the level of processing elements of each macroprocessor. The suggested structure of pipelining increases flexibility and simplifies programming of the system. The group of systems with programmable architecture contains the following ones: ES-2703 (1985), Trace (1991) and Module-8 (1998).

In spite of all advantages of reconfigurable MCSs and rather numerous researches of their design, they were not wide spread until recently, because of specific and complicated structural programming, that requires skills of circuit engineering and non-standard thinking. As a rule, programming of reconfigurable MCSs takes plenty of time and hence, reconfigurable MCSs are not very promising for solving volatile problems with a single vector of input data.

However, so-called data flow problems may be perfectly solved on reconfigurable multipipeline MCSs. Data flow problems consist in processing of huge data arrays (flows) according to the algorithm, which remains unchangeable. The examples of data flow problems are problems of mathematical physics, digital signal processing, etc.

1.2. Data Flow Problems and Methods of Their Solution

In general case, a data flow problem may be formulated as follows. Let us assume that there is an ordered set of data vectors $D_i < d_1^i, d_2^i, ..., d_k^i, ... >$, $i = 1, 2, ..., N$, each element of which is to be processed according to a fixed algorithm, represented as an algorithm graph $G(Q, X)$ (see Fig. 1.7).

Each vertex $q_j \in Q$ of $G(Q, X)$ is put in correspondence with a certain operation O_j, contained in a set of feasible operations O. The edge $x(q_j, q_{j+1}) \in X$ defines that the result of the operation O_j is the input data of the operation O_{j+1}, which corresponds to the vertex q_{j+1}.

The data flow problem consists in transformation of the cortege (flow) of the input data vectors D_i, $i = 1, 2, ..., N$ into the cortege (flow) of the output data vectors $B_i < b_1^i, b_2^i, ..., b_l^i, ... >$, $i = 1, 2, ..., N$ according to $G(Q, X)$.

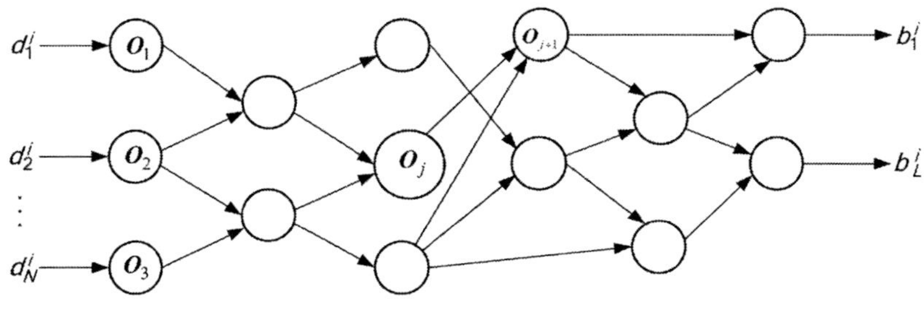

Figure 1.7. A graph of an algorithm.

In the simplest case the problem, formulated above, may be solved on a sequential von Neumann computer (see Fig. 1.8) with a processor P as the main element.

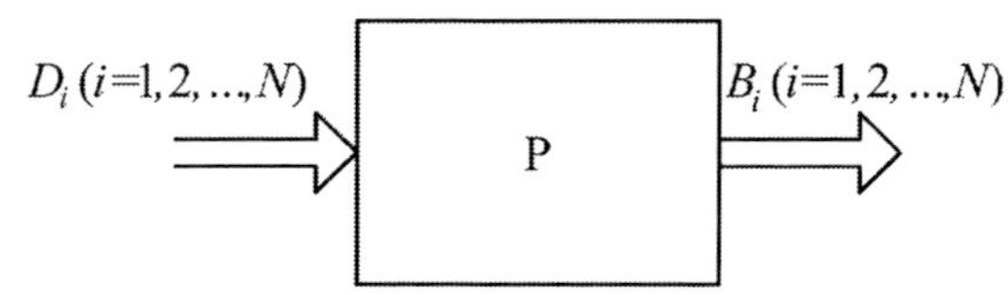

Figure 1.8. Von Neumann's computer.

Firstly, the processor P is to be preliminarily programmed to sequential performing of all operations O_j, $j = 1, 2, ..., M$, corresponding to the vertices of $G(Q, X)$. Then the sequence of the vectors D_i, $i = 1, 2, ..., N$ is to be supplied to the von Neumann computer.

The time of processing of *N* input data vectors is obtained according to the equation

$$T_{pr} = N\sum_{i=1}^{M} f(O_i)\tau , \tag{1.1}$$

where $M = |Q|$ is the number of the vertices of $G(Q,X)$;

$f(O_i)$ is the number of cycles, required to perform the operation O_i of the corresponding vertex of $G(Q,X)$;

τ is the cycle duration.

The condition $T_{pr} < T_{MAX}$ defines effectiveness of processing.

Here T_{MAX} is the time needed for processing of all input data vectors. If the inequality is true, then the processing method is effective, and is ought to be used as the one with the most simple implementation. However, if the main parameters of the data flow problem, such as the number *N* of input vectors, the size of each vector, the number of vertices in the algorithm graph, are rather big, then the abovementioned condition will not be fulfilled.

The time T_{pr} of processing of the cortege of the input vectors may be minimized by parallelization of computations [25, 40-42]. In the simplest case, parallelization requires C processors P_i, $i = 1,2,...,C$, each of which can operate independently. It is evident, that if each processor P_i, $i = 1,2,...,C$ is programmed to perform the algorithm according to $G(Q,X)$, and the set of the input vectors D_i, $i = 1,2,...,N$ is split into $]N/C[$ of disjoint subsets, then each subset of the input vectors may be independently processed by the processor P_i, i.e. concurrently with other subsets (see Fig. 1.9).

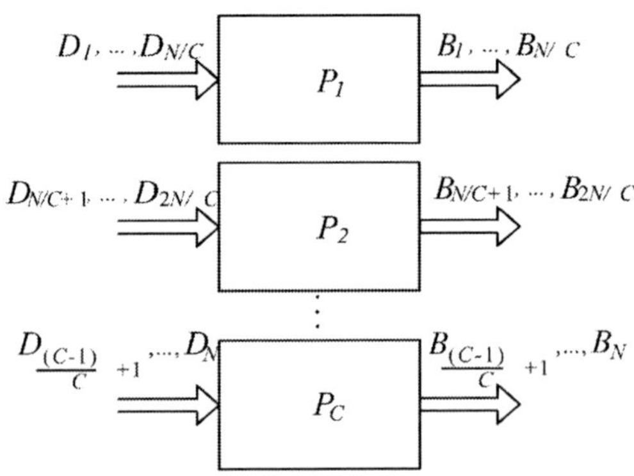

Figure 1.9. A parallel method of processing of the input data vectors.

In this case, the problem time, i.e. the processing time of the set of the input data vectors, is obtained according to the equation

$$T_{pr} = \left]\frac{N}{C}\right[\sum_{i=1}^{M} f(O_i)\tau . \tag{1.2}$$

Evidently, the problem time is reduced in C times in comparison with the case, when all input vectors are sequentially processed by a single processor.

Another method of reducing of the problem time is usually called *pipeline processing*. The following example explains the sense of the method. The graph $G(Q,X)$ may be partitioned into disjoint subgraphs $G_1(Q_1,X_1)$, $G_2(Q_2,X_2)$, ..., $G_H(Q_H,X_H)$ (see Fig. 1.10).

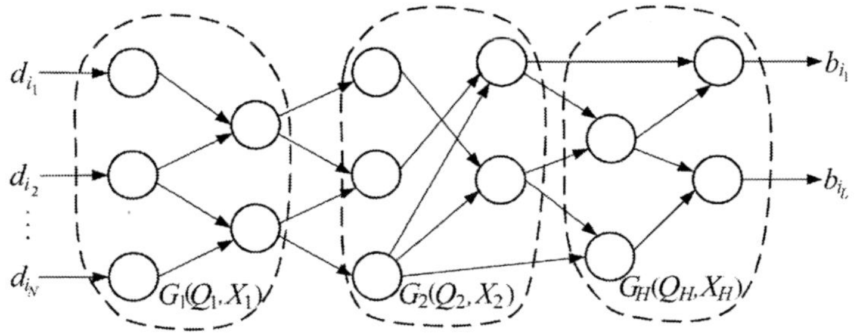

Figure 1.10. Partition of $G(Q,X)$ into disjoint subgraphs.

Let us assume, that the system contains H sequentially connected processors P_i, $i=1,2,...,H$, as it is shown in Fig. 1.11, and each processor P_i is programmed to perform the corresponding subgraph $G_i(Q_i,X_i)$, $i=1,2,...,H$ of the algorithm graph $G(Q,X)$.

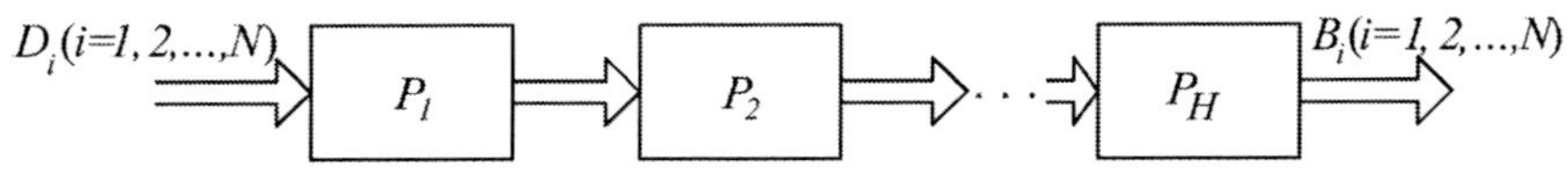

Figure 1.11. A chain of processors.

If the single input vector D_i enters such chain, then the processing time, i.e. the time needed to define the output vector B_i, is obtained according to the equation:

$$T_{pr}=\sum_{i=1}^{H}\sum_{j=1}^{M_i}f(O_j^i)\tau, \qquad (1.3)$$

where M_i is the number of the vertices of $G_i(Q_i,X_i)$;

O_j^i is the operation which corresponds to the vertex j of $G_i(Q_i,X_i)$.

Taking into account that $H\times M_i=M$, the equation (1.3) is obtained as follows:

$$T_{pr}=\sum_{i=1}^{M}f(O_i)\tau. \qquad (1.4)$$

Hence, the processing time of a single data vector, processed by the processor chain, is equal to the processing time of a single data vector processed by a single processor. Nevertheless, if the sequence of N input data vectors D_i, $i=1,2,...,N$ enters the same processor chain, then the processing time is considerably less than that of processing of vectors by a single processor. When the vector D_1 is processed by P_1 according to $G_1(Q_1, X_1)$ and the results enter P_2, the processor P_1 starts to process the input vector D_2. When D_2 is processed, P_1 starts to process D_3 according to $G_1(Q_1, X_1)$. When P_2 has performed all operations, specified by $G_2(Q_2, X_2)$, on the results of processing of D_1 in P_1, it can perform the same operations on D_2. If it is accepted that the time of input vector processing by P_i according to $G_i(Q_i, X_i)$, $i = 1,2,...,H$ is $T^i_{pr} = \sum_{j=1}^{M_i} f(O^i_j)\tau$, then it will take $T_1 = \sum_{i=1}^{H} T^i_{pr}$ for the first vector of output data B_1 to leave the processor chain. All the rest vectors $B_2,..., B_N$ will leave the processor chain with a period $\Delta T = \max\limits_{i=1,...,F}(T^i_{pr})$.

As a result, the total time of processing of *N* data vectors by the processor chain is the following:

$$T_{pr} = \sum_{i=1}^{H} T^i_{pr} + (N-1)\cdot \max_{i=1,...,H}(T^i_{pr}) \approx H \cdot \max_{i=1,...,H}(T^i_{pr}) + (N-1)\cdot \max_{i=1,...,H}(T^i_{pr}) = (N+H-1)\cdot \Delta T \text{ , (1.5)}$$

where $\Delta T = \max\limits_{i=1,...,H}(T^i_{pr}) = \max\limits_{i=1,...,H}\left(\sum_{j=1}^{M_i} f(O^i_j)\tau\right)$.

It is possible to neglect $(H-1)$, if N is rather large. Hence, T_{pr} is obtained according to the equation

$$T_{pr} \approx N \cdot \Delta T \approx N \cdot \frac{\sum_{j=1}^{M} f(O_j)\tau}{H}, \tag{1.6}$$

i.e. in comparison with processing of N data vectors by a single processor (see (1.1.)), in this case the processing time reduces in about H times, where H is the number of processors in the chain.

Usually such processor chain is called simply a *pipeline*, each processor is called a *pipeline stage*, and the number H of the chain processors is called a *pipeline length.*

As it may be seen from the equation (1.5), the time needed to process an array, which consists of N elements, by the processor pipeline is $\Delta T = \max\limits_{i=1,...,H}(T^i_{pr})$, i.e. it depends on the "slowest" stage of the pipeline. Hereinafter, the value, reciprocal to ΔT, will be named the processing speed of the pipeline. The higher is the processing speed of the pipeline, the less time is needed to process the array of input data vectors. It is clear, that the highest speed of processing may be achieved, if $G(Q, X)$ is partitioned into the subgraphs $G_i(Q_i, X_i)$ and $T^i_{pr} \approx T^j_{pr}$, $i = 1,2,...,H$, $j = 1,2,...,H$.

In comparison with parallel processing, pipeline processing has a considerable advantage, which consists in reducing of the number of input and output data channels. As a result, pipeline methods of processing became widespread in certain systems. However, in the case of parallel processing the granularity of parallelism cannot exceed N, which is equal to the size of vector cortege; pipeline processing is limited by the number M of operations of the algorithm graph, which is, as a rule, much less than N.

Parallel and pipeline methods of processing are to be combined for further minimization of the processing time of the input vector flow. It is necessary to organize C pipelines of the length H, operating concurrently and processing a certain subset of the input vectors D_i each (see Fig. 1.12).

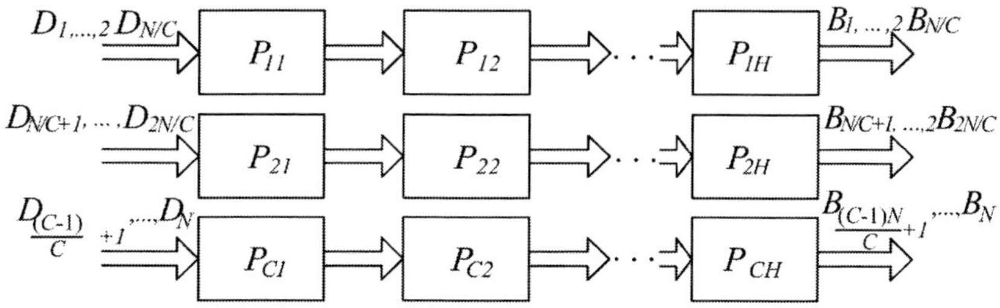

Figure 1.12. Combination of parallel and pipeline processing.

For such method of organization of computations, the processing time of the flow, that consists of N vectors, is the following:

$$T_{pr} = \left(\left]\frac{N}{C}\right[+ H - 1 \right) \Delta T . \tag{1.7}$$

If N is large, then $(H-1)$ may be neglected, and therefore

$$T_{pr} \approx \frac{N}{C} \cdot \Delta T \approx \frac{N}{C} \cdot \max_{i=1,...,H} \left(\sum_{j=1}^{M_i} f(O_j) \cdot \tau \right) \approx \frac{N}{C \cdot H} \cdot \sum_{j=1}^{M} f(O_j) \cdot \tau . \tag{1.8}$$

1.3. Principles of Multipipeline Processing of Data Flow

Earlier, when the principles of pipeline processing were described, it was assumed, that the processor P_i, included into the pipeline stage i, sequentially performed the operations O_j^i, $j = 1,2,...,M_i$, which corresponded to the vertices of the subgraph $G_i(Q_i, X_i)$. As long as the speed of data processing in the pipeline depends on the slowest pipeline stage, sequential data processing, performed by the pipeline stages, leads to reducing of the processing speed and, hence, to increasing of the total processing time T_{pr}, needed to process all vectors of the input data.

Parallelization of data processing, performed on each pipeline stage, is the evident method of increasing of the data processing speed of the pipeline. In extreme case, the pipeline stage i may be represented as M_i processors P_{ij}, $j = 1,2,...,M_i$, operating concurrently. Each processor performs the operation that corresponds to the operational vertex $q_j \in Q_i$ of $G_i(Q_i, X_i)$ (see Fig. 1.13).

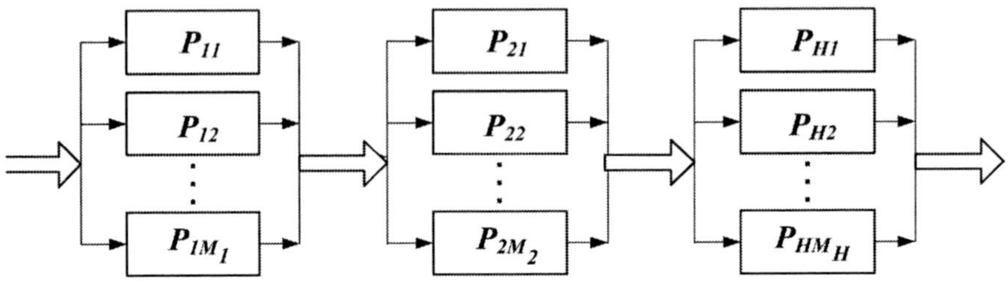

Figure 1.13. Representation of the pipeline stage i as M_i processors P_{ij}, $j = 1,2,...,M_i$, operating concurrently.

The processing time of the flow, that consists of N vectors, for such pipeline, is obtained according to the equation

$$T_{pr} = (N + H - 1) \cdot \Delta T, \tag{1.9}$$

and in this case

$$\Delta T = \max_{i=1,...,H}\left(T^i_{pr}\right) = \max_{i=1,...,H}\left(\max_{j=1,...,M_i} f(O^i_j) \cdot \tau\right) = \max_{i=1,...,M} f(O_i) \cdot \tau,$$

where $\max_{i=1,...,M} f(O_i)$ is the most lasting operation, corresponding to the vertices of $G(Q, X)$.

If N is large and $M_i \approx \frac{M}{H}$, then the equation (1.9) may be represented in the approximate form:

$$T_{pr} = N\frac{\sum_{j=1}^{M} f(O_j) \cdot \tau}{H \cdot M / H} = \frac{N}{M}\sum_{j=1}^{M} f(O_j) \cdot \tau. \tag{1.10}$$

Hence, in this case the time T_{pr}, in comparison with the standard pipeline (1.6), is reducing in about M / H times, where M is the number of the vertices of $G(Q, X)$, and H is the number of the pipeline stages.

However, the structure chart, shown in Fig. 1.13, is effective, if all vertices of $G_i(Q_i, X_i)$, $i = 1, 2, ..., H$ are information-independent and may be performed completely concurrently (at the same time). The subgraphs $G_i(Q_i, X_i)$ may be partitioned into information-independent subgraphs, but the granularities of parallelism of the newly obtained subgraphs will differ. As a result, the processing speed of the pipeline stages will decrease, and hence, the total processing speed of the pipeline will also decrease.

In general case, the operational vertices of $G_i(Q_i, X_i)$ are information-dependent, i.e. the operation, corresponding to the vertex $q_j \in Q_i$, cannot be performed until the operation, corresponding to the vertex $q_{j-1} \in Q_i$, is performed (see Fig. 1.10). Therefore, the highest speed of pipeline processing and the minimum total time T_{pr} of all input data processing may be provided, if the structure of connections between the pipeline stages is completely similar to the structure of connections between the vertices of the algorithm graph $G(Q, X)$. Such organization of pipeline computations is called *structural* [11].

Indeed, the highest processing speed may be provided, if each operational vertex q_i, $i = 1, 2, ..., M$ of $G(Q, X)$ is put in correspondence with the processing element P_i, $i = 1, 2, ..., M$, and all P_i are connected according to the information edges of $G(Q, X)$ (see Fig. 1.14). If the vectors D_i, $i = 1, 2, ..., N$ of the input data set successively enter such computing structure, then the processing time is obtained according to the equation

$$T_{pr} = (N + H_{\max} - 1) \cdot \Delta T ,$$

where $H_{\max}$ is the number of the vertices of the critical path of $G(Q, X)$, i.e. of the path, connecting the input and output vertices, with maximum value $\sum_{j=1}^{K} f(O_j) \cdot \tau$;

O_j, $j = 1, 2, ..., K$ are the operations that correspond to the vertices of the critical path;

$\Delta T = \max\limits_{i=1,...,H_{\max}} \left(f(O_j) \right) \cdot \tau$ is the time needed to perform the most lasting operation, corresponding to the vertices of the critical path.

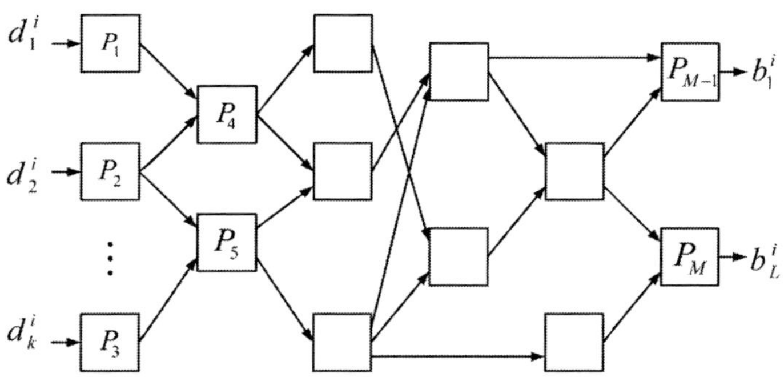

Figure 1.14. The structure chart that provides the highest speed of pipeline data processing.

If N is large and ΔT is obtained according to the equation

$$\Delta T = \max_{i=1,2,\ldots,H_m} f(O_j) \approx \frac{\sum_{j=1}^{M} f(O_j^i)}{M},$$

then we get

$$T_{pr} = \frac{N}{M}\sum_{j=1}^{M} f(O_j)\cdot\tau . \qquad (1.11)$$

Hence, in comparison with all methods of computer system organization (see (1.1), (1.2), (1.6), (1.8), (1.10)) that were considered above, structural organization of computations provides the minimum time of processing of the data vector array.

It may be noted, that the computing structure, shown in Fig. 1.14, combines both parallel and pipeline methods of data processing, because the input data is processed concurrently in different pipeline processor chains. Hereinafter, the pipeline structure of such type is called a *multipipeline computing structure (MPCS)*.

If the algorithm graph $G(Q,X)$ remains unchangeable during all period of functioning of the computing structure, then a specialized multipipeline computer system is to be created to solve the problem.

It is clear, that such approach is unacceptable, if it is necessary to design a computer for solving various data flow problems with different algorithm graphs $G(Q,X)$. In such case, it is necessary to provide possibility of reconfiguration of the information connections between the processors, and hence, to provide possibility of mapping of algorithm graphs with different structures. Hereinafter, the multipipeline computer with such capabilities will be called a multipipeline structure with reconfigurable architecture or a *reconfigurable multipipeline computing structure*. To provide possibility of reconfiguration of multipipeline computer architecture, a certain communication system (CS) is to be organized within the multipipeline computer. The information connections between the processors P_i, $i = 1,2,...,M$ are created by means of this communication system (see Fig. 1.15). Let us consider several methods of communication system creation.

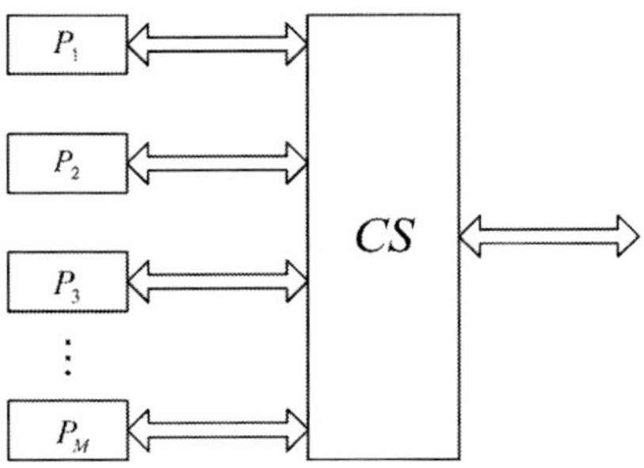

Figure 1.15. The structure chart of the reconfigurable multipipeline computer.

The simplest method is to use a full-accessible switch that provides all-to-all connections between the processors of the structure [7]. The main advantage of the method is possibility of organization of arbitrary multipipeline structures. However, hardware resources, needed to design the full-accessible switch, are proportional to M^2, where M is the number of the processors of the structure. Implementation of the full-accessible switch faces difficulties, if M is large.

In contrast to the previous one, another method does not require large hardware resources. The sense of this method consists in creation of a multipipeline with "hard" architecture, and in provision of communication capabilities by means of auxiliary data transit functions of processors. For example, Fig. 1.16 shows the multipipeline computer with "hard" matrix architecture.

In addition to the operations from the set O, each processor P_i, $i = 1, 2, ..., M$ of the structure is to perform an auxiliary data transit function, according to which data enters the processor and leaves it without any processing.

Theoretically, any algorithm graph $G(Q, X)$ may be mapped on the matrix multipipeline structure (see Fig. 1.17), but a substantial number of processors will be used for data transit between other processors of the structure, not for carrying out calculations.

It is clear, that mapping process of $G(Q, X)$ on the multipipeline structure with hard architecture is sufficiently complicated. Placement of the operational vertices of $G(Q, X)$ in the processors must provide their most effective use, i.e. the number of the processors, used for data transit, must be minimum.

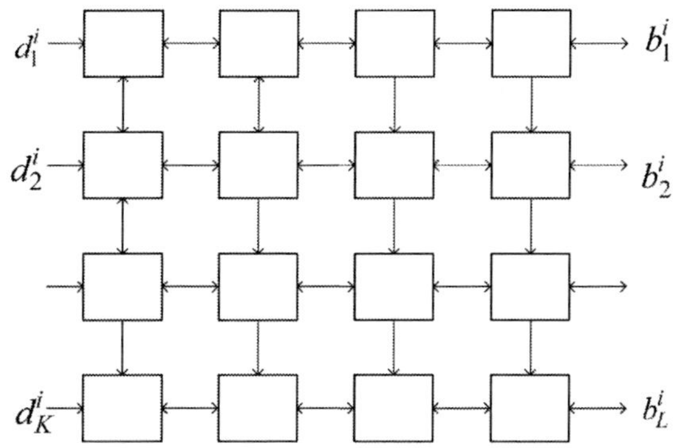

Figure 1.16. The multipipeline computer with matrix architecture.

Though the suggested method does not require a complex switch, it, however, has two substantial disadvantages. Firstly, switch functions are distributed between the processors, and, hence, the part of the processors is used as switch nodes. As a result, effectiveness of use of hardware system resources is decreasing. Secondly, data transit through the processors leads to additional time costs, when data passes through the multipipeline, and reduces the speed of pipeline processing, increasing the total time of data flow processing.

It is possible to suggest a combined method, which contains both of the variants, analysed before. The main idea of the method is a multipipeline cascade structure (see Fig. 1.18). Each cascade is a subset of processors P, and a switch SW_i, which provides all-to-all communication between the processors of the cascades i and $i+1$, $i=1,2,...,H$, and direct data transference to the inputs of the switch SW_{i+1} of the next cascade.

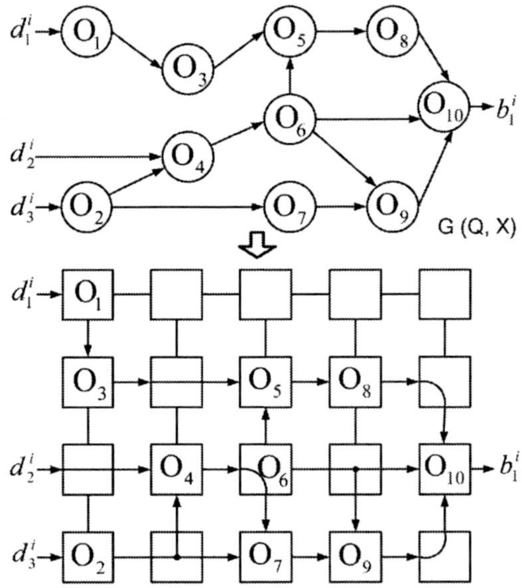

Figure 1.17. Mapping of $G(Q,X)$ on the matrix multipipeline structure.

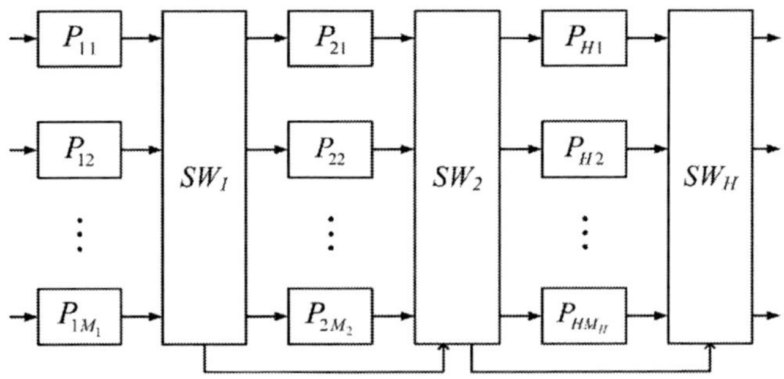

Figure 1.18. A multipipeline cascade structure.

In comparison with the first analyzed method, the combined method considerably reduces hardware resources, required for implementation of the switches. Indeed, hardware resources, used for creation of H switch cascades, are proportional to the value

$$H \cdot M_i^2 \approx H \cdot \left(\frac{M}{H}\right)^2 = \frac{M^2}{H},$$

where M_i is the number of the processors P of the multipipeline cascade i, i.e. hardware resources will be in H times less, than resources that are required for the full-accessible switch, which provides all-to-all connections between the processors of the structure. At the same time, in comparison with the second method, the cascade structure of multipipeline provides considerably greater useful loading of the processors, because all switching functions are performed by the switch cascades.

1.4. MULTIMICRO- AND MULTIMACROPIPELINE COMPUTING STRUCTURES

Earlier we considered the fundamental principles of multipipeline computing structure design. However, the definition of an operation $O_j \in O$, implemented by a pipeline stage, was not given. Operation complexity determines requirements of computational power of the multipipeline stage, the multipipeline processing speed, and effectiveness of communication between the processor stages. Let us make a detailed analysis of this problem.

Let us assume that a certain algorithm graph $G(Q,X)$ is given. Each vertex q_j of the graph defines a certain operation O_j, contained in a certain set of feasible operations O (see Fig. 1.19).

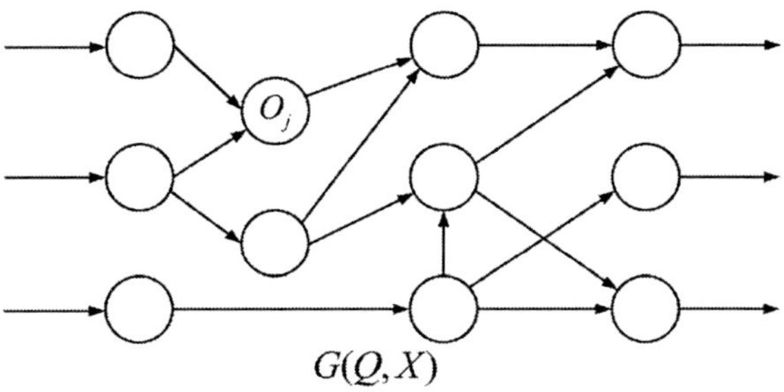

Figure 1.19. An initial algorithm graph $G(Q,X)$.

Let us assume that the operation $O_j \in O$ may be decomposed into a number of more simple operations. In general case, O_j may be represented as an algorithm graph $G_j^{-1}(Q_j^{-1},X_j^{-1})$, whose vertices correspond to more simple operations than O_j (see Fig. 1.20). The vertices of G_j^{-1} may be decomposed in the similar way, when each vertex of the graph is represented as an algorithm graph $G_l^{-2}(Q_l^{-2},X_l^{-2})$ of the lower level.

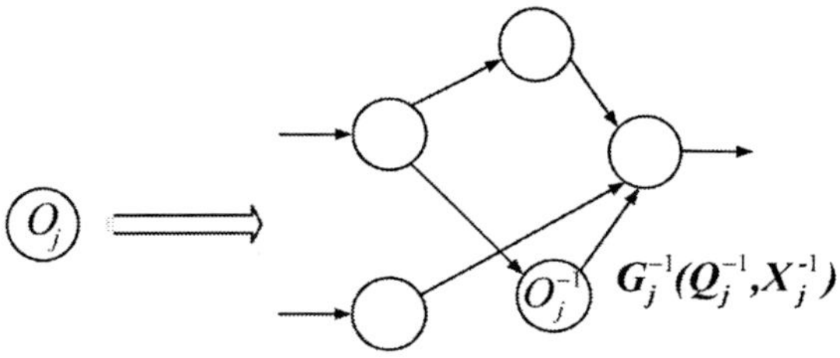

Figure 1.20. Representation of the operation O_j in the form of an algorithm graph $G_j^{-1}(Q_j^{-1},X_j^{-1})$.

Decomposition may be continued until a certain level of indivisible primitive operations is achieved. Hereinafter, such primitive operations will be called *microoperations*. As a result, the initial graph $G(Q,X)$ may be replaced by a graph of microoperations $G^{-n}(Q^{-n},X^{-n})$, where n is the number of partition levels. The graph of microoperations is considerably more complex than $G(Q,X)$.

The obtained graph of microoperations may be executed by a multipipeline computer, which contains processors that perform microoperations, corresponding to the vertices of $G^{-n}(Q^{-n},X^{-n})$. It is obvious, that implementation of such microoperations does not need complex processors; only elementary single-digit processing elements, called *cells of a homogeneous computing environment*, are sufficient to be used. A multipipeline that consists of the computing cells, which perform a certain set of elementary microoperations, will be hereinafter called a *multimicropipeline computing structure*.

The advantage of the multimicropipeline structure is the maximum possible pipelining of data flow processing, and hence, the maximum possible speed of pipeline processing and the minimum possible processing time of a large array of data vectors. Indeed, as it was shown above, the speed of multipipeline processing is in inverse proportion to the time ΔT, which is needed to perform the most lasting operation of the stage, included into the critical path. The simpler is the operation, the smaller is the time ΔT. The total processing time of the array of N input vectors is obtained according to the equation

$$T_{pr} = (N + H_{\max} - 1) \cdot \Delta T , \tag{1.12}$$

where $H_{\max}$ is the number of the multipipeline stages, included into the critical path.

The simpler are the microoperations, corresponding to the vertices of $G^{-n}(Q^{-n},X^{-n})$, the more complex is the graph, and hence, the greater is $H_{\max}$. However, if N is large, then $H_{\max}$ may be neglected. As a result, T_{pr} is obtained as follows

$$T_{pr} \approx N \cdot \Delta T ,$$

i.e. the smaller is ΔT, the smaller is the total processing time of the array of data vectors.

However, practical implementation of the multimicropipeline method of organization of computations comes across numerous technical difficulties. Firstly, microitemization of the algorithm is a complicated and laborious process, which gives a graph of a very large size. In turn, multimicropipeline implementation of such a graph of microoperations will require a very large number of computing cells. Though each cell of the environment may be rather simple, the huge number of cells will lead to an unacceptable size of hardware resources, needed for multimicropipeline organization.

The problem of sharp increase of hardware resources, that are needed for multimicropipeline implementation, aggravates, if it is necessary to create a general-purpose multimicropipeline for solving various data flow problems with different graphs of data processing. In this case, as it was shown in the previous subsection, the multipipeline must contain a full-accessible switch, which must provide all-to-all communication between all the processors of the multipipeline. Since the number of the computing cells of the multimicropipeline is to be sufficiently large, then hardware costs for the full-accessible switch, which are proportional to square of the number of processors, will be still inadmissibly high. Therefore, the general-purpose multimicropipeline must be designed solely with the help of the abovementioned principle of communication, based of additional data transit functions of the processors. The general-purpose multimicropipeline may be represented as a matrix of processing elements (see Fig. 1.21). In addition to the capabilities of carrying out microoperations, each computing cell of the multimicropipeline is to provide data transit, transferring information that enters its inputs to any of its outputs and further to the inputs of the adjacent cells, including branching of the transferred transit data simultaneously to several outputs.

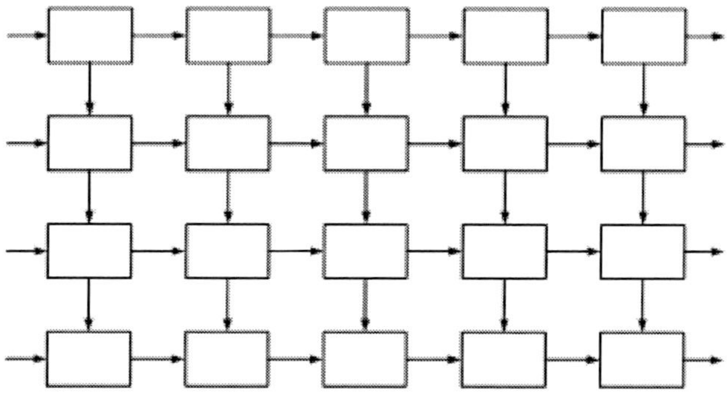

Figure 1.21. The general-purpose multimicropipeline represented as a matrix of computing cells.

Though mapping of the algorithm of microoperations on the matrix structure is a nontrivial problem, its solution is easier, than hardware implementation of the full-accessible communication between the computing cells of the multimicropipeline.

More detailed consideration of the principles of organization, functioning and programming of multimicropipeline computing structures will be given in Chapter 2.

The variant of partition of the initial algorithm graph into microoperations was considered above. Enlargement of the vertices of the algorithm graph is an alternative to this method. Let $G(Q,X)$ be an algorithm graph of a data flow problem. Let us partition the graph into the subgraphs $G_j(Q_j,X_j)$, $i=1,2,...,M$, and put each subgraph into correspondence with some vertex q_j^1 of $G^1(Q^1,X^1)$. The edges of $G^1(Q^1,X^1)$ connect the vertices q_j^1 and q_l^1, which correspond to the incident subgraphs $G_j(Q_j,X_j)$ and $G_l(Q_l,X_l)$ of $G(Q,X)$ (see Fig. 1.22).

Let us assign a “large” operation O_j^i, whose algorithm is described by $G_j(Q_j,X_j)$, to each q_j^1 of $G^1(Q^1,X^1)$. As a result we obtain the enlarged graph $G^1(Q^1,X^1)$ of the algorithm of the problem.

Enlargement of the algorithm graph may be continued. The graph $G^1(Q^1,X^1)$ is being partitioned into subgraphs, and a larger algorithm graph $G^2(Q^2,X^2)$ is obtained as a result.

The process is being continued until the algorithm graph is diminished into $G^n(Q^n, X^n)$, containing the only vertex q^n.

The graph $G^i(Q^i, X^i)$, $i = 1,2,...,n$, obtained as a result of enlargement of the initial algorithm graph $G(Q,X)$, will be hereinafter called *an algorithm macrograph.* The operations, assigned to the vertices of the algorithm macrograph will be called *macrooperations.*

To create a multipipeline structure, which performs the algorithm macrograph $G^i(Q^i, X^i)$, it is necessary to put each vertex of the graph into correspondence with some processor, and to connect the processors according to the macrograph topology. Since some "large" macrooperation is assigned to each vertex of the macrograph, then the processor of the multipipeline is to perform it, i.e. the processor is to have sufficient computational capabilities. The multipipeline, which contains processors capable to perform large macrooperations, will be hereinafter called *a multimacropipeline computing structure.*

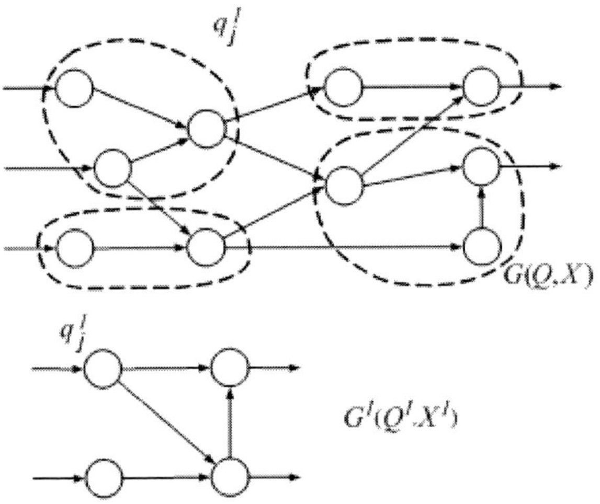

Figure 1.22. Partition of the algorithm graph $G(Q,X)$ into subgraphs.

The main advantage of the multimacropipeline architecture in comparison with the multimicropipeline one is considerable reduction of the number of processors owing to processor "enlargement". Owing to this circumstance, it is possible to implement the multimacropipeline computer with the full-accessible switch, and hence, the multimacropipeline structure may be adapted to topologies of various algorithm graphs. However, the advantage is attained owing to considerable complication of the processors, and especially owing to reduction of the multipipeline processing speed and, consequently, owing to increasing of the total processing time of the whole input data array. The more complicated are the macrooperations, the larger time is needed to perform them in the multimicropipeline stages and, hence, the larger is ΔT, which defines the rate of data output from the multipipeline structure. For instance, if the multipipeline stages are implemented by means of a standard sequential processor with von Neumann architecture, then the time ΔT will be in K times larger, than that for the multipipeline structure. Here K is the number of the vertices of the algorithm micrograph, which correspond to one vertex of the algorithm macrograph.

Another suggested method of multimacropipeline organization [1] may eliminate the disadvantage. The main idea of the method is the following. Let us put each vertex q_j^i of the macrograph $G^i(Q^i, X^i)$ (see Fig. 1.23) in correspondence with a special processor, introduced as a set of processing elements (PE), and with a switch, providing fully-accessible communications between the processing elements. Each PE is to perform the operation set, assigned to the vertices of the initial algorithm graph $G(Q, X)$. The number of the PEs is to be sufficient for mapping of all vertices of the subgraph $G_j(Q_j, X_j) \in G(Q, X)$, corresponding to the vertex q_j^i of the macrograph $G^i(Q^i, X^i)$.

Hereinafter, the processor will be called *a macroprocessor*. The macroprocessor may be used as a base element of a local multipipeline that implements $G_j(Q_j, X_j)$, corresponding to the vertex q_j^i of $G^i(Q^i, X^i)$ (see Fig. 1.24).

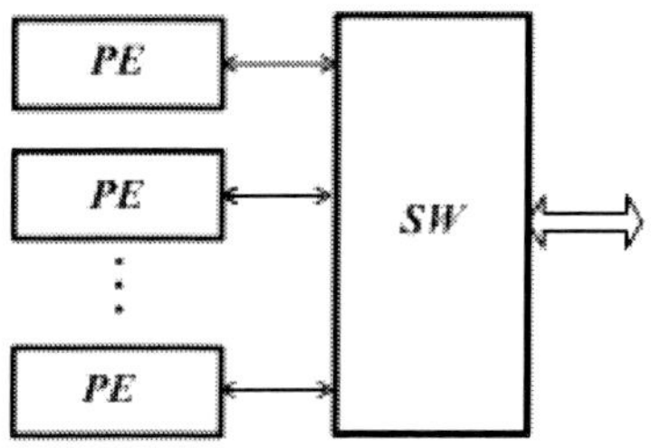

Figure 1.23. A structure chart of a macroprocessor.

The multipipeline macroprocessor implements the local multipipeline, which maps the corresponding subgraph $G_j(Q_j, X_j)$ of $G(Q, X)$. The upper level system switch (SSW) implements information connections between the macroprocessors according to the topology of $G^i(Q^i, X^i)$. The data processing speed of such multipipeline computing structure will depend on the time needed to perform the most lasting operation of the operation set O_j, carried out by the processing element.

In comparison with the multimicropipeline method of computing structure organization, the last considered method has a number of advantages.

First of all, communication capabilities of the processing elements are more flexible. Besides, since the switch is divided into two levels (the local switch of the macroprocessor and the upper level system switch), then hardware resources, needed for their implementation, are not too large. In addition, programming of the structures is considerably easier, because it is based on large macrooperations, but not on elementary microoperations. However, on the other hand, the speed of multimacropipeline data processing will be lower than in multimicropipeline.

A detailed analysis of principles of organization, functioning and programming of multimacropipeline structures will be given in Chapter 3.

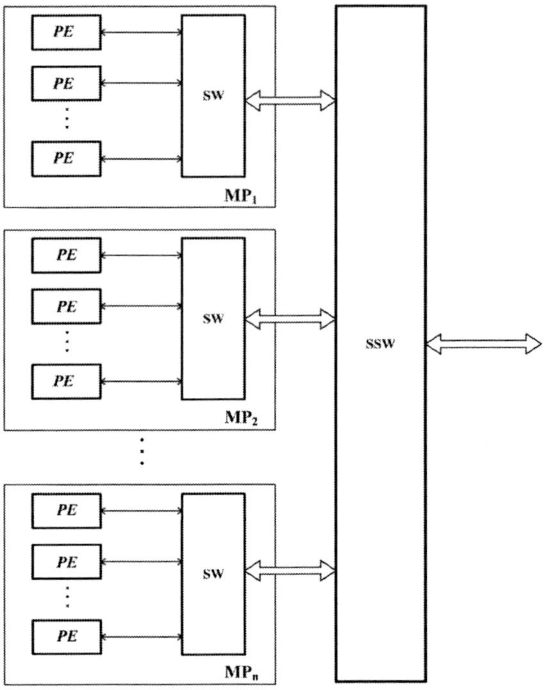

Figure 1.24. A structure chart of a multimacropipeline.

1.5. The Structural-Procedural Method of Organization of Multipipeline Computations

In the previous subsections we considered the methods of organization of multimicropipeline and multimacropipeline computations. The result of use of both methods is a multipipeline structure with some level of itemization of a mapping graph of a data flow algorithm. It was assumed that hardware resources of the multipipeline structure (processors and communication capabilities) are sufficient to map the algorithm graph. However, this condition may not be fulfilled in reality. Usually, the number of operational vertices of the algorithm graph of the solving problem is much greater than the number of processing elements in the multipipeline. Therefore, mapping of a large algorithm graph on a multipipeline structure with limited hardware resources comes across several problems [11, 22], that may be solved with the help of an evident method given in [11]. According to the method, the initial algorithm graph $G(Q,X)$ is to be partitioned to the subgraphs $G_i(Q_i,X_i)$, $i=1,2,...,F$, so that each subgraph $G_i(Q_i,X_i)$ could be implemented in the given multipipeline structure. The initial algorithm graph is to be partitioned according to the following conditions.

1) The number of the vertices of $G_i(Q_i,X_i)$, $i=1,2,...,F$ is to be equal to (but cannot exceed) the number of the multipipeline processors.

2) The number of the internal edges of $G_i(Q_i,X_i)$, $i=1,2,...,F$ cannot exceed the number of connections provided by the switch of the multipipeline structure.
3) The number of the input and output edges of the subgraph cannot exceed the number of communication channels between the multipipeline and external memory.

Figure 1.25 shows an example of $G(Q,X)$ of a digital signal processing algorithm and its partition into the set of more simple subgraphs $G_i(Q_i,X_i)$, $i=1,2,...,F$.

The multicriterion partition of $G(Q,X)$ is a sufficiently complicated process that is analysed in details in [17]. Here, in order not to obscure the general understanding of the idea, it is assumed that $G(Q,X)$ is somehow partitioned into the subgraphs $G_i(Q_i,X_i)$, $i=1,2,...,F$ according to the abovementioned requirements.

Then, to implement the data flow problem, which is described by $G(Q,X)$, the subgraphs $G_i(Q_i,X_i)$, $i=1,2,...,F$ are to be successively mapped on the multipipeline structure. Besides, the following considerations are to be taken into account.

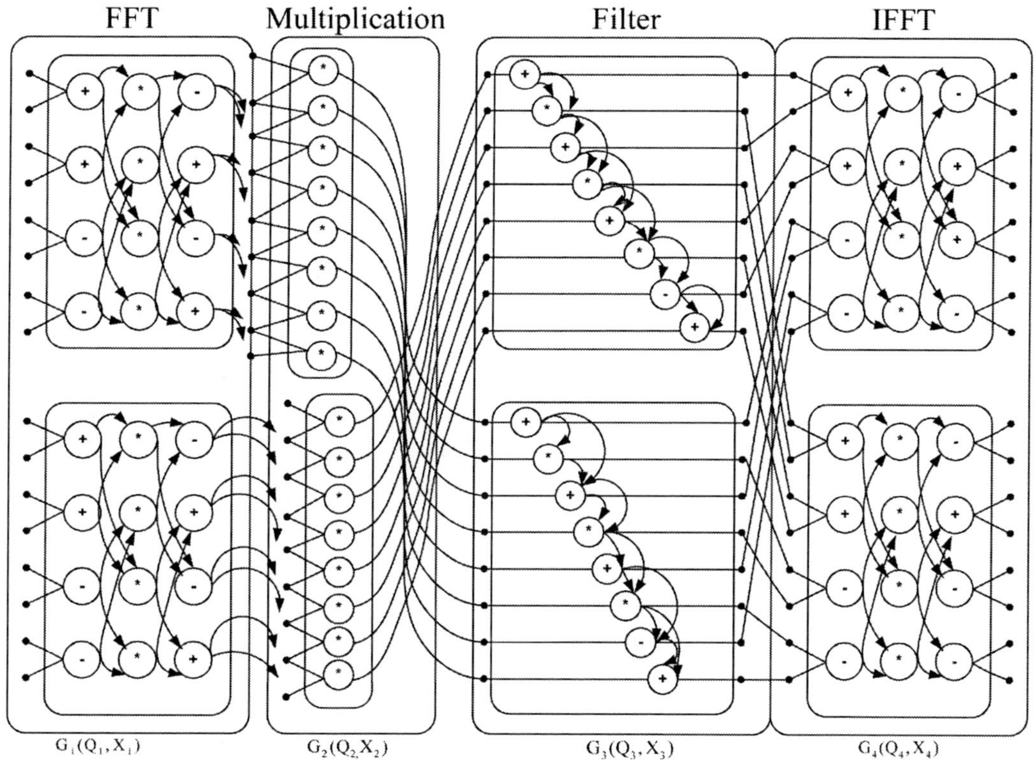

Figure 1.25. An information graph of a DSP algorithm.

Since mapping of $G_i(Q_i,X_i)$ on the multipipeline is a sufficiently complicated and lasting procedure, then each $G_i(Q_i,X_i)$, $i=1,2,...,F$ is to be mapped only once to reduce the time of data flow problem implementation. In this case, the data flow of all input vectors is to pass through the multipipeline, implementing $G_i(Q_i,X_i)$.

All vectors of input data, which are to pass through the multipipeline, must be ready before mapping of $G_i(Q_i, X_i)$ on the multipipeline. In other words, before implementing $G_i(Q_i, X_i)$ in the multipipeline, it is necessary to implement all subgraphs connected with $G_i(Q_i, X_i)$ by outgoing edges. The results of data flow processing, obtained in the multipipeline of $G_i(Q_i, X_i)$, are to be stored in an intermediate memory, and are to be available for further processing when the subgraphs $G_{i+1}(Q_{i+1}, X_{i+1}), ..., G_F(Q_F, X_F)$, incident to $G_i(Q_i, X_i)$, are implemented in the multipipeline. It is desirable to provide the rate of writing/reading of data vectors in/from memory not less, than the processing speed of the multipipeline stages. Otherwise, the speed of multipipeline data processing will depend on the data access speed.

Taking into account the abovementioned considerations, the generalized structure chart of the multipipeline computing structure, implementing large algorithm graphs, may be represented as it is shown in Fig. 1.26.

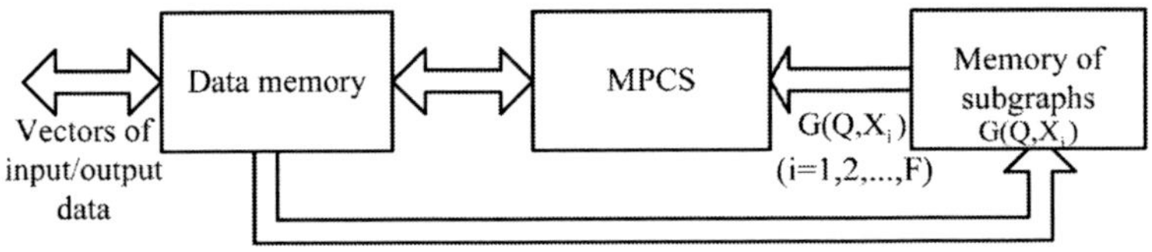

Figure 1.26. The structure chart of the multipipeline computing structure (MPCS), implementing large information graphs.

Figure 1.27 shows the structure chart of functioning of the computing structure.

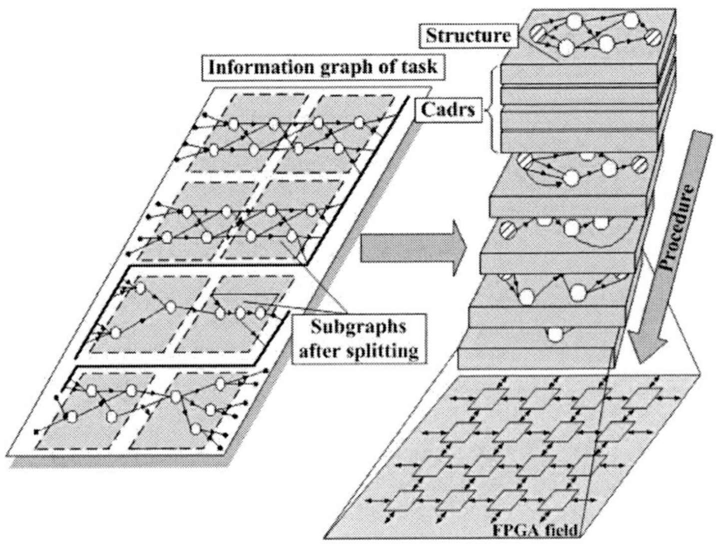

Figure 1.27. Structural-procedural organization of computations.

The initial graph $G(Q,X)$ of the algorithm of the data flow problem is partitioned into the subgraphs $G_i(Q_i,X_i)$, $i=1,2,...,F$ according to the abovementioned requirements. Then the subgraphs are ordered in a queue according to the condition: the indices of the subgraphs, which precede $G_i(Q_i,X_i)$, are to be less than the index i of $G_i(Q_i,X_i)$. The subgraphs $G_i(Q_i,X_i)$, $i=1,2,...,F$ are mapped on the multipipeline structure successively, according to the subgraph sequence. The stream of input data vectors follows from memory to the inputs of the multipipeline, which implements $G_i(Q_i,X_i)$. The output data vectors are being written into memory. The process continues until all $G_i(Q_i,X_i)$, $i=1,2,...,F$ are implemented in the multipipeline and all needed data vectors are processed. After execution of the last subgraph $G_F(Q_F,X_F)$, memory contains the set of result data vectors, which then leave the computing structure. The procedure, described above, will hereinafter be called *a structural-procedural method of organization of multipipeline computations*, because each subgraph is structurally implemented in the multipipeline. At the same time, the subgraphs are mapped on the multipipeline according to the sequential procedure [43].

Figure 1.28 shows the example of structural-procedural implementation of the DSP algorithm with partition of its information graph into the subgraphs according to Fig. 1.25.

The time, needed to process a data array of N elements in a multipipeline structure with structural-procedural organization of computations, is obtained as

$$T_{pr}=\sum_{i=1}^{F}((N+H^{i}_{\max}-1)\cdot\Delta T_i+T^{i}_{reconf}),$$

where $H^{i}_{\max}$ is the length of the critical path of $G_i(Q_i,X_i)$;

ΔT_i is the time, needed to carry out the most lasting operation, corresponding to the vertices of $G_i(Q_i,X_i)$;

T^{i}_{reconf} is the time of adjustment (reconfiguration) of the multipipeline, which implements $G_i(Q_i,X_i)$;

F is the number of the subgraphs, obtained as a partition result of the initial graph of the algorithm.

If $\Delta T_i=\Delta T$, $i=1,2,...,F$, $T^{i}_{reconf}=T_{reconf}$, $i=1,2,...,F$ and $H^{i}_{\max}=H$, $i=1,2,...,F$, then the last equation may be represented as

$$T_{pr}=F((N+H_{\max}-1)\cdot\Delta T+T_{reconf}), \tag{1.13}$$

or as

$$T_{pr}=F(N\Delta T+T_{reconf}), \tag{1.14}$$

when N is large.

As it may be seen from the last equation, a considerable part of time, needed to process the N size data array in the multipipeline computer system with structural-procedural organization of computations, is taken by the reconfiguration procedure, when calculations are not performed. To reduce the total time T_{pr} of the problem execution, it is necessary to diminish the time, which is taken by the system reconfiguration. The simplest solution is to use a multipipeline computing structure, that consists of several reconfigurable subsystems. Let us make a detailed consideration of the method.

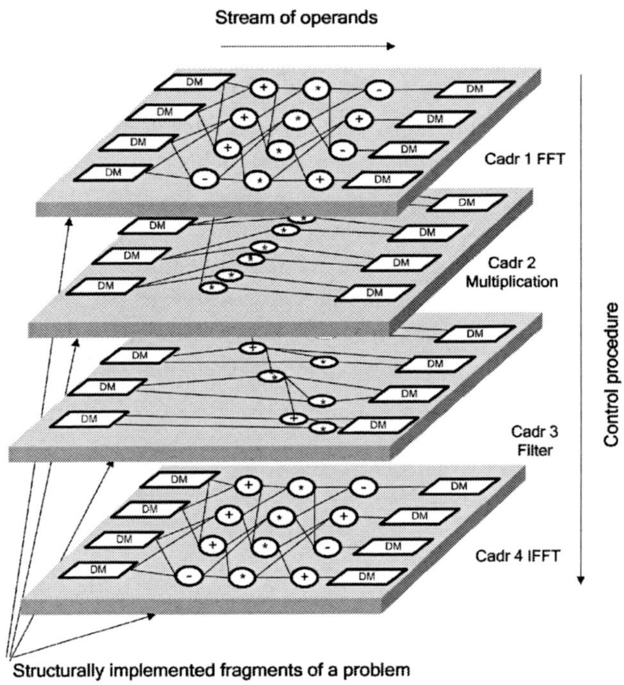

Figure 1.28. Structural-procedural implementation of the DSP algorithm.

It is assumed that the computer system consists of two MPCSs, each of which structurally implements its own subgraph $G_i(Q_i, X_i)$ of the algorithm graph $G(Q,X)$ of the solving problem (see Fig. 1.29).

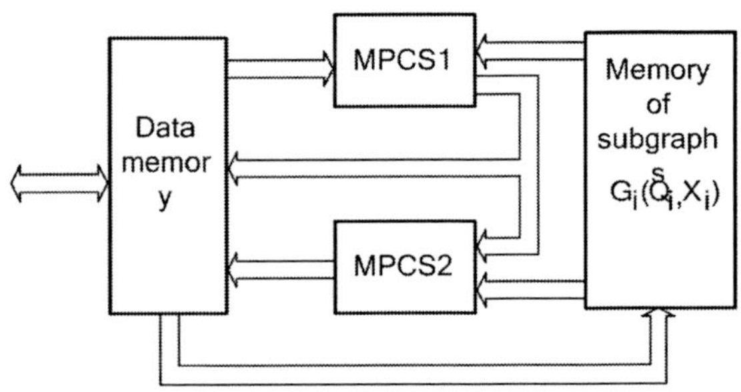

Figure 1.29. A two-stage MPCS with parallel reconfiguration.

So, if the two MPCSs are concurrently loaded with the two sequential subgraphs $G_i(Q_i, X_i)$ and $G_{i+1}(Q_{i+1}, X_{i+1})$, $i = 1,2,...,F/2$ (see Fig. 1.30) of $G(Q,X)$, and the array of the input data vectors of $G_i(Q_i, X_i)$ enters the MPCS1, then the time, needed to obtain the first processed vector in the MPCS1 outputs, is $T_{cal}^{i,1} = H_{\max}^i \cdot \Delta T_i$.

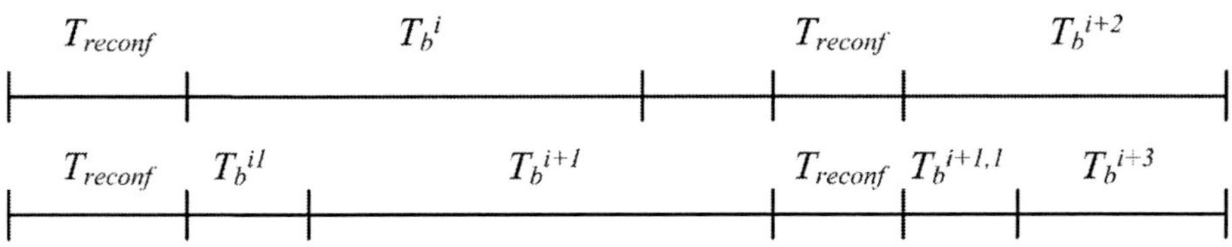

$$T_{cal}^i = (N + H_{\max}^i - 1) \cdot \Delta T_i$$

Figure 1.30. The timing diagram of the computational process of the two-stage MPCS with parallel reconfiguration.

The obtained data vector directly enters the MPCS2 and is processed according to $G_{i+1}(Q_{i+1}, X_{i+1})$. As a result, processing of the whole data array by the computing structures, implemented according to the subgraphs $G_i(Q_i, X_i)$ and $G_{i+1}(Q_{i+1}, X_{i+1})$, takes the time $(T_{cal}^{i,1} + T_{cal}^{i+1})$, where T_{cal}^i is the time, needed to process the array of data vectors by the MPCS according to $G_i(Q_i, X_i)$ (see Fig. 1.30). Then the MPCS1 and the MPCS2 are loaded with the next two subgraphs $G_{i+3}(Q_{i+3}, X_{i+3})$ and $G_{i+4}(Q_{i+4}, X_{i+4})$ of the algorithm graph $G(Q,X)$, and data processing is repeated. With such organization of computational process, the total time of solution is obtained according to the equation

$$T_{pr} = \sum_{i=1}^{F/2} (T_{cal}^{i,1} + T_{cal}^{i+1} + T_{reconf}) = \sum_{i=1}^{F/2} ((H_{\max}^i + H_{\max}^{i+1} + N - 1) \cdot \Delta T + T_{reconf}),$$

where $\Delta T = \max(\Delta T_i, \Delta T_{i+1})$;

$T_{reconf} = \max(\Delta T_{reconf}^i, \Delta T_{reconf}^{i+1})$;

F is the number of the subgraphs, obtained after partition of $G(Q,X)$.

Assuming that $H_{\max}^i \approx H_{\max}^{i+1}$, we obtain

$$T_{pr} \approx F/2((2H_{\max} + N - 1) \cdot \Delta T + T_{reconf}), \qquad (1.15)$$

i.e. the time of solution is reducing almost twice in comparison with the variant, when the only one MPCS is used (see Fig. 1.13).

If the number of the MPCS stages is increasing, then the total time of structural-procedural implementation of the problem is reducing even more. But saving of the time of the problem implementation will be achieved owing to proportional increasing of hardware cost.

Organization of structural-procedural computations may lead to another possible difficulty. Let us assume that an endless (or extremely large) stream of data vectors is to be processed in a multipipeline computer. A data source generates streams of data vectors, that enter the multipipeline structure in constant equal time periods ΔT_{in}. If the stream is processed by a standard multipipeline computer, then the speed of data processing will simply depend on ΔT_{in}, and the result data vectors will leave the computer outputs in every ΔT_{in} period. However, such method is unacceptable with structural-procedural organization of multipipeline computations. Usually the time of multipipeline computer reconfiguration is rather large, so it may turn out, that (see (1.13))

$$\Delta T_{in} < T_{pr} = F((1 + H - 1) \cdot \Delta T + T_{reconf}) = F(H \cdot \Delta T + T_{reconf}),$$

i.e. when the next vector of data enters the multipipeline, processing of the previous data vector is still not finished. An evident solution to this difficulty is to accumulate a subset of input data vectors, and then to transfer it to the inputs of the multipipeline with structural-procedural organization of computations. If the number of data vectors in the subset is large, then the multipipeline may have no time to process them before the moment, when a new accumulated subset enters the multipipeline.

If the data in the subset is few, then the multipipeline will finish processing before the next subset is accumulated and will remain idle. Therefore, the optimal size of the vector subset must fulfill the condition: the time of the subset processing in the multipipeline must be equal to the time of accumulation of the subset, i.e.:

$$N \cdot \Delta T_{in} = T_{pr},$$

or, taking into account (1.13),

$$N \cdot \Delta T_{in} = F((N + H - 1) \cdot \Delta T + T_{reconf}), \qquad (1.16)$$

where N is the number of data vectors in the subset;

ΔT_{in} is the time period (rate) between input data entry;

T_{pr} is the time of the subset processing in the multipipeline with structural-procedural organization of computations.

The optimal number N of data vectors in the subset may be obtained, taking into account the equation (1.16)

$$N = \frac{F((H-1)\Delta T + T_{reconf})}{T_{in} - F\Delta T}. \qquad (1.17)$$

Hence, to process a large amount (stream) of input data, that enter in $1/\Delta T_{in}$ periods, according to the complex algorithm graph $G(Q, X)$, it is necessary to accumulate the vector subsets with the number of elements (1.17), and to process each subset in the multipipeline with structural-procedural organization of computations.

As it was shown above, one of the basic principles of structural-procedural organization of computations is creation of the ordered sequence of the subgraphs $G_i(Q_i, X_i)$, $i = 1,2,...,F$ of the initial algorithm graph $G(Q,X)$ according to the following rule: the set of input data vectors, needed for implementation of $G_i(Q_i, X_i)$, must be completely ready by the moment when $G_i(Q_i, X_i)$ is mapped on the multipipeline.

It is easy enough to fulfill such requirement, if the data d_l^i, $l = 1,2,...,k$ of the input vector $D_i =< d_1^i, d_2^i, ..., d_k^i >$ enters the multipipeline concurrently or according to the pre-defined sequence. However, if the data d_l^i, $l = 1,2,...,k$ is obtained in different non-synchronized data sources, then it may enter the multipipeline in random, a priori undefined moments of time.

In this case, the precise sequence of the multipipeline implementation of the subgraphs $G_i(Q_i, X_i)$, $i = 1,2,...,F$ of the algorithm graph $G(Q,X)$ cannot be generated. Under existing conditions, the subgraph $G_i(Q_i, X_i)$ with the ready-to-use set of input data vectors should be mapped on the multipipeline (see Fig. 1.31).

On the whole, the system is to operate as follows. Data storage of each subgraph $G_i(Q_i, X_i)$ contains a bank, whose capacity is sufficient to store the set of N data vectors.

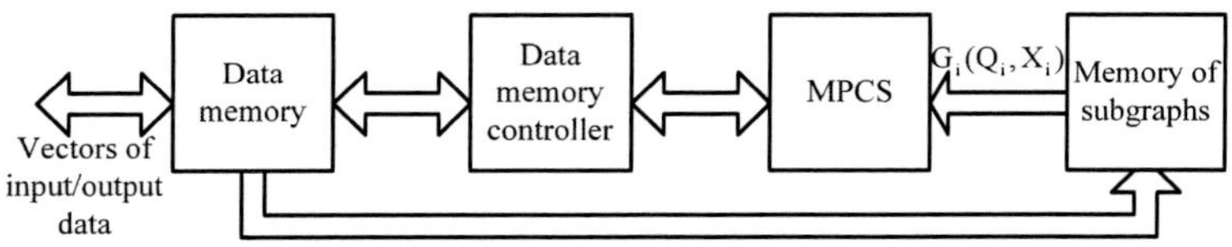

Figure 1.31. The structure of the multipipeline computer with a priori undefined sequence of the implemented subgraphs $G_i(Q_i, X_i)$.

The input data, which is being transferred from the data sources, is being written into correspondent cells of data memory. The data memory controller (DMC) is continuously scanning banks of data memory of the subgraphs $G_i(Q_i, X_i)$, $i = 1,2,...,F$. As soon as the bank of $G_i(Q_i, X_i)$ is filled with the set of input data vectors, the DMC yields a corresponding instruction and transfers it to the memory of subgraphs, according to which $G_i(Q_i, X_i)$ is mapped on the multipipeline computing structure. If several subgraphs are ready to be executed, then the DMC queues them. When the MPCS is programmed to execute $G_i(Q_i, X_i)$, the DMC defines the bank with the set of input data of the subgraph in the data memory and yields data with the rate $1/\Delta T$. The output data, obtained after processing in the MPCS, is transferred into the DMC, which writes it into the corresponding banks of data memory.

Then, the DMC defines the next $G_i(Q_i, X_i)$, that is ready to be executed, i.e. the subgraph with the generated set of input data, and yields an instruction into the memory of subgraphs for the subgraph mapping on the MPCS. The process is going on until all

$G_i(Q_i, X_i)$, $i = 1,2,...,F$ of $G(Q,X)$ have been implemented on the MPCS. The obtained vectors of output data are transferred to the system outputs.

The detailed description of the principles of structural-procedural organization of computations in multipipeline structures will be given in Chapter 3.

So, generalizing everything that was considered above, it is possible to formulate the main principles of organization and functioning of reconfigurable multipipeline computing structures:

1) An MPCS is a computation field, formed by uniting a set of single-type processing elements (cells, processing elements, macroprocessors, etc.). The architecture of connections between the united elements may be changed one way or another (see Fig. 1.32).
2) The hardware (structural) analog of the information graph of some solving problem, represented as the multipipeline structure, topologically similar to the modeling graph, is created by means of specialized tools of structural programming within the computation field (see Fig. 1.32).

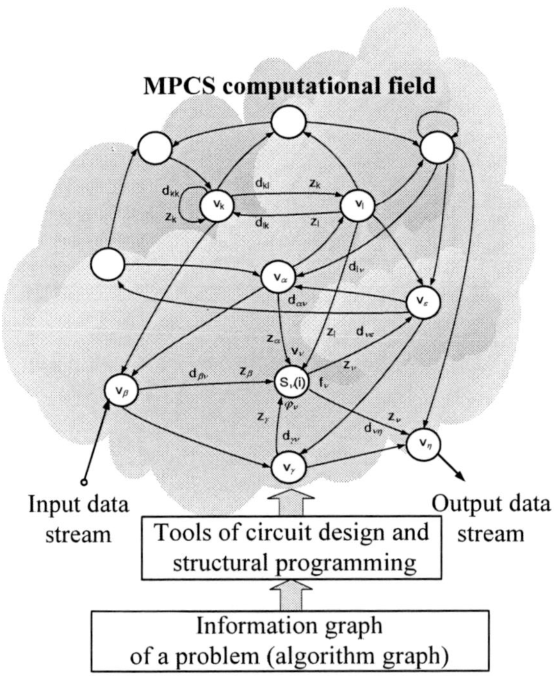

Figure 1.32. Mapping of the problem graph on the computation field of the MPCS.

3) The simpler is the computing element of the MPCS computation field, the higher speed of data processing may be provided by the multipipeline. However, simplification of the computing element leads to higher complexity of procedures that map the information graph of the solving problem on the MPCS computation field.
4) If hardware constraints of the computation field do not permit "placement" of the whole information graph of the problem, then the graph is beforehand to be

partitioned into disjoint subgraphs, each of which may be structurally implemented in the computation field. Besides, the sequential procedure of structural implementation of the obtained graphs in the MPCS computation filed is organized.

5) The larger is the MPCS computation field, the more complex information graph of the problem may be structurally implemented within the field with no partition into subgraphs. Hence, it is possible to reduce a part of non-productive burden, needed for reconfiguration of the MPCS computation field according to structural implementation of various subgraphs, in the total time of the problem solution.
6) The cortege (stream) of input data enters the multipipeline, which is organized within the MPCS computation field and implements the information graph (subgraph) of the solving problem. The cortege (stream) of output data leaves the multipipeline (see Fig. 1.32) in $T = (H_{max} - 1)\Delta T$, where H_{max} is the length (the number of computing elements) of the critical path of the multipipeline; ΔT is the time of data processing by the computing element.

Chapter 2

Multimicropipeline Computing Structures Based on Homogeneous Environments

2.1. Architecture of Homogeneous Computing Environment

As it was shown in the previous chapter, the method of multipipeline computing structure design is based on creation of structures that consist of processing elements, which perform a set of elementary single-bit operations, and which are placed in lattice-points of a lattice; adjacent processing elements are connected by direct links. In literature such multimicropipeline structures are usually called *homogeneous computing environments (HCE)*, consisting of the processing elements called *the HCE cells* [5]. Advantages of organization of the HCE-based multipipeline computing structures are simple engineering implementation and capability to provide the highest speed of pipeline processing. Disadvantages of the method are complicated generation of a detailed description of an algorithm graph and difficulty of effective mapping of the graph on a homogeneous environment.

For the first time a notation "homogeneous computing environment" was suggested by E.V. Evreinov in 1960s [44]. The theory of homogeneous computing environments and peculiarities of their application were considered in [2-4, 10, 12, 45-47]. These works and conferences, dedicated to homogeneous computing systems, structures and environments, attracted a great interest to this scientific domain both in Russia and abroad. Transputers were rather close to homogeneous computing structures; systolic arrays, analyzed in the well-known paper by H.T. Kung and C.E. Leiserson "Systolic Arrays" (1978) [37], were close to computing environments.

Figure 2.1 shows a general organization of an HCE-based multimicropipeline computing structure.

According to the definition, given in [2], an HCE, in general case, is an n-dimensional lattice structure of computing cells, each of which performs a set of elementary (usually single-bit) operations.

Each HCE cell is connected with adjacent cells by channels which are intended for transfer of processed data and adjusting information. Figure 2.2 shows the structure of the HCE computing cell.

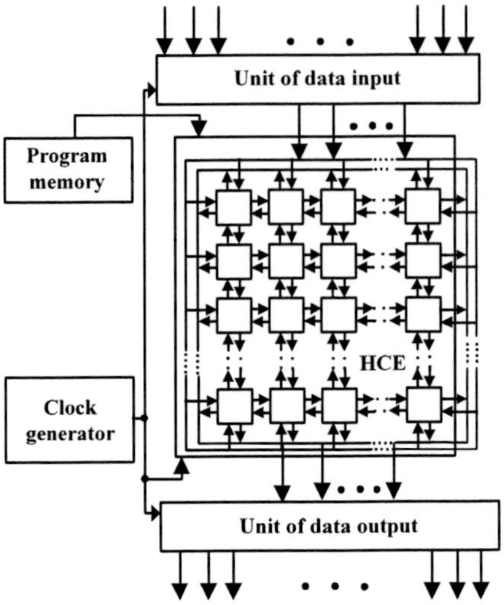

Figure 2.1. Organization of an HCE-based computing structure.

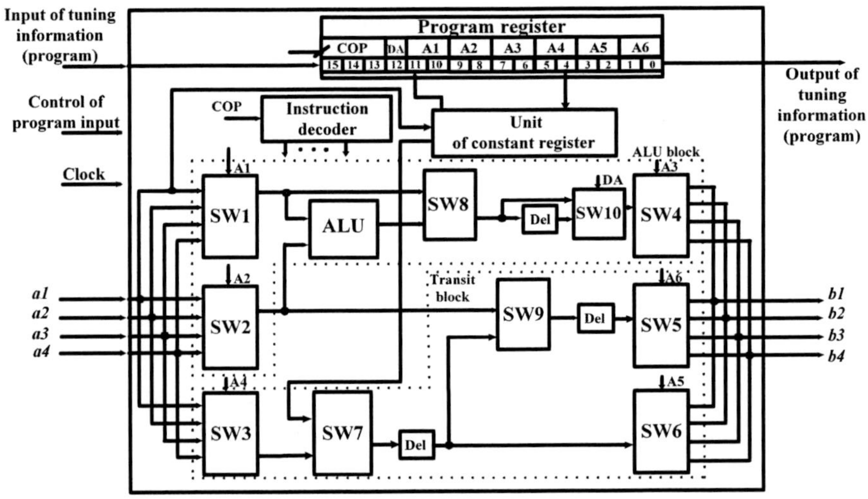

Figure 2.2. The structure of the HCE computing cell.

The computing cell has 4 data inputs *(a1,...,a4)* and outputs *(b1,...,b4),* an input of program input control, an input and an output of a programming channel, clock inputs and a "Reset" input. The cell contains functional units such as a program register, an instruction decoder, an arithmetic-logical unit (ALU), 3 input and output switches (SW1, SW2, SW3 and SW4, SW5, SW6), switches of the ALU channel (SW8 and SW10), a switch of a constant (SW7), a switch of transit (SW9) and a unit of a constant register.

The main control element of the cell is the program register, which consists of 16 digits, united into fields:

- digits 0 and 1 are the address of the output of the second transit channel;
- digits 2 and 3 are the address of the output of the first transit channel;
- digits 4 and 5 are the address of the output of the ALU channel;

- digits 6 and 7 are the address of the input of the first transit channel;
- digits 8 and 9 are the address of the input of the second operand;
- digits 10 and 11 are the address of the input of the first operand;
- digit 12 is a delay attribute;
- digits 13, 14, 15 are the code of an ALU operation.

The computing cell may operate in two modes: adjustment and operation. If a signal of program input control is equal to logic "1", then the computing cell switches over to the adjusting mode according to which data is written in the program register in each cycle.

If the signal of program input control is equal to logic "0", then the computing cell switches over to the operation mode and performs an arithmetic-logical operation on single-digit operands, passing from the input switches SW1, SW2. The operation is performed in the ALU channel according to the code of the program register. In the next cycle the cell transfers results of processing through the switches SW8, SW10 to the output switch SW4. Simultaneously, the cell may redirect input data streams from its inputs to outputs through the two transit channels, which transfer data through the switches SW2, SW9, SW5 and SW3, SW7, SW6 to the output of the cell with one and two cycles delay, respectively. If the delay attribute DA=1, then the cell performs additional one cycle delay of data, processed in the ALU block.

According to the code of operation (COP), the computing cell performs the following arithmetic-logical operations:

- carry storage addition (COP = 001, "+");
- logical multiplication with inverse (COP = 010);
- modulo 2 addition (COP = 101, "$\oplus$");
- memorizing of "1" (COP = 110, "MEM1");
- generation of a constant (COP = 111, "GC");
- extended transit (COP = 011, "ET");
- no operation (COP = 000, "NOP").

Hence, an instruction set of the computing cell contains arithmetic, logical and transit operations.

The operation "Memorizing of 1" is performed as follows:

- if the second operand is equal to logical "0", then the cell transfers the first operand to the output;
- if the second operand is equal to logical "1", then the cell keeps the previous output value.

The operation "Memorizing of 1" takes two cycles before the result leaves the cell.

The operation "Generation of a constant" is performed as follows:

- if DA=0, then an 8-digit constant, contained in digits 4-11 of the program register, is being cyclically transferred to the cell output;
- if DA=1, then an 8-digit data, that enters the input "a1" and is stored in digits 4-11 of the program register, is being transferred to the cell output with 8 cycles delay.

When performing operations "Extended transit" and "No operation", the cell redirects data streams from its inputs to its outputs through the ALU channel and two transit channels.

Let us consider the main principles of organization of multimicropipeline computations on the HCE (see Fig. 2.3).

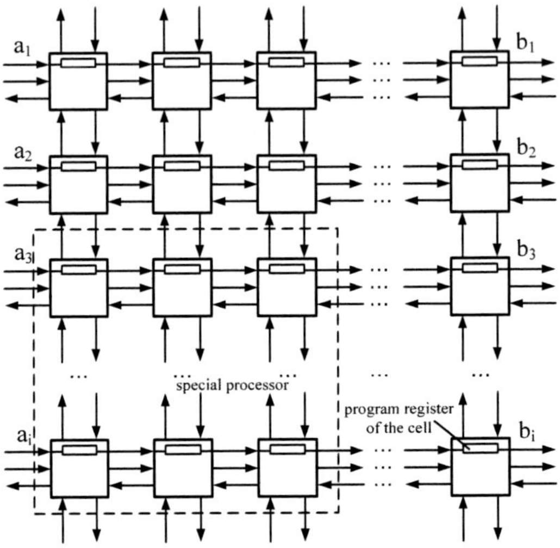

Figure 2.3. A fragment of the HCE field with adjusting chains of program registers.

As it was shown above, adjusting results of the 16-digit cell register defines the operation, performed by the cell on two digits, and also connects channels of adjacent cells.

A group of adjusted cells of the environment may be considered as *a specially designed processor* or *a special processor* (see Fig. 2.3), which performs an arithmetic or logical operation, for instance, multiplication of 32-digit numbers.

Let us assume, that there is a graph of a certain data flow problem (see Fig. 2.4), which is to be solved on the HCE. A special processor, performing an operation that corresponds to a vertex, is formed for each vertex of the graph in the HCE field. The communication channels between the special processors, which correspond to the edges of the information graph of the problem, are also adjusted by the HCE cells, which perform data transit. As a result, a hardware structure, similar to the information graph of the problem, is formed in the HCE and may be used for data stream processing (see Fig. 2.5).

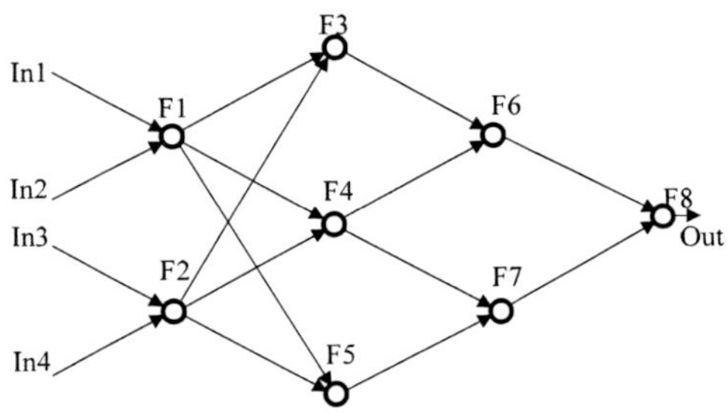

Figure 2.4. A graph of a problem.

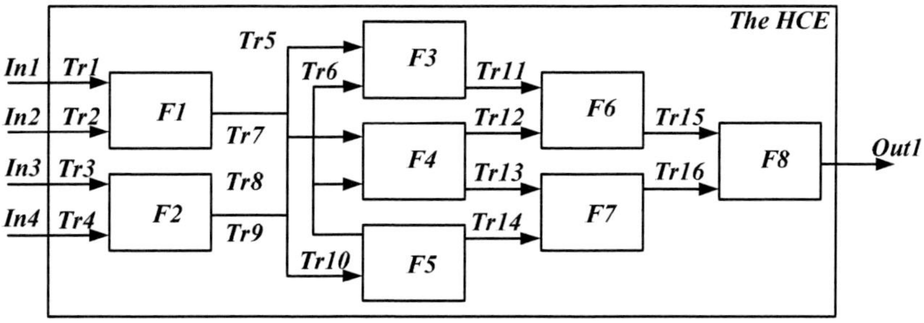

Figure 2.5. Mapping of the graph of the problem on the HCE.

Figure 2.5 shows the special processors F_i, adjusted in the homogeneous environment. In reality, special processors which perform different operations, contain different numbers of the HCE cells, i.e. parts of the computation field, occupied by different special processors, have different sizes.

It is evident, that the HCE programming takes plenty of time. Programming process may be considerably simplified by creating libraries of *microprogram modules*, i.e. the adjusting programs of special processors that perform operations of various levels. In particular, there are libraries of the microprogram modules, used for adjustment of the special processors, that perform logical and arithmetic operations on fixed-point and floating-point numbers, operations of integer and complex arithmetic, matrix operations, algorithms of digital signal and image processing in the HCE [50-55].

The program that generates the result array of adjusting codes of the HCE cells, uses the library of the special processor microprogram modules, that are to be used according to the problem graph.

Special translating programs create register traces for data transfer between the special processors and synchronize all data streams in the HCE field.

The processes of HCE adjustment and operating are separate. At first, a stream of adjusting instructions is generated. Then the adjusting data is loaded into the instruction registers of each computing cell, and a data stream enters the cell. The loaded program analyses all variants of calculation branching.

During execution the program is being stored in the instruction registers of the HCE cells. Operating of the HCE elements is synchronized by a single external clock generator. Data is processed in a pipeline. The input data stream enters the computing cells of the matrix, passes through the pipeline synchronously with clock pulses and is processed according to the program [56].

The following subsections contain a detailed analysis of organization of computational processes in the homogeneous computing environment.

2.2. Organization of Computational Processes in the HCE

Let us define notions of an HCE microoperation, a microinstruction and a microprogram module.

A microoperation is an elementary operation, that is performed in operational or transit channels of a cell according to a code, stored in a microinstruction field.

A microinstruction is a set of codes of operations, simultaneously performed in a computing cell during one cycle. The microinstruction is stored in the 16-digit adjusting register of the cell. The microinstruction of the computing cell consists of 8 fields.

A microprogram module (MPM) is a set of microinstructions, i.e. adjusting instructions of a certain rectangular domain of the homogeneous computing environment. As a result of the microprogram module execution, *a special processor*, intended for a certain arithmetic, logical or any other so-called large operation, is synthesized within the HCE.

Figure 2.6 shows the example of the structure chart of the special processor, that performs subtraction of *n*-digit numbers in modified radix complement representation.

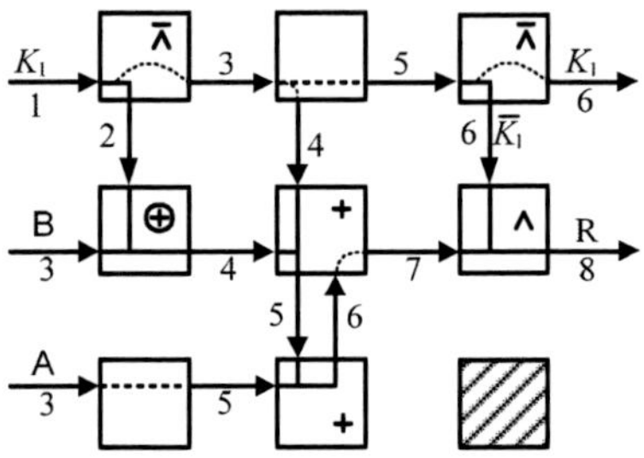

Figure 2.6. The structure chart of the special processor, that performs subtraction.

Structure charts of special processors are formed according to the following rules: traces of operands, entering the ALU of the cell, are marked with solid lines; transit traces are marked with dot lines. The structure charts also contain labels of operations, performed in the cell, and numbers of cycles of operand entries.

The special processor structure chart, shown in Fig. 2.6, corresponds to the microprogram module of subtraction. The module is represented as a table 2.1 and defines adjusting codes of each cell that is contained in the module of subtraction.

Hence, a certain domain of the HCE, adjusted by microinstructions of a microprogram module, is a special processor, adjusted to performing of a certain operation on operands of an arbitrary but fixed capacity.

Table 2.1. The microprogram module of subtraction

A number of a cell	COP	DA	A1	A2	A4	A3	A6	A5
1.1	100	0	00	00	00	11	10	01
1.2	001	0	00	00	00	01	10	11
1.3	100	0	00	00	00	11	01	10
2.1	101	0	01	00	00	10	01	11
2.2	001	0	01	00	11	11	00	10
2.3	010	0	00	01	00	10	01	11
3.1	000	0	00	00	00	01	10	11
3.2	001	0	01	00	00	01	10	11
3.3	000	0	00	00	00	00	00	00

In most cases, operations that are performed by special processors, synthesized in the HCE field, are binary, and hence, two vectors A and B of operands continuously enter the special processor.

With the delay of S_k cycles, that depends on the operation carried out by the special processor, the operation results, represented in sequential code, continuously leave the special processor, starting with the less significant digit. As a rule, the sizes of input and result data are similar.

So, according to data entry, the special processor is the unit of sequential data processing. However, there are considerable distinctions between the HCE special processor and the arithmetic-logical unit. The HCE special processor hardwarily performs every operation in contrast to a traditional PC, that provides software implementation of operations. According to hardware implementation, an algorithm of an operation is performed concurrently. If input data is entering continuously, then different time slices are being continuously processed, and all phases of the algorithm are being executed concurrently. The HCE contains N special processors, each of which consists of m_i concurrently operating computing cells. The computing cells hardwarily perform operations, and the set of concurrently functioning special processors hardwarily implement the algorithm.

Data exchange between the HCE special processors is realized according to a principle of register communication. The register channels, that are used for data transfer between the special processors and between the HCE peripheral cells considered as inputs of the field, are serial shift registers that can shift or branch data.

Data, represented in binary code, is moving at the clock rate through communication registers, adjusted within the HCE. In every cycle binary numbers successively, digit-to-digit enter the special processor, starting with less significant digits.

In every n_i cycles, where n_i depends on capacity of processed data, a new machine word enters the special processor. It should be noted, that owing to the principle of register communication the time of data passing through the special processors may be defined very precisely.

Thus, the HCE is adjusted to the algorithm of the solving problem on two levels. The first level is adjustment of a special processor hardware structure to an operation; the second level is structural mapping of the algorithm on the HCE by creating the register communication channels between the special processors SP_i (see Fig. 2.7).

2.3. A Mathematical Model of an HCE

An HCE structure consists of two levels:

- a level of a computing cell;
- a level of a special processor.

The structure chart of the HCE computing cell was described above. The HCE cell is connected with four adjacent cells. Let us designate inputs and outputs of the cell as a_{ij} and b_{ij}, and write the following formulas:

$$a_1(i,j) \leftarrow b_3(i,j-1);$$
$$a_2(i,j) \leftarrow b_4(i-1,j);$$
$$a_3(i,j) \leftarrow b_1(i,j+1);$$
$$a_4(i,j) \leftarrow b_2(i+1,j).$$

Data transit, i.e. data transfer to the outputs of the cell without processing, is performed concurrently with the operations, carried out in the ALU. Processing results, likewise data transit, may be transferred to one of four outputs of the cell. The 16-digit code, stored in the program register (see Fig. 2.2), adjusts the cell to performing of one or another operation and transit functions.

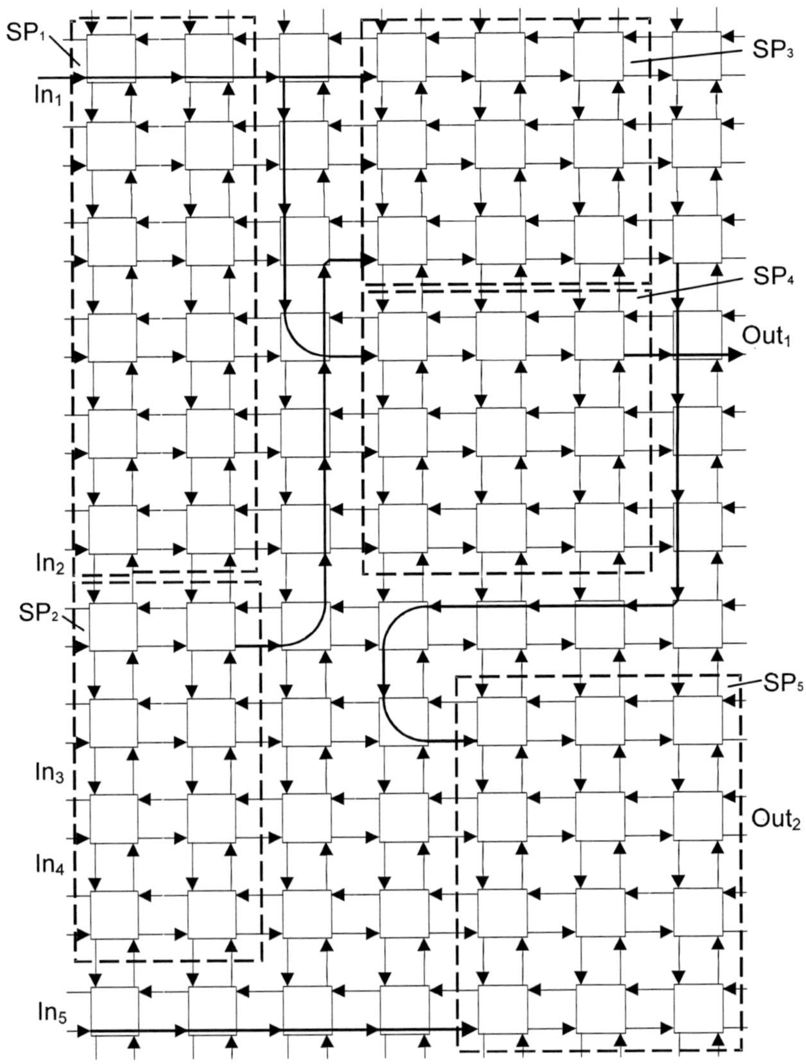

Figure 2.7. Implementation of a certain problem within the HCE field.

Digits 0–2 define the operation code; digit 3 defines the data delay in the ALU block, digits 4–15 define addressing of the input and output data streams. Data, stored in the program register of the cell, is defined during adjustment and remain unchangeable during the HCE operating, i.e. the cell is adjusted by the code $< f, A_1, A_2, A_3, B_1, B_2, B_3 >$, where f is the code of the operation of the cell, A_i are the adjusting codes of the inputs, B_i are the adjusting codes of the outputs.

It is necessary to distinguish internal and external inputs of the computing cell. Each cell is connected with four adjacent cells, i.e. it has four external inputs/outputs. There are three internal inputs/outputs in the cell: two inputs are used for the ALU operand entries and one input is used as a transit channel. The number of the cell outputs is quite the same: one ALU output and two outputs, used as transit channels.

A set $K_i = \{K_1, ..., K_n\}$ defines selection of data switching and data transit functions. The address code of the external input/output of the cell is defined by a 12-digit code, which is stored in the instruction register (digits 4-15):

$$\varphi(k) = < A_1, A_2, A_3, B_1, B_2, B_3 > .$$

The adjusting code consists of the adjusting codes of the ALU and the transit block.

Development of a microprogram module of any special processor starts from development of the algorithm of some operation that must be performed. The algorithm may be represented as a logical net or a graph, whose vertices correspond to the operations, included into the instruction set of the HCE cell. Then the weighted graph is mapped on the HCE field. The number of cells, required for the graph mapping, is minimized and hence a matrix of the HCE cells is formed. When the matrix is adjusted, it can carry out a certain arithmetic or logical operation. Consequently, the special processor that is adjusted within the HCE, may be described by a rectangular matrix, which consists of the elements $P_{i,j}$, represented as 16-digit binary codes. Each binary code defines the operation, performed by the cell, and adjusts the input and output switches of the cell:

$$P = \begin{pmatrix} P_{11} & \dots & P_{1m} \\ \dots & \dots & \dots \\ P_{n1} & \dots & P_{nm} \end{pmatrix},$$

$$P_{i,j} < f, A_1, A_2, A_3, B_1, B_2, B_3 >,$$

where f is the operation code;

A_1, A_2, A_3 are the codes, which define connections of the inputs of the operational and transit channels to the outputs of adjacent cells;

B_1, B_2, B_3 are the codes, which define connections of the outputs of the operational and transit channels to the inputs of adjacent cells of the HCE.

The structure chart of the special processor may be described by matrices F, Q_{in}, Q_{out}:

$$F = \begin{pmatrix} f_{11} & \cdots & f_{1m} \\ \cdots & \cdots & \cdots \\ f_{n1} & \cdots & f_{nm} \end{pmatrix},$$

where $f_{i,j}$ is the adjusting code of the cell with coordinates i, j,

$$Q_{in} = \begin{pmatrix} q_{11}(A_1, A_2, A_3) & \cdots & q_{1m}(A_1, A_2, A_3) \\ \cdots & \cdots & \cdots \\ q_{n1}(A_1, A_2, A_3) & \cdots & q_{nm}(A_1, A_2, A_3) \end{pmatrix},$$

$$Q_{out} = \begin{pmatrix} q_{11}(B_1, B_2, B_3) & \cdots & q_{1m}(B_1, B_2, B_3) \\ \cdots & \cdots & \cdots \\ q_{n1}(B_1, B_2, B_3) & \cdots & q_{nm}(B_1, B_2, B_3) \end{pmatrix},$$

and $q_{i,j}$ (A_1, A_2, A_3) and $q_{i,j}$ (B_1, B_2, B_3) are the codes, which define connections of the "internal" inputs and outputs of the cell.

The special processor, adjusted within the HCE, operates correctly, if operands enter precisely in time. The coordinates of the cells, used as the inputs of the special processor, and cycles of data entry are registered in *a special processor passport*, which is considered below. In addition, the passport contains the coordinates of the cells, that transfer data to the outputs of the special processor, and numbers of output cycles.

The HCE consists of several special processors, functioning concurrently. The information graph of the problem is hardwarily implemented in the HCE.

The graph vertices correspond to the special processors; the edges correspond to connections between the processors. The connections are implemented by means of the register channels, adjusted within the HCE.

To execute the program, developed according to the graph of the problem, the HCE is to be programmed with the help of two types of microprograms. The microprograms of the first type consist of adjusting instructions of a rectangular domain of the HCE cells, that perform some arithmetic, logical or "large" operation. These microprograms adjust the special processors that carry out certain operations. In addition, the microprogram set of the first type contains microprograms that adjust the cells to a data delay operation. The first type microprograms contain macroinstructions for adjustment of rectangular (in most cases) environment fragments, and are called *microprogram modules*.

The microprograms of the second type consist of microinstructions that adjust communication traces. The HCE uses register communication, according to which not buses, but register structures serve the purpose of data transfer from one special processor to another [13].

The register shift structures (or the register channels) consist of the same HCE cells, that are included in the special processor. All data, which is being processed in the HCE and is represented in sequential code, passes through the register channels at the clock rate. In addition to the processed data, auxiliary information such as service constants and an overflow digit may be transferred by the same register channels, adjusted in the HCE. In contrast to the microprograms of the first type, which adjust certain connected domains of the HCE, the microprograms of the second type contain adjusting microinstructions of the cell shift registers, used as the communication channels with turns and branching. The second type microprograms are called *trace microprograms*. In contrast to the first type microprograms (or microprogram modules), that may be stored in libraries, the second type microprograms are to be generated each time by special programs.

As long as the HCE adjustment in the multipipeline structure is performed line-by-line, i.e. the instruction registers of the cells, placed in one line, are sequentially connected with each other and form a "long" register, then it is necessary to write the correspondent adjusting information to the registers of the cells that are not included in any special processor, but may be used as data transit cells.

Adjustment of the "free" cells requires a special microprogram, which is developed exactly like other microprograms that adjust the HCE according to the problem algorithm. Then a special software tool analyses the adjusting microprogram modules and generates a file of adjusting information according to the adjusting rules of the HCE cells [87].

Figure 2.8 shows the structure of the HCE microprogram software.

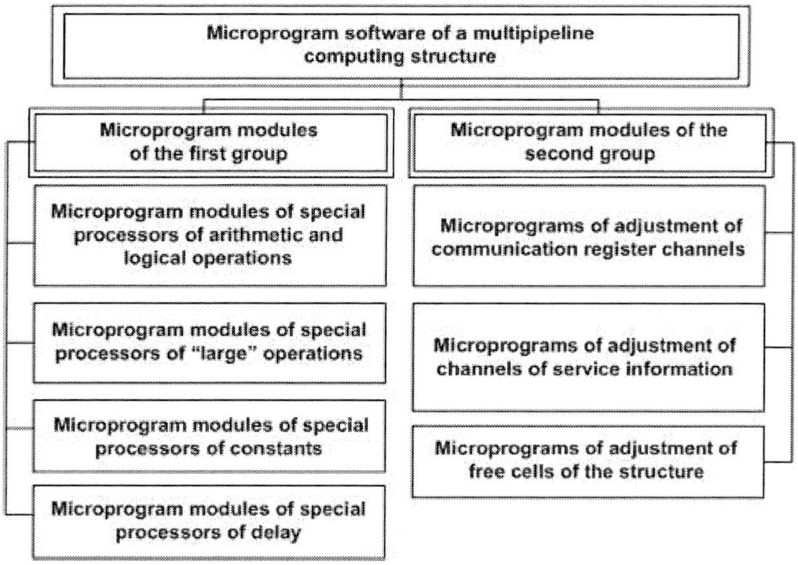

Figure 2.8. The structure of the HCE microprogram software.

2.4. Synthesis of Special Processors in HCEs

Synthesis of any special processor in the HCE consists of two steps. On the first step a microprogram module of the special processor or a so-called matrix of microinstructions is developed. The microinstructions of the microprogram module, that are placed in registers of the HCE computing cells, adjust a connected domain to a certain arithmetic or logical operation. The geometric configuration of the domain is uniquely specified by the microprogram module.

Let us analyse a development process of the microprogram modules of the special processors (see Fig. 2.9).

The structure chart of the special processor consists of blocks. A graph of a logical net is developed for each block according to communication capabilities of the cells. The graphs of the logical nets of the blocks are placed in the HCE. The debugged blocks of the special processor are assembled into the microprogram module, which adjusts the rectangular domain of the HCE to the operation.

The microprogram module of the special processor is added into the library of microprogram modules, that adjust the HCE to implementation of the solving problem.

One of approaches of optimal development of the HCE special processors is a method of principal axes balancing [46], according to which an information graph of an operation is mapped on a coordinate grid of the HCE. The graph edges are directed along the coordinate axes of the HCE, and the vertices match certain cells. According to the number of the graph edges, connected to the left and to the right sides of the cells that are placed on the principal axes, notions of an axis and a balanced axis are defined. For optimization the mapped graph fragments are replaced to balance the principal axes. As a result, the total length of the communication channels within the HCE is minimized, and the number of cells, needed for the operation, is also reduced.

The synthesized special processor is described by a passport of a special processor that usually contains four tables such as the table of special processor parameters, the table of inputs, the table of outputs, the table of transits and partition.

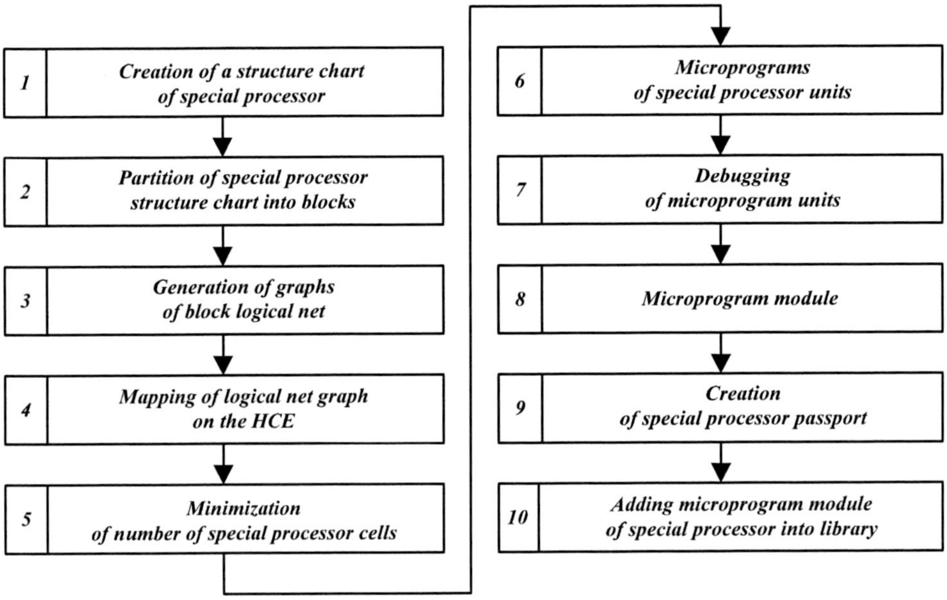

Figure 2.9. Development of the microprogram module of the special processor.

Let us make a detailed analysis of the tables. The table of parameters contains the following information.

1) *Mnemonics.* It specifies the operation, performed by the special processor, which is synthesized within the HCE.
2) *Data type.* It defines an encoding and a format of data, that may be processed by the special processor, synthesized within the HCE.
3) *Data size.* One of the important advantages of the HCE is the capability to process data of the size, required to obtain result of the needed accuracy. According to the attribute "Data accuracy", contained in the passport table, the required module is selected from the library of microprogram modules or is synthesized by a special program.
4) *Commutative property of inputs.* The special processor of binary operations has two inputs, which receive operands A and B. In the special processors of subtraction, division, comparison and some others, the operands A and B enter the inputs which are specially fixed for these operands.
5) *"Height" of the special processor.* Structural realization of the problem algorithm within the HCE is premised on the capability to choose special processors according to the operations and the information connections of the problem. The selected special processors of a certain geometric configuration provide the most effective use of computational resources, taking into account design features and restraints of the HCE.
6) *"Length" of the special processor.* The special processors that perform arithmetic or logical operations, contain different numbers of the HCE cells. The number of cells in certain special processors, for instance, in those that perform multiplication or division, is proportional to the size of processed data.
7) *Period of data entry.* The homogeneous computing structure provides the maximum performance, when operands continuously, digit-to-digit enter the inputs of the special processors at the clock rate. However, in certain cases, when computational resources are insufficient for a structural implementation of the algorithm with continuous data entry, the mode of decimated data entry may be used.

The table of inputs contains two values such as the coordinates of the cells, that receive operands or a service constant, and a number of cycles of operand entries in the special processor according to the unit digit, contained in the constant $K1$. The table of outputs contains the coordinates of the cells, that are used for data stream output, and the number of cycles needed for a certain operation to be performed.

Owing to a significant "length" of the special processors that perform "large" operations, or process floating-point data, or perform fixed-point multiplication and division, or process 32-digit data or 64-digit data, creation of the register communication channels within the HCE according to the information graph of the problem is complicated. Therefore, the transit channels are to be allowed to cross the special processor domains, placed within the HCE.

To allow crossing, a column of computing cells (so-called "transit cells") is added into the matrix of the special processor. The computing cells of the column are adjusted to transfer of data, obtained in the special processor, and to data transit. It is clear, that the "transit cells" cannot be arbitrary added into the structure of the HCE special processor without breaking synchronization of data streams. The "transit cells" may be inserted into such a part of the special processor domain, which has no feedbacks, i.e. no right-to-left data transfer is allowed. The coordinates of the special processor fragments, between which the columns of "transit cells" for extraneous data transit may be inserted, are obtained during development of the microprogram module of the special processor.

Information, contained in the tables, completely specifies functioning of the special processors and is used as a base for structural implementation of the algorithm within the HCE.

Table 2.2 shows an example of the passport of the special processor of subtraction, viewed in the subsection 2.2.

Table 2.2. The passport of the special processor of subtraction

Parameter		Value	
Format		F2	
Data size		32	
Commutative property of inputs		0	
"Height" of special processor		3	
"Width" of special processor		3	
Parameters of inputs			
Channels	Coordinates of cells	Addresses of inputs	Input cycles
A	3.1	00	3
B	2.1	00	3
K1	1.1	00	1
Parameters of outputs			
Channels	Coordinates of cells	Addresses of outputs	Output cycles
R	2.3	10	8
K1	1.3	10	6

2.5. The Structure of Microprogram Software

Microprogram software contains microprogram modules (MPM) of three levels:

- the MPMs of the special processors of arithmetic and logical operations;
- the MPMs of the special processors of "large" operations;
- the MPMs of standard problems.

The first level microprogram modules will be considered below.

The set of "large" operations $O = \{O_1, O_2, ..., O_m\}$ is to be universal enough to solve various problems. It is advisable for the operations to correspond with mathematical operations. At last, the algorithms of "large" operations are to be as simple as possible and must provide the required accuracy of calculations.

As it is noted in [34], it is impossible to select a suitable universal set of "large" operations. In contrast to the set of arithmetic and logical operations, which is not numerous, the number of "large" operations of a hypothetical universal set is about hundreds and thousands. Therefore each set of "large" operations depends on the problem domain such as equations of mathematical physics, linear algebra, digital signal processing, etc. In [1] it is suggested to add the following base operators into the set of "large" operations:

- functional operators that derive functional dependences;
- differential operators, applied to dependent and dependent variables;
- Riemann and Stilties integral operators;
- matrix operators that perform operations of addition, multiplication, transposing, etc.;
- vector and tensor operations.

The special processors of "large" operations may be synthesized on the base of microprogram modules of the special processors of arithmetic and logical operations.

A library of microprogram modules of standard problems, that are effectively solved on the HCE, is to be created just as the library of base microprogram modules of the special processors of "large" operations. If the library of microprogram modules contains a standard problem, it, by all means, also contains the base microprogram modules of the special processors of "large" operations, the information graph of the problem and its mapping on the HCE. The time of structural implementation of the problem of some class in the HCE may be enormously reduced, if a standard solution of problems from such class is already implemented.

All HCE microinstructions are connected not only by temporal relations, but by spatial relations as well. Each microinstruction of the microprogram specifies adjustment of a certain cell with certain coordinates in the module. When the microprogram module is placed on the HCE, the microinstruction gets an absolute address. When the HCE is adjusted, each cell gets a microinstruction, which specifies functions of the ALU block and functions of the transit block, that are to be performed by the cell during structural implementation of the algorithm. The functions of the cells remain unchangeable, until the HCE is adjusted to performing of a new algorithm.

2.6. HCE ADJUSTMENT

Figure 2.10 shows the logic diagram of HCE adjustment to structural implementation of an algorithm.

A program, written on the HCE programming language, is translated by a special software suit. One of the first and the most important steps of translation is creation of a weighted information graph of the solving problem, which gives a complete description of the problem, a base for structural implementation of the algorithm in the HCE. The vertices of the information graph correspond to the operators, and the edges – to connections between the vertices.

The next step is mapping of the weighted information graph on the HCE. The graph is to be mapped taking into account design parameters of the HCE such as the width of the environment field, the number of inputs, inhomogeneity of connections between its circuit boards, fast transit channels and other architectural features.

When the vertices of the information graph are specified and each vertex is put in correspondence with a certain arithmetic, logical or "large" operation, a mapping tool analyses the passport tables of the special processors, that contain all parameters of each special processor, adjusted in the HCE. The passport tables contain such parameters as geometric configuration of the special processor, the coordinates of its inputs of operands and a service constant, the coordinates of its output cells, etc.

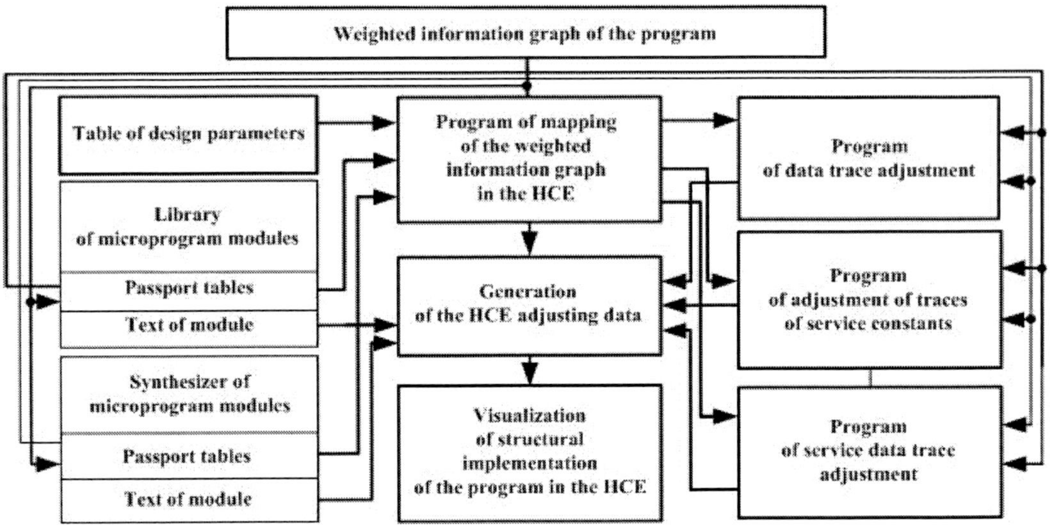

Figure 2.10. The logic diagram of HCE adjustment.

If implementation of the problem requires use of modules, generated by special programs, then the passport tables of correspondent special processors are also mapped on the HCE. Initial mapping of the information graph is performed according to the size of the special processors, adjusted within the HCE, and according to the number and distribution of the information graph edges, which specify data streams between the special processors.

After placement of the special processors, register communication channels are created and synchronized by adjusting and synchronization tools. These tools also use the passport tables of the special processors, that contain numbers of cycles in which operands enter the special processors. For each special processor the number of cycles is different and depends on the cycle, when the unit digit of the service constant enters the special processor.

Data synchronization tools can use delay units, that may require replacement of the special processors in the HCE [58]. After optimization, that provides effective use of the HCE computational resources, an array of the HCE adjusting data is generated. The general adjusting tool analyses the microprogram modules of the special processors, the microprogram modules of register communication trace adjustment, and the microprograms of the service constant and register channels, created for control of data processing.

The array of the HCE adjusting data is to be generated according to peculiarities of the writing mode of each special processor. For visual control of structural implementation of the algorithm and for control of effectiveness of computational resource use, graphic information is generated. Graphic information may be created in a printed form or displayed on the screen. Generally, it contains marked domains, adjusted to certain operations.

Problems of computer-aided programming of the HCE were analysed in [58-63].

2.7. HCE PROGRAMMING

2.7.1. Generation and Use of Constants

Constants, that are stored in a cell register or in registers of a group of cells, are widely used in development of the special processors, that perform arithmetic and logical operations. If GC0=1, then the constant is written into the cell register. If GC1=1, then the constant is written into registers of the group of cells. The constants are needed for obtaining a sign and an absolute value of a number, for extraction, nullification or digit propagation, etc.

According to microprogramming experience, rather limited variety of constants is usually used in the special processors, for instance: the constants with the only binary unit, or the constants, whose number of units depends on the size or format of data [64]. Typical constants are used in various special processors, that perform arithmetic and logical operations.

Let us analyze certain groups of 16-digit constants.

The constants, used for sign extraction and absolute value determination:

0 0 0 1 1 1 1 1 1 1 1 1 1 1 1 1
1 1 1.0 0 0 0 0 0 0 0 0 0 0 0 0

The constants with the only unit for extraction of a digit:

0 0 0 0 0 0 0 0 0 0 0 0 0 0 0 1
0 0 0 0 0 0 0 0 0 0 0 0 0 0 1 0
....
1 0 0 0 0 0 0 0 0 0 0 0 0 0 0 0

The gate constants:

0 0 0 0 0 0 0 0 0 0 0 0 0 0 0 1
0 0 0 0 0 0 0 0 0 0 0 0 0 0 1 1
....
0 1 1 1 1 1 1 1 1 1 1 1 1 1 1 1

The constants with the only zero for nullification of a digit:

1 1 1 1 1 1 1 1 1 1 1 1 1 1 1 0
1 1 1 1 1 1 1 1 1 1 1 1 1 1 0 1
....
0 1 1 1 1 1 1 1 1 1 1 1 1 1 1 1

The constants for digit propagation, containing two units, separated with an arbitrary number of zeros:

0 0 0 0 0 0 0 0 0 0 0 0 0 1 0 1
1 0 0 0 0 0 0 0 0 0 0 0 0 0 0 1

Special constants. A set of special constants contains all the rest 16-digit constants.

In the HCE it is possible to organize processing of data of an arbitrary size.

When synthesizing the special processors of arithmetic and logical operations, there are two variants of placement of the constants: inside and outside the special processor. As an example let us consider the special processor with a built-in constant, that performs comparison of two numbers in true representation.

The algorithm of comparison $|A|<|B|$:

1) negation of A;
2) determination of $|B|$;
3) determination of $|\overline{A}|$;
4) addition $|\overline{A}|+|B|$;

If a carry digit, appeared in the most significant digit is equal to unity, then the inequality $|A|<|B|$ is true. The carry digit is used as an attribute digit.

Figure 2.11 shows the structure chart of the special processor of “Comparison $|A|<|B|$” for the data size n<80. The attribute digit leaves the special processor in the cycle s, where s=17+n. Starting from the cycle s, the attribute digit is generated by the special processor with a period, equal to the size of every input number. Values of intermediate digits are arbitrary.

When the size of operands grows, the number of computing cells, whose registers store the constant needed for functioning of the special processor, also grows.

For automatic generation of the special processor which will process data of an arbitrary size (n≤80), the special processor of comparison $|A|<|B|$ contains a reserved group of cells, intended for storage of the constant of the required size. In the structure chart of the special processor (see Fig. 2.11) the reserved group of cells is marked with a dot line.

Addition of the group of cells for storage of constants in the structure of the special processor has considerable disadvantages. First of all, a significant number of the cells, needed for constants in each special processor, is lost. Besides, the greater is the size of processed data, the greater is the number of the lost cells.

2.7.2. External Supply of Constants into a Special Processor

The disadvantages of the special processors with built-in constants were mentioned above. They may be eliminated, if the special processor receives only one external constant which is used as the base for generation of all other required constants by the special processor itself [50]. The base constant, generated in a special processor of constants, is transferred through the HCE register channels to all special processors, placed in the HCE.

Use of the external base constant involves considerable modifications of special processor design. The special processor needs an auxiliary input for the base constant entry. The base constant is transferred unchanged to an auxiliary output of the special processor along with operation results.

Generation of the base constant and constants, needed for calculations, requires the individual constant processors, placed in the HCE along with the special processors of operations. The constant processor is regular and may generate any constant.

Figure 2.12 shows the structure chart of the special processor of comparison $|A|\leq|B|$ for n-digit fixed-point numbers in true representation.

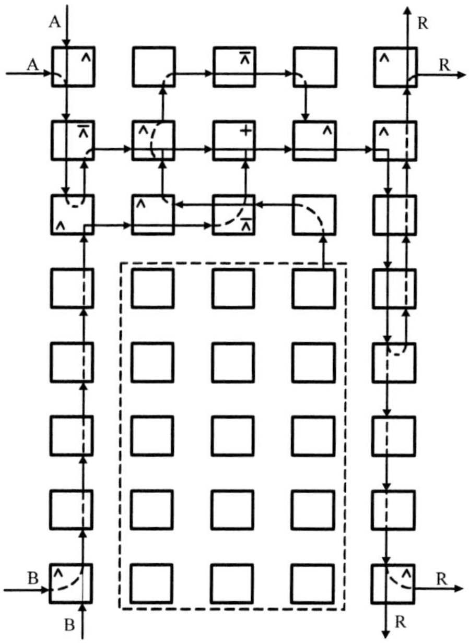

Figure 2.11. The special processor with a built-in constant.

In contrast to the special processor of comparison $|A| < |B|$ with a built-in constant, the special processor, shown in Fig. 2.11, receives the base constant $K1 = 0...01$ from the outside. As a result of external constant use, the special processor of $|A| \leq |B|$ |contains only 9 computing cells, and the size of the special processor does not depend on the size of processed data. The important property of independence of the special processor size from the size of processed data is also kept by the special processors, that perform other operations, for instance, addition.

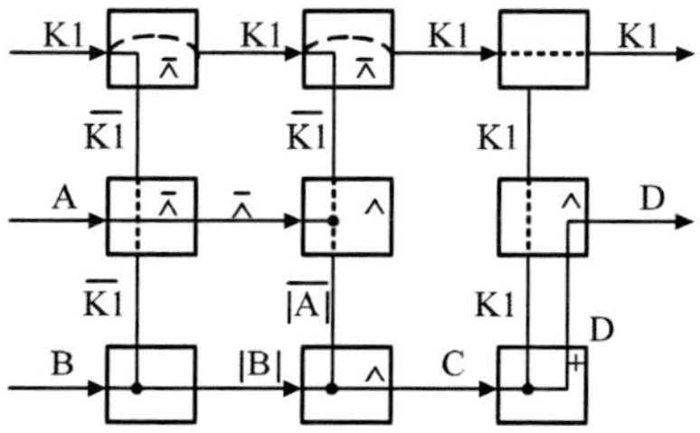

Figure 2.12. The special processor of comparison.

2.7.3. Formats of Operands

Data, represented in sequential code, enters the HCE starting from the less significant digit.

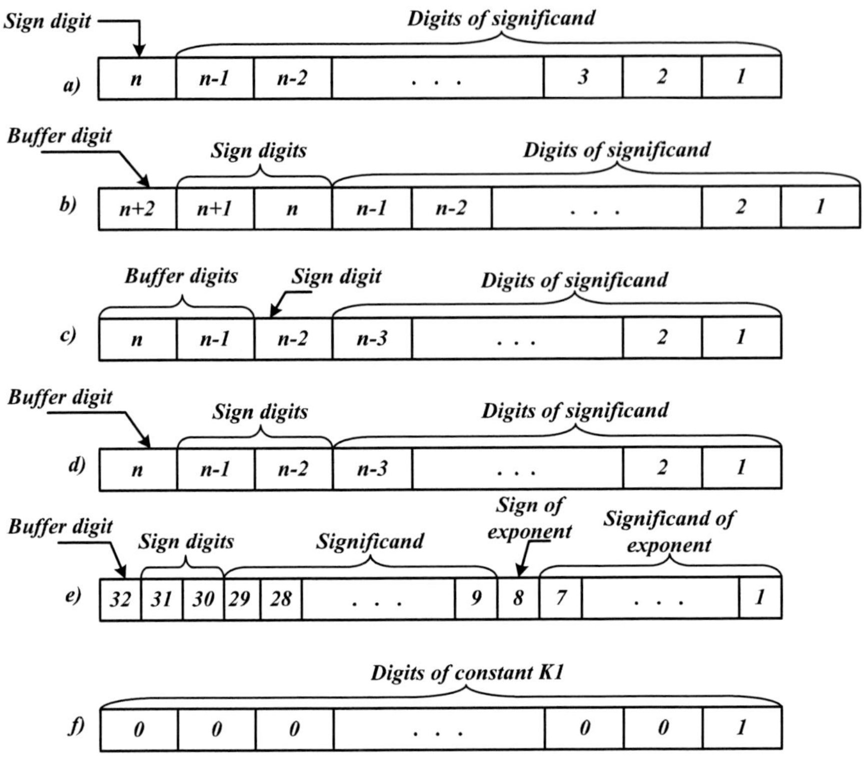

Figure 2.13. Formats of operands.

Figure 2.13*a* shows a format, that is usually used for binary fixed-point *n*-digit numbers in true representation. However, use of data, represented in the abovementioned format, meets significant difficulties during development of the special processors of arithmetic operations with continuous entry of data, represented in sequential code. Figure 2.13*b* shows a data format, that is effective for algebraic addition. To provide such data format with continuous entry of data, represented in the format shown in Fig. 2.13*a*, a machine word is to be paralleled inside the special processor. Before leaving the special processor, the paralleled streams are to be united into one data stream. However, the method of input data paralleling inside the special processor leads to rapid, more than double, increasing of the number of cells [60, 65].

It is possible to suggest another method, that does not lead to increasing of the number of the special processor cells. The method consists in addition of two buffer digits to the data format (see Fig. 2.13*c*). Hence, a number in true representation (format F1) is a sequence of the digits, in which two the most left digits are buffer digits; the next digit is a sign digit; all the rest digits contain a fractional part of the number. The point is placed between the digits 3 and 4. Numbers in modified radix complement representation (format F2) have one buffer digit, the next two digits are sign digits, and all the rest digits contain a fractional part of the number (see Fig. 2.13*d*).

Floating-point numbers (format FP) are represented in a 32-digit sequence: the most left digit is a buffer digit, the next two digits are sign digits; the next 21 digits contain a significand, the 20th digit is a sign of an exponent and 7 digits contain a significand of the exponent (see Fig. 2.13*e*). The formats F1 and F2 may contain an arbitrary number of digits, but base formats contain 8, 16 and 32 digits.

In addition to the formats F1, F2 and FP, the HCE may process other formats, for instance, the format of the constant $K1$, shown in Figure 2.13*f*.

2.7.4. The Special Processors of Arithmetic and Logical Operations

Owing to hardware implementation of operations in the HCE, use of codes and data formats, which provide minimum computational burden during performing of various operations, is crucial. In other words, the microprogram modules of the special processors of arithmetic and logical operations are to contain a minimum number of microinstructions. Each microinstruction, included in the module and identified by the coordinates (i, j), is the adjusting program of the cell with the same relative coordinates (i, j) in a certain HCE domain. The microinstructions of the microprogram module adjust the domain to hardware implementation of the special processor of some specific arithmetic or logical operation.

The absolute address of the adjusted cell, contained in the special processor *k* of the HCE, is determined by the coordinates $(p_k + i, q_k + j)$, where (p_k, q_k) are the coordinates of the left upper cell of the special processor *k*. The application programmer may define the coordinates (p_k, q_k), placing the special processor in the HCE according to the structurally implemented problem algorithm. However, the coordinates may be modified during automatic creation of register communication channels, synchronization of data streams and optimization.

Let us analyse the structures of the special processors, that perform arithmetic and logical operations in the HCE cells. The structure of development of microprogram modules of special processors was considered above.

The Special Processors of Addition and Subtraction

The special processors process data in modified radix complement representation (MRCR) in the whole range of legitimate values. Numbers enter the special processor in different cycles, and hence, optimal realization of an operation is provided. The cycles are numerated, starting from the cycle, when the constant $K1$ represented as 0...01, enters the special processor. The size of the constant is equal to the size of operands. The constant is transferred unchanged to the output of the special processor.

In some cases an overflow digit, obtained in the special processor, is to be transferred through register channels into peripheral cells of the HCE. However, variants of the special processors of addition and subtraction with no segment of overflow indication were developed. The special processors of addition and subtraction may process data of an arbitrary size, that is determined by the size of the constant $K1$.

The Operation of Addition

The numbers A and B, represented in the format F2 (modified radix complement representation $ozza_{n-3}...a_1$), enter the special processor of addition on cycles 1 and 2, respectively. A carry digit is added to the sum of A and B. Then a buffer digit is cleared by the negated constant $K1$. Figure 2.14 shows the structure chart of the special processor of addition with no overflow indication.

The Operation of Sign Modification

The operation negates a sign of a number. Figure 2.15 shows the structure chart of the special processor of sign modification.

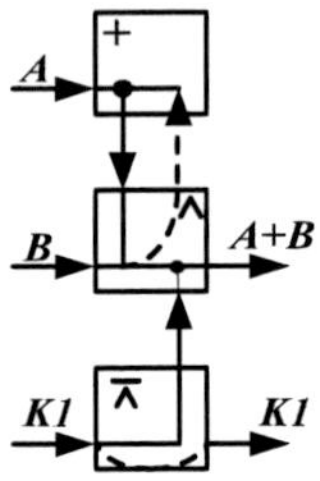

Figure 2.14. The special processor of addition.

The Operation of Subtraction

The operand B is modified according to the algorithm of sign modification. Then a carry digit is added to the sum of A and $(-B)$, and a buffer digit is cleared. The structure chart of the special processor of subtraction of data in modified radix complement representation is shown in the subsection 2.2.

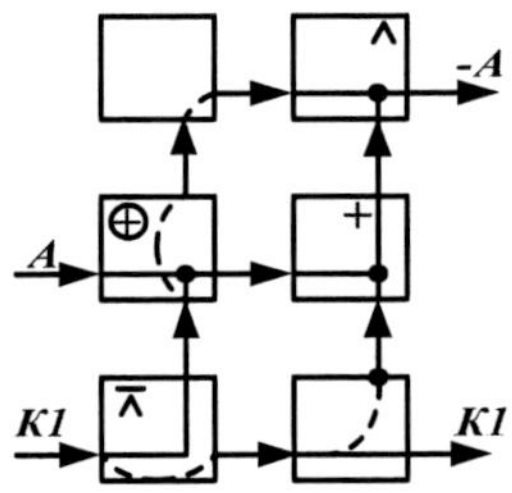

Figure 2.15. The special processor of sign modification.

The Operation of Overflow Addition

The special processor of addition with an overflow, processing RC data, performs the following algorithm:

1) bitwise overflow addition of operands;
2) extraction of the overflow;
3) generation of the two-digit sign of the addition result;
4) nullification of the buffer digit of the addition result.

Figure 2.16 shows the structure chart of the special processor of addition. The operand A enters the matrix cell with the coordinates (3, 1) on the second cycle; the operand B enters the cell (4, 1) on the first cycle. The constant $K1$ enters the cell (2, 1). The result of addition is obtained in the cell (2.3) on cycle 10.

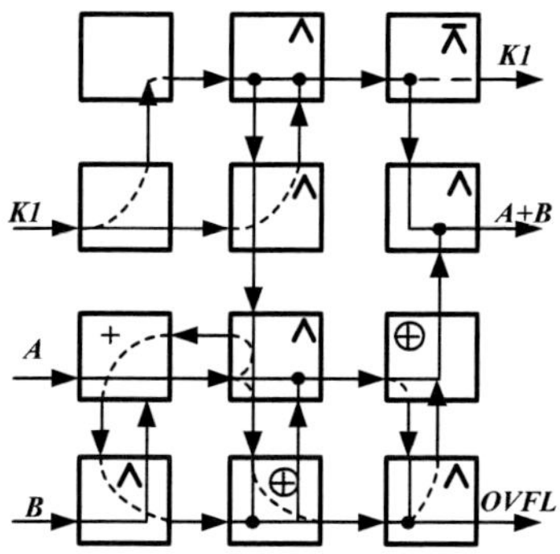

Figure 2.16. The modified special processor of addition.

The Special Processor of Multiplication

Data may enter the HCE continuously, one-by-one $(\rho = 1)$ and with intervals $(\rho < 1)$. If $\rho < 1$, then some special processors, for instance, the processor of multiplication, require a considerably less number of computing cells to perform the operation.

Let us consider the special processor of multiplication of fixed-point data in true representation with $\rho = 0.5$. The special processor of multiplication performs the following algorithm:

1) the sign of the product is obtained as a modulo 2 sum of the multiplicand sign and the multiplier sign. The sign of the product is written into the sign digit of the multiplier;
2) multiplication starts from the less significant digits. The digits of the multiplicand are logically multiplied by each extracted digit of the multiplier. Then the results of multiplication are summarized. After three additions the low-order digits of the accumulative sum are cleared by two buffer digits and one sign digit;
3) adjunct of the result sign to the product, accumulated in the block of addition.

Figure 2.17 shows the structure chart of the special processor of multiplication of 8-digit numbers in true representation with $\rho = 0.5$.

The size of the special processor is 4×11 cells. Operands enter the cells (4, 1) and (2, 1) on cycles 3 and 5. The service constant $K1 = 0...01$ enters the cell (3, 1) on the first cycle. Results of multiplication leave the special processor from the cell (4, 11) on cycle 26.

Let us consider the special processors of multiplication with continuous, one-by-one data entry $(\rho = 1)$. Multiplication is performed starting from the less significant digits. All digits of the multiplicand are multiplied by a fixed digit of the multiplier. Each digit of a partial product is a result of logical multiplication of the correspondent digits of the multipliers. The obtained partial products are summed up, taking into account a shift of every partial product.

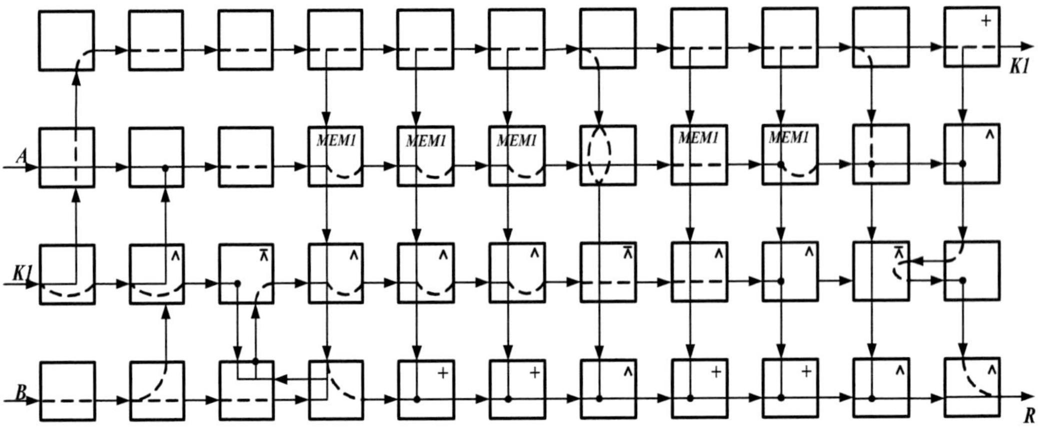

Figure 2.17. The special processor of multiplication $(\rho = 0.5)$.

Binary multiplication is effectively performed on numbers in true representation. The sign of the product is obtained as modulo 2 sum of the signs of the multipliers. To provide equality of sizes of input and output data, the result of the special processor of multiplication is rounded. The size of the matrix of the HCE cells, occupied by the special processor of multiplication of fixed-point numbers in true representation (F1) with continuous data entry, is 8×23 cells.

Multiplication of Fixed-Point Numbers, Represented in Modified Radix Complement Representation

When performing multiplication of data in modified radix complement representation (MRCR), input data is transformed in true representation, and output data is transformed in the RCR. Figure 2.18 shows the structure chart of the special processor of multiplication of 8-digit RCR numbers.

The multipliers *A* and *B* enter the special processor through the cells with the coordinates (3, 1) and (6, 1) on cycle 5. On the first cycle the constant $K1 = 00000001$ enters the cell (4, 1). On cycle 31 the constant $K1$ leaves the special processor of multiplication.

If multipliers continuously enter the special processor of multiplication, then on cycle 50 8-digit products, represented in format F2, appear on the output of the cell (4, 15).

The special processor of multiplication of 8-digit RCR numbers is a matrix of 8×15 computing cells.

The special processors of multiplication of 16-digit and 32-digit RCR numbers take 8×23 and 8×39 cells, respectively.

Analysis of the structure charts of the special processors, which perform multiplication of numbers of various sizes, shows that the size of the special processor is proportional to the size of input data.

An additional group of cells is to be inserted into the structure of the special processor to process one additional digit (see Fig. 2.19).

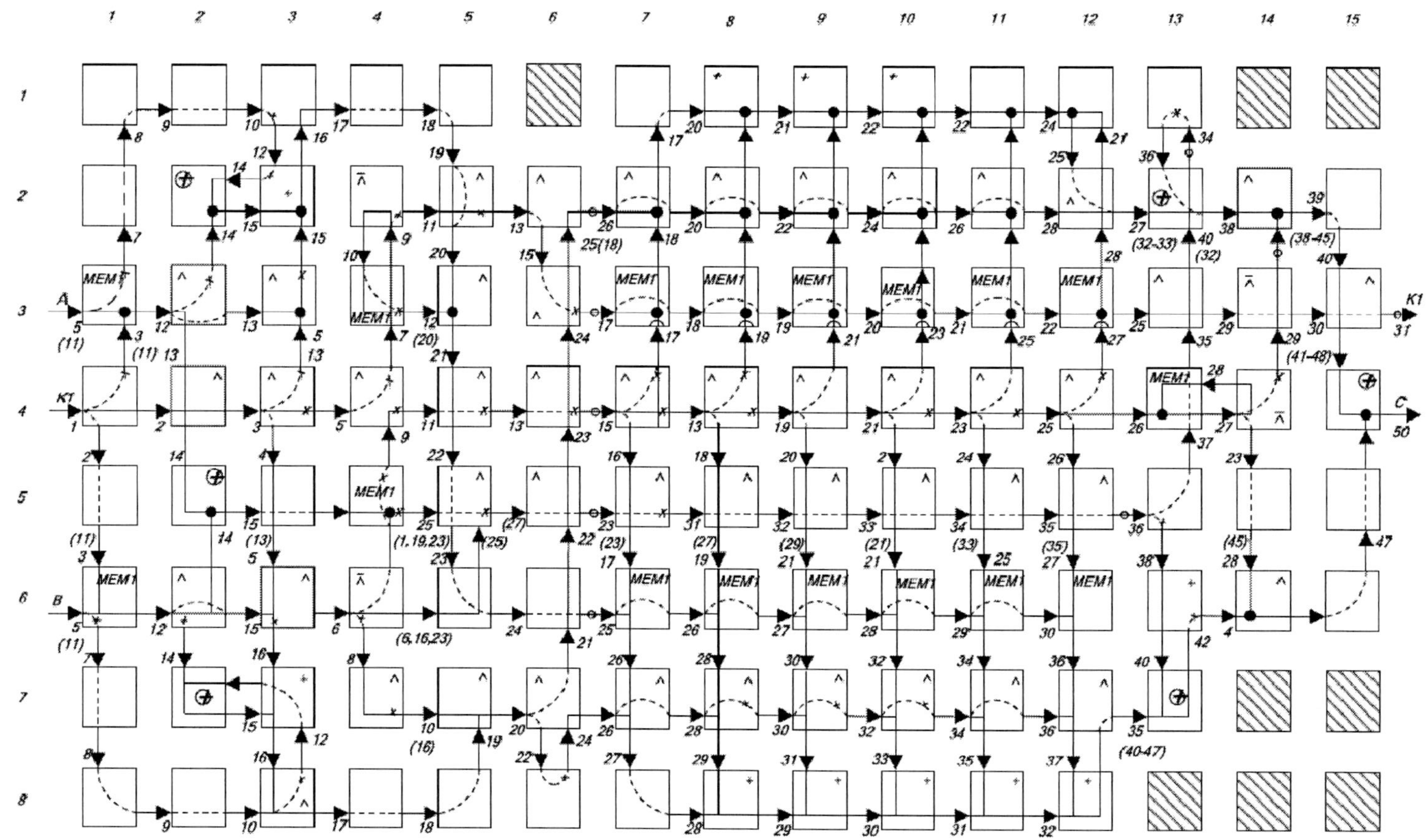

Figure 2.18. The structure chart of the special processor of multiplication.

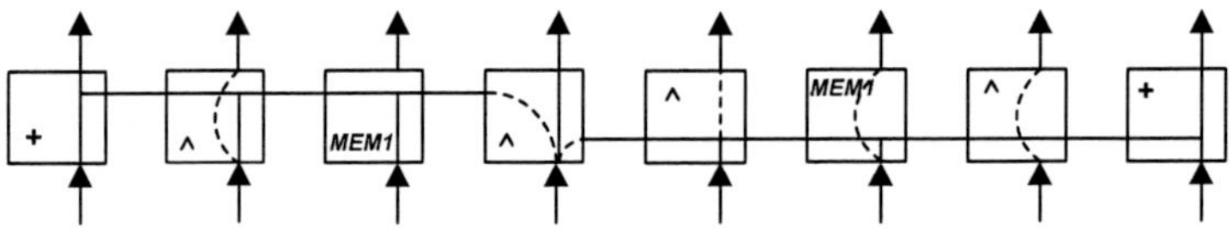

Figure 2.19. Processing of one digit in the special processor of multiplication.

As a result, the microprogram of the special processor of multiplication, processing numbers of an arbitrary size, may be synthesized automatically.

The Special Processors of Division

Division is the most complicated arithmetic operation and it requires much more time to be performed in comparison with other arithmetic operations. Therefore hardware implementation of division in the HCE requires considerable computational resources.

Modern computers usually perform division as a sequence of operations of subtraction and shift. Such algorithms as "Restoring division" and "Non-restoring division" became the most widely used. It turned out that implementation of the "Restoring division" algorithm in the HCE is the most effective, with low hardware and time costs.

On the first step of division of binary numbers, a divisor is subtracted from a dividend. If an obtained residue is not negative, then the first digit of a quotient becomes a unity. After that the obtained residue is shifted to the left by one digit, and the whole sequence of operations is repeated again.

If the residue is negative, then the first digit of the quotient becomes a zero. The divisor is added to the obtained residue and then the retrieved residue is shifted to the left by one digit, and the whole sequence of operations is repeated again.

The structure chart of the special processor of division of numbers in true representation may be conditionally split into two parts: constant and variable. The function of the constant part is to test the condition $|A| \geq |B|$, where A is the dividend and B is the divisor, and to transform the divisor in true representation into the RCR. The sign of the quotient is obtained as modulo 2 sum of the dividend sign and the divisor sign.

The variable part consists of a number of units which generate digits of the quotient. The number of units depends on required accuracy of the quotient.

Figure 2.20 shows the structure chart of the special processor of division of 8-digit numbers. A brief description of the structure chart is the following. The input data streams $\overline{A}$ and $\overline{B}$ consist of numbers in true representation in the format F1 $(ooza_{n-3}...a_1)$. A constant $K1 = 00000001$ enters the computing cell (5, 1). The overflow digit leaves the cell (8, 2) on cycle 23. The quotient appears in the outputs of the cells (1, 20) and (8, 20) on cycle 102. The constant $K1$ leaves the special processor on cycle 52.

The matrix of the special processor of division of 16-digit numbers takes 8x44 cells of the HCE. The first digit of the quotient leaves the cells (1, 44) and (8, 44) on cycle 326.

To obtain one digit of the quotient, a group of cells, shown in Figure 2.21, is required. The special processor of division of numbers of various sizes may be generated by adding groups of cells into the structure of the special processor. The number of the groups is determined by the size of processed data.

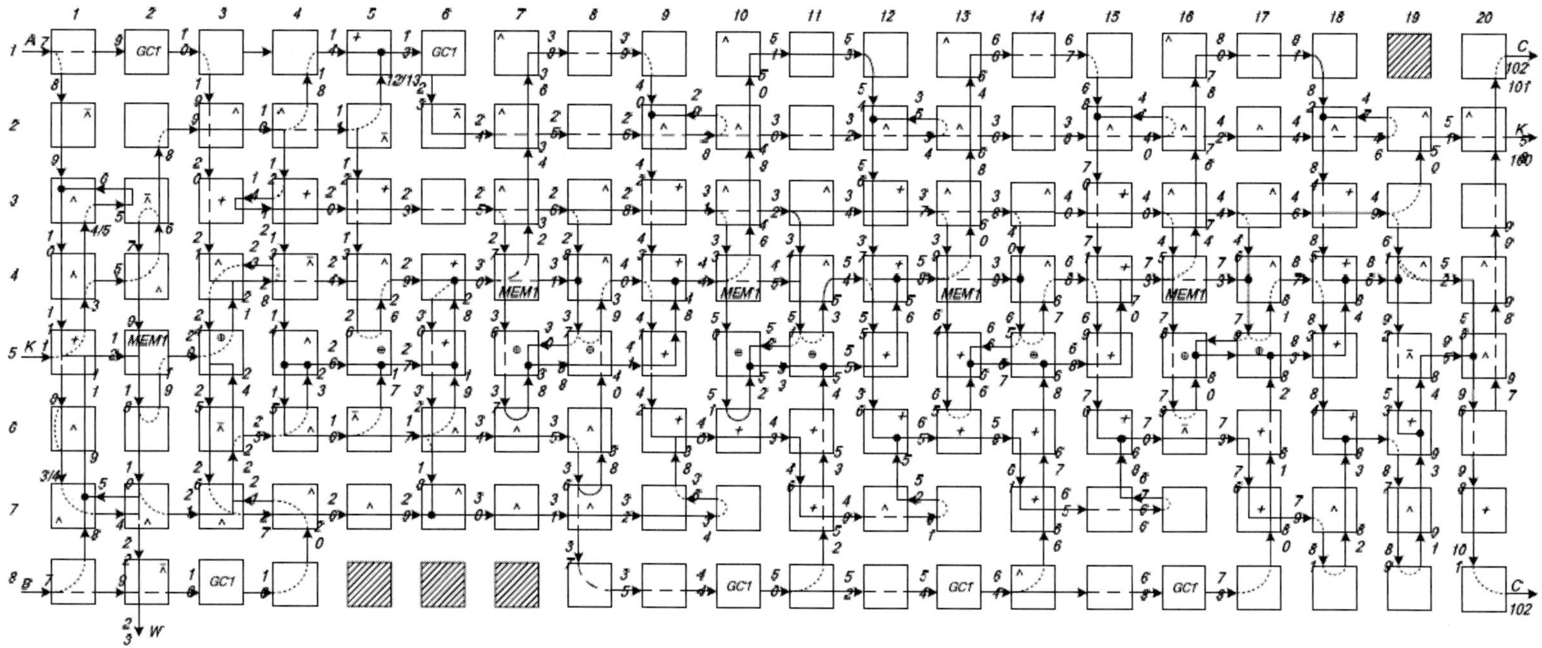

Figure 2.20. The special processor of division.

The special processors of comparison, logical and arithmetic shifts, code transformations and other operations are created in a similar way.

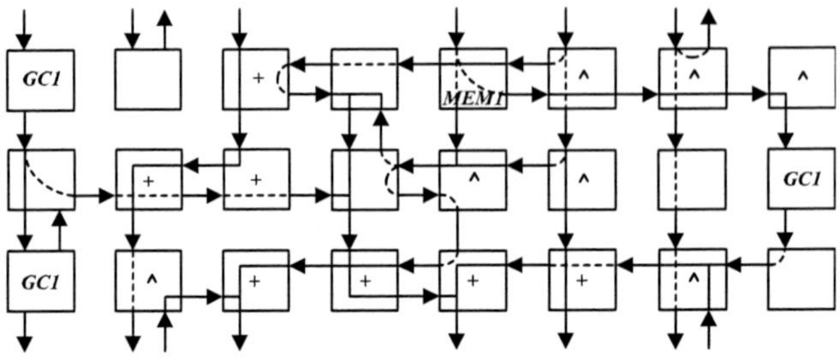

Figure 2.21. Obtaining of one digit in the special processor of division.

Table 2.3 contains parameters of the special processors of arithmetic and logical operations for fixed-point number processing. The third column contains the number of the HCE cells required to perform the operations. The last column of the table contains the number of cycles, needed to obtain operation results.

It should be noted that some operations, for instance, addition of two numbers, are carried out by the special processor, whose size does not depend on the size of processed data. The sizes of the special processors of multiplication, division and shift is proportional to the size of input data.

Table 2.3. Parameters of the special processors of arithmetic and logical operations for fixed-point data

№	Name of microprogram module	Size of module	Number of cycles
1	Addition	3×1	5
2	Addition with overflow	4×3	10
3	Subtraction	3×3	8
4	Subtraction with overflow	4×3	15
5	Multiplication of 8-digit numbers	4×10	27
6	Multiplication of 16-digit numbers	4×21	46
7	Multiplication of 32-digit numbers	4×44	87
8	Division of 8-digit numbers	8×23	93
9	Division of 16-digit numbers	8×46	308
10	Division of 32-digit numbers	10×94	1073
11	Comparison $A = B$	3×1	4
12	Comparison $A \geq B$	3×2	N+13
13	Comparison $A > B$	3×2	N+13
14	Comparison $A \leq B$	3×2	N+13
15	Comparison $A < B$	3×2	N+13
16	Comparison $A = 0$	3×1	N+13
17	Addition of absolute values	3×2	9
18	Subtraction of absolute values	4×5	24
19	Comparison of absolute values $\|A\| = \|B\|$	3×1	5

№	Name of microprogram module	Size of module	Number of cycles
20	Comparison of absolute values $\|A\| \geq \|B\|$	3×2	10
21	Comparison of absolute values $\|A\| > \|B\|$	3×3	10
22	Comparison of absolute values $\|A\| \leq \|B\|$	3×3	10
23	Comparison of absolute values $\|A\| < \|B\|$	3×3	10
24	Comparison of absolute values $\|A\| = 0$	2×2	N +6
25	Logical left shift of 8-digit numbers	6×7	17
26	Logical left shift of 16-digit numbers	6×9	20
27	Logical left shift of 32-digit numbers	6×11	23
28	Logical right shift of 8-digit numbers	6×7	27
29	Logical right shift of 16-digit numbers	6×9	40
30	Logical right shift of 32-digit numbers	6×12	61
31	Arithmetic left shift of 8-digit numbers	6×7	18
32	Arithmetic left shift of 16-digit numbers	6×9	21
33	Arithmetic left shift of 32-digit numbers	6×11	24
34	Arithmetic right shift of 8-digit numbers	6×7	27
35	Arithmetic right shift of 16-digit numbers	6×9	39
36	Arithmetic right shift of 32-digit numbers	6×13	61
37	Codes transformation	4×4	25

The Special Processors of Arithmetic Operations for Floating-Point Numbers

In the HCE a binary floating-point number is represented as follows: a significand of an exponent takes digits 1–7; digit 8 is a sign of the exponent; a significand of the number takes digits 9–29; a sign of the number takes digits 30-31; digit 32 is a buffer digit.

The algorithm of addition (subtraction) of floating-point numbers consists of the following steps: subtraction of exponents; generation of a shift constant for the number with the less exponent; arithmetic right shift of the significand with the less exponent; addition (subtraction) of the significands; normalization of the obtained significand; generation of the exponent of the result.

Implementation of floating-point arithmetic operations requires splitting of input data in two streams – a significand and an exponent – by a parallelization unit. Two operands enter the parallelization unit and four numbers are obtained on its outputs. The result numbers are the following: the first operand significand and the sign of the significand take digits 9-32, digits 1-8 are zero digits; the first operand exponent and the sign of the exponent take digits 1-8, digits 9-32 are zero digits; the significand of the second operand and the exponent of the second operand. The digits, which contain zeros, are the buffer digits. Let us consider the special processor of multiplication of floating-point numbers. Let us have two floating-point numbers: $A = M_1 \cdot 2^{P_1}$ and $B = M_2 \cdot 2^{P_2}$, where M_1 and M_2 are significands; P_1 and P_2 are exponents. The product of the numbers is $AB = M_1 M_2 \cdot 2^{P_1+P_2} = M_3 \cdot 2^{P_3}$.

Data enters the special processor of multiplication of floating-point numbers continuously, digit-to-digit, hence, the data stream is split as in the special processor of multiplication of fixed-point numbers. Implementation of multiplication of significands in the special processor of multiplication of floating-point numbers is similar to that in the special processor of multiplication of fixed-point data. Then overflow of the obtained exponent of the product is tested. If the exponent does not exceed the specific limit, then it is appended to the normalized significand. The size of the special processor of multiplication of floating-point numbers is 8×42 cells. When implementing division of floating-point numbers, the significand of the quotient is the result of division of the dividend significand by the divisor significand. The exponent of the quotient is the result of subtraction of the divisor exponent from the dividend exponent. The size of the special processor of division of floating-point numbers is 8×73 cells.

2.8. Implementation of Complex Arithmetic in HCEs

Plenty of important applied problems can be solved solely by means of complex arithmetic, which is constantly used for solving problems of such problem domains as theory of elasticity, ballistics, cartography, hydrodynamics and air dynamics, etc. It is impossible to solve problems of electrical engineering and radio engineering, digital signal processing without complex arithmetic. Therefore, the subsection is dedicated to implementation of base operations of complex arithmetic in the HCE. The following equations define operations, performed on complex numbers:

- addition $(A+Bi)+(C+Di)=(A+C)+(B+D)i$;
- subtraction $(A+Bi)-(C+Di)=(A-C)+(B-D)i$;
- multiplication $(A+Bi)(C+Di)=(AC-BD)+(BC+AD)i$;
- division $\frac{(A+Bi)}{(C+Di)}=\left(\frac{AC+BD}{C^2+D^2}\right)+\left(\frac{BC-AD}{C^2+D^2}\right)i$;
- comparison $A+Bi=C+Di$, if $A+C$ and $B=D$.

The algorithm of addition of complex numbers in the HCE contains the following steps:

1. bitwise addition $A+C$, and $B+D$;
2. extraction of a carry digit;
3. generation of a result with a sign digit;
4. clearing buffer digits.

Figure 2.22 shows the structure chart of the special processor of overflow addition of complex numbers.

The algorithm of subtraction of complex numbers contains the following steps:

1. negation of the constant $K1$;
2. transformation of the operands C and D in radix complement representation;
3. bitwise addition $A+C_{add}$, $B+D_{add}$;
4. extraction of a carry digit;
5. generation of a result with a sign digit;
6. clearing a buffer digit.

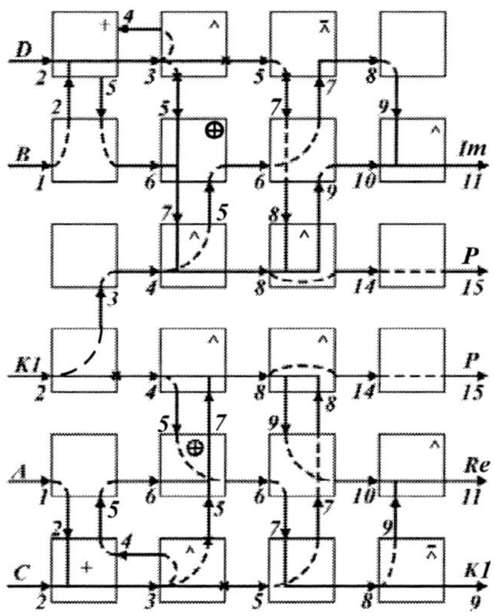

Figure 2.22. The structure chart of the special processor of overflow addition of complex numbers.

Figure 2.23 shows the structure chart of the special processor of subtraction of complex numbers.

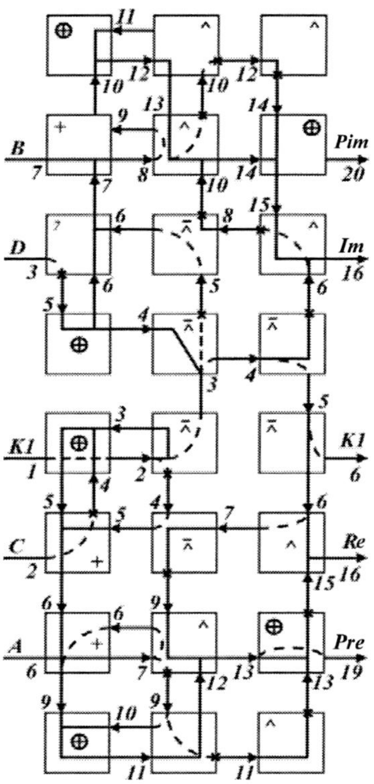

Figure 2.23. The structure chart of the special processor of subtraction of complex numbers.

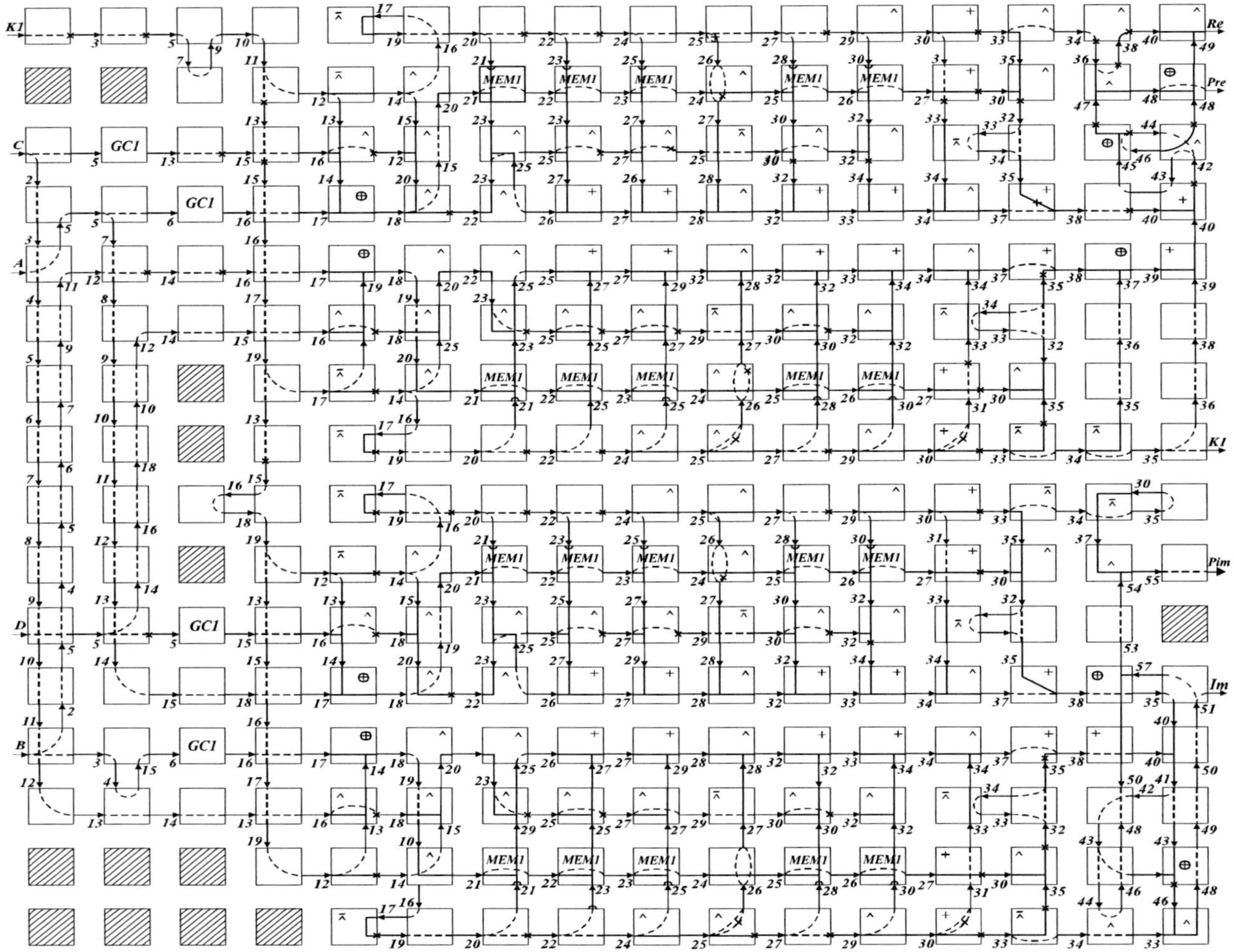

Figure 2.24. The structure chart of the special processor of complex multiplication.

The algorithm of multiplication of complex numbers is more complicated:

1. extraction of the sign of the operands C and B;
2. determination of the product sign;
3. obtaining absolute values of C and D;
4. obtaining partial products;
5. generation of the result with the sign digit;
6. transformation of R_{BD} into radix complement representation;
7. bitwise addition $R_{AD} + R_{BC}$;
8. bitwise addition $R_{AC} + R_{BD_{add}}$;
9. extraction of carry digits of imaginary parts;
10. clearing buffer digits.

Figure 2.24 shows the structure chart of the special processor of complex multiplication.

2.9. Implementation of Image Processing Algorithms in HCEs

Implementation of problems of image processing demonstrates all advantages of multimicropipeline processing on the HCE.

Majority of image processing algorithms uses apertures or windows with the size from 2×2 to 9×9 pixels. It is known, that HCEs may provide extremely effective "window" image processing [66]. Videoprocessors [67] and highly-performance image processing complex for VLSI control [70] were designed on the base of the HCE. The works [52, 66] give the detailed description of the library of adjusting microprogram modules of special processors, that perform basic algorithms of image processing, including algorithms of edge definition and contrast increasing, algorithms of threshold noise reduction, linear and non-linear filtration within the HCE.

Let us analyze examples of HCE-based implementation of tradition algorithms of image processing such as an algorithm of contrast increasing, algorithms of noise reduction by means of various filters, algorithms of spectral transformations, image measure of image similarity, edge definition, etc.

2.9.1. Pseudomedian Filtration in the HCE

Pseudomedian filtration is a method of non-linear image processing, based on obtaining of order statistics. As it is known, the method of median filtration requires generation of variational series of N^2 elements. The method of pseudomedian filtration requires generation of variational series of N elements, and hence, considerably reduces hardware cost. The algorithm of pseudomedian filtration consists of the following steps. At first, a median of every row of sliding window is obtained. Then, a new median is obtained on the base of N values of the previously obtained median.

Figure 2.25 shows the structure chart of pseudomedian filtration.

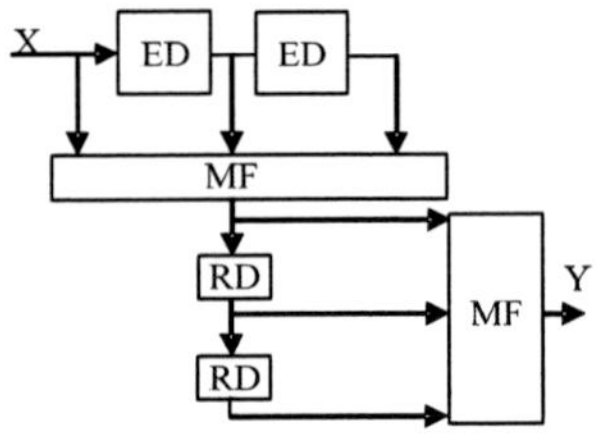

Figure 2.25. The structure chart of pseudomedian filtration.

The structure chart has the following designations:

- *ED* is the delay of one image element;
- *RD* is the delay of a row of image elements;
- *MF* is the unit, which obtains the median of three image elements.

Figure 2.26 shows the structure chart of the unit, which determines the median of three image elements, based on a sorting net.

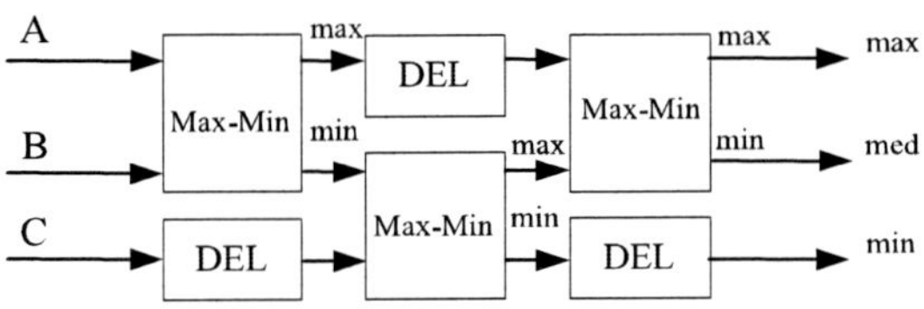

Figure 2.26. The structure chart of the median unit.

The structure chart has the following designations:

- *Max-Min* is the unit of two numbers ordering;
- *DEL* is the unit of delay on a number of cycles, required to perform ordering.

The special processors, which carry out the operation of search for the maximum of two numbers, may be used for integer and fixed-point data processing. The input data of the special processors are two operands A and B, and the constant $K1 = 00000001$, which enters on the first cycle. Figure 2.27 shows the structure chart of the special processor of $\max(A, B)$ for 8-digit operands.

The algorithm of $\max(A, B)$:

1. synchronization of $K1$ with B;
2. negation of $K1$;
3. negation of B;
4. addition of unity and the less significant digit of B;
5. addition $A + B_{add}$.

The condition $A \geq B$ is true, if the most significant digit of Z is equal to zero. If $Z = 1$, then $A < B$.

Figures 2.28*a* and 2.28*b* show the structure charts of the special processors of max(A, B) for 16-digit operands and operands of an arbitrary size. The algorithm of the operation max(A, B) for data with an arbitrary size is similar to that for 16-digit data, but the operands enter the microprogram module twice. The operand A enters the module on cycles 6 and *n*, and the operand B – on cycles 3 and *n*. The special processor of search for the maximum of two numbers takes 20 cells of the HCE.

The algorithm of search for the minimum of two numbers slightly differs from the algorithm of search for the maximum.

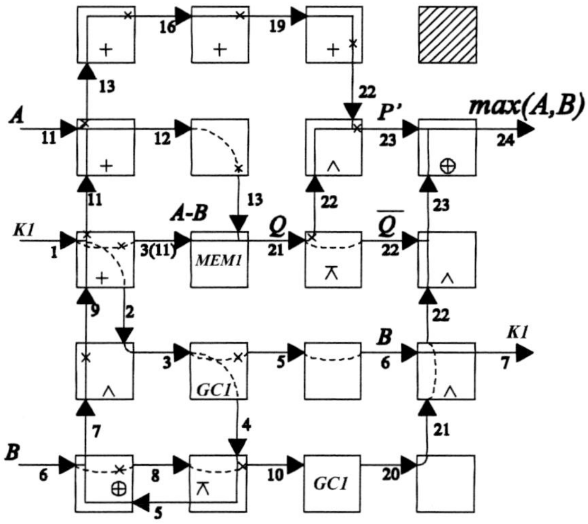

Figure 2.27. The special processor of max(A, B).

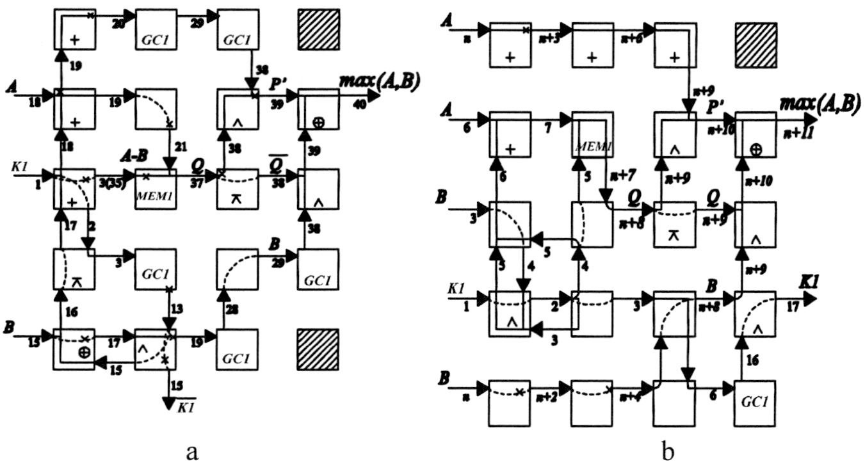

Figure 2.28. The structure charts of the special processors of max(A, B).

Figures 2.29*a* and 2.29*b* show the structure charts of the special processors of $\min(A, B)$ for 8-digit and 16-digit operands.

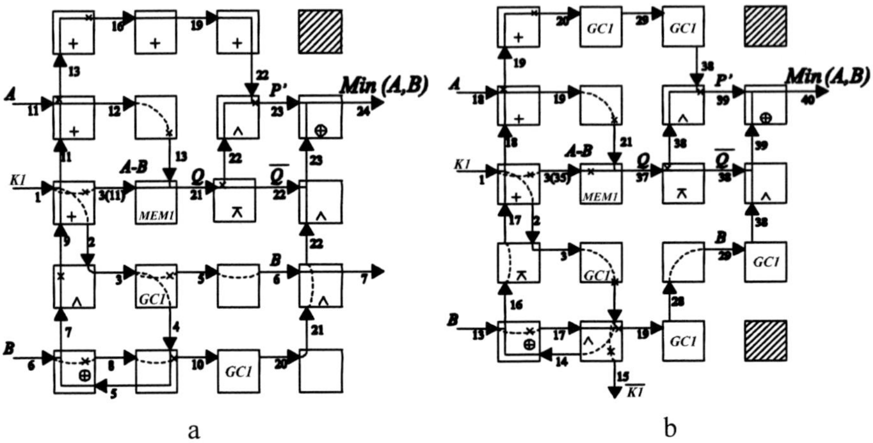

Figure 2.29. The structure charts of the special processors of $\min(A, B)$.

The modules, that calculate both $\max(A, B)$ and $\min(A, B)$, are used in many cases of HCE-based implementation of image processing algorithms. Figures 2.30*a* and 2.30*b* show the structure charts of the special processors of the ordering operation for 8-digit and 16-digit operands.

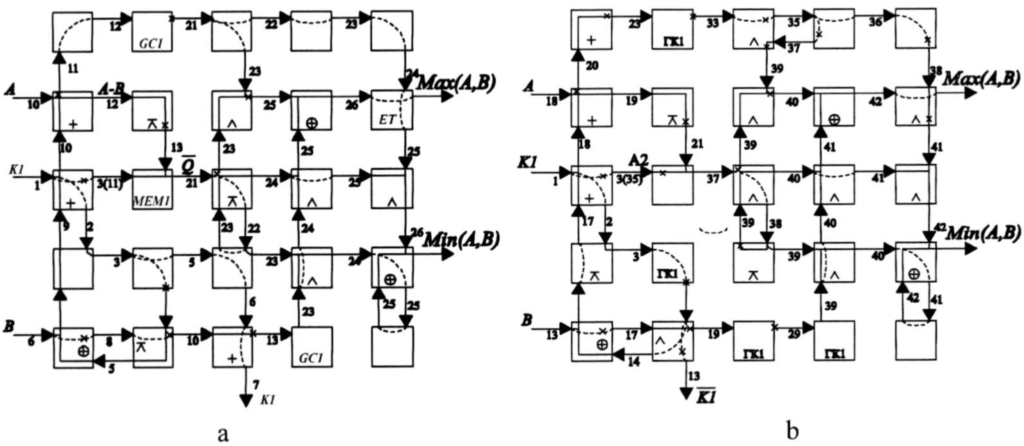

Figure 2.30. The structure charts of the special processors of $\max(A, B)$ and $\min(A, B)$.

2.9.2. Algorithms of Image Contrast Increasing in the HCE

Rather often methods of image processing use simplified object models, that do not involve a considerable information loss. One of such methods consists in processing of image edges, not image points. The edges contain plenty of information, weakly depend on colour and brightness, and do not depend on image generator type or on its frequency range. The edge completely defines an image form and contains complete information needed for image recognition according to its form. Owing to the edge methods it is possible to avoid analysis of inner points of a two-dimensional image and to considerably reduce amount of processing data. Hence, computer systems may provide practically a real-time problem solution. In contrast to bitmap methods, the methods of edge analysis, in majority of cases, allow use of models, that are invariant to image moving, rotations and zooming. The algorithms of edge definition and image contrast increasing widely use differential operators of Roberts, Laplace, Prewitt, Sobel, Kirsch, etc. [52,70-73]. For creation of modules, which perform these operators in the HCE, the operational processors, considered in subsection 2.7.4, are used.

The Roberts Operator

Edges and brightness jumps are determined by means of non-linear filters. The Roberts operator is the most simple:

$$G_R = |A_2 - A_3| + |A_4 - A_1|,$$

where A_1, A_2, A_3, A_4 are the image elements of the sliding window of 2×2 size.

The functional diagram of the Roberts operator, shown in Figure 2.31, consists of the following units:

- a sorting unit;
- a unit of conjugation;
- a unit of integer numbers subtraction;
- a unit of addition of absolute values of two integer numbers.

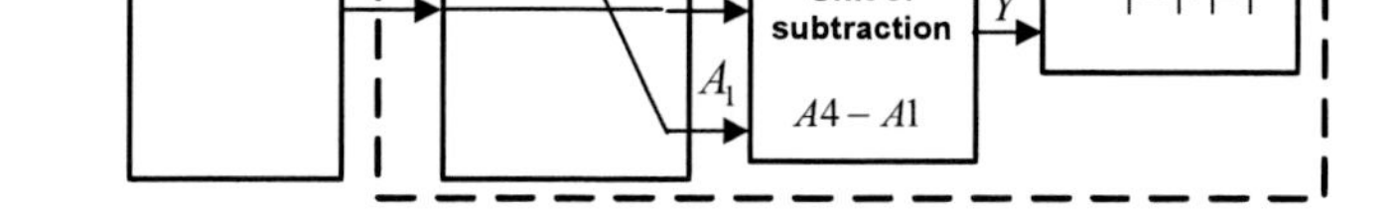

Figure 2.31. The functional diagram of the Roberts operator.

Figure 2.32 shows the structure chart of the special processor, which performs the Roberts operator.

The sorting unit selects the image elements A_1, A_2, A_3, A_4 of the sliding window from a data stream X_i and transfers them into the unit of conjugation, which synchronizes the streams of the image elements with the constant $K1$. The units of subtraction calculate differences of the pair values A_2, A_3, and A_1, A_4. The unit of addition sums absolute values of the data streams X, Y, i.e. carries out the operation $G_R = |A_2 - A_3| + |A_4 - A_1|$.

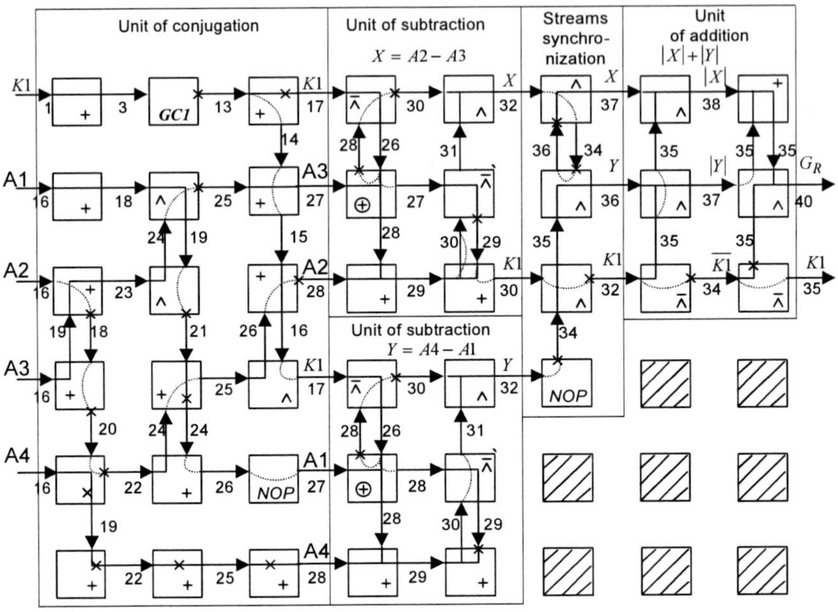

Figure 2.32. The structure chart of the special processor, which performs the Roberts operator.

Implementation of the special processor of the Roberts operator takes 40 computing cells of the HCE.

The Laplace Operator

The Laplace operator is written as follows

$$G_L = \left| A_5 - \frac{1}{5}(A_1 + A_2 + A_3 + A_4 + A_5 + A_6 + A_7 + A_8 + A_9) \right|,$$

where $A_1 - A_9$ are the elements of the sliding window of 3×3 size.

Figure 2.33 shows the functional diagram of the Laplace operator.

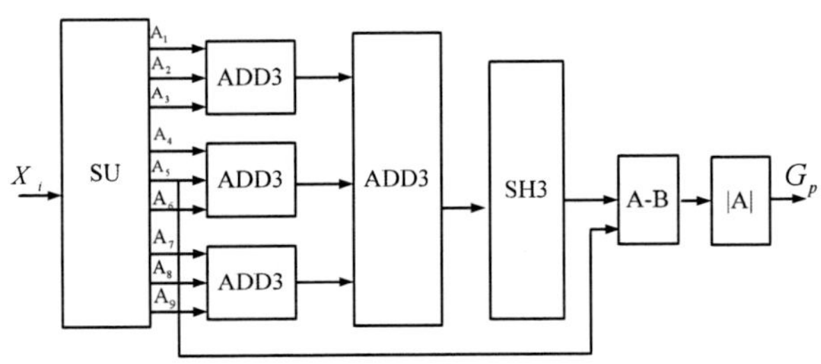

Figure 2.33. The functional diagram of the Laplace operator.

The functional diagram contains the following units:

- *SU* is a sorting unit;
- *ADD3* is a three-input adder;
- *SH3* is a unit of three cycles right shift;
- *A-B* is a unit of subtraction of two numbers;
- $|A|$ is a unit of the absolute value of *A*.

Figure 2.34 shows the structure chart of the special processor of the Laplace operator, which takes 4×14 computing cells of the HCE.

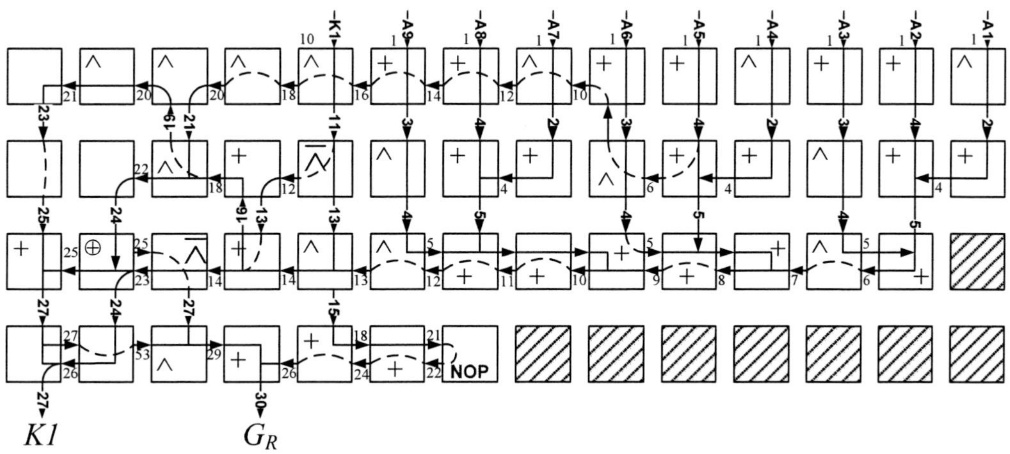

Figure 2.34. The structure chart of the special processor of the Laplace operator.

As may be seen from Fig. 2.34, 8 cells remain wasted and may be used for transit of data streams in the HCE.

The Prewitt Operator

The Prewitt operator is written as follows

$$G_P = |X| + |Y|,$$

where $X = (A_1 + A_4 + A_7) - (A_3 + A_6 + A_9)$, $Y = (A_3 + A_2 + A_1) - (A_9 + A_8 + A_7)$;
$A_1 - A_9$ are the elements of the sliding window of 3×3 size.

Figure 2.35 shows the structure chart of the special processor of the Prewitt operator. As may be seen from the figure, the special processor of the Prewitt operator takes 9×11 computing cells of the HCE.

The constant $K1$ is used in the special processor of the Prewitt operator. The data $A_1 - A_9$, represented in sequential code, enters the special processor on cycle 10. The result G_P leaves the special processor on cycle 56.

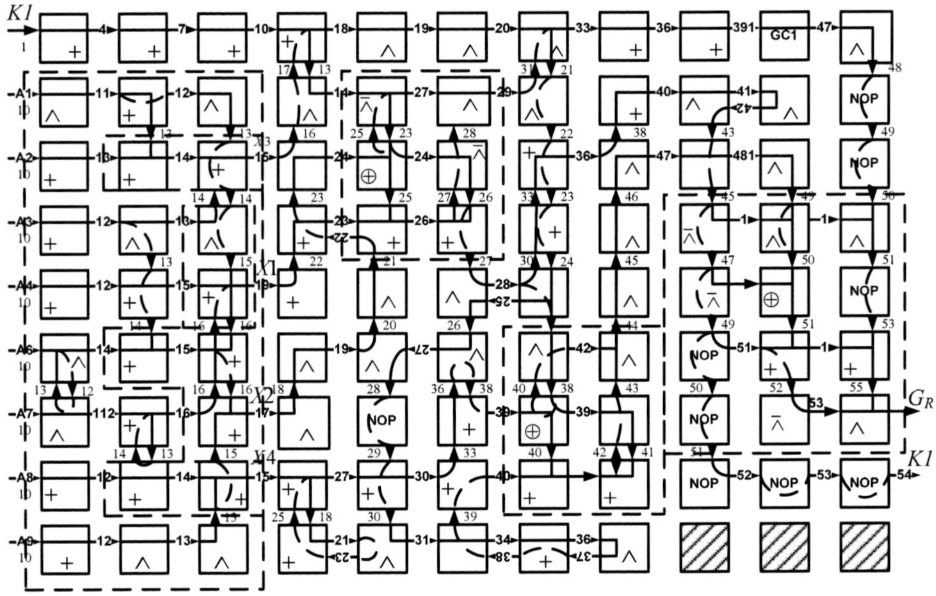

Figure 2.35. The structure chart of the special processor of the Prewitt operator.

The Sobel Operator

The Sobel operator is written as $G_S = |X| + |Y|$, where

$$X = (A_1 + 2A_4 + A_7) - (A_3 + 2A_6 + A_9),$$
$$Y = (A_3 + 2A_2 + A_1) - (A_9 + 2A_8 + A_7).$$

$A_1 - A_9$ are the elements of the sliding window of 3×3 size.

The distinctive feature of the algorithm is calculation of the image elements of the sliding window. In addition, data streams are branched and delayed to synchronize input data streams and calculations.

Figure 2.36 shows the functional diagram of the Sobel operator. Tables 2.4 and 2.5 contain the passport of the special processor of the Sobel operator.

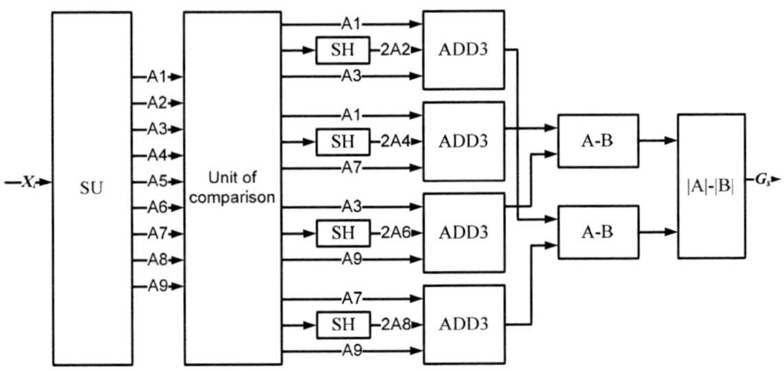

Figure 2.36. The functional diagram of the Sobel operator.

Table 2.4. The passport of the special processor of the Sobel operator

Parameter	Value
Format	F2
Data size	8
"Height"	14
"Width"	12

Table 2.5. The passport of the special processor of the Sobel operator

Channels	Cells	Input/output addresses	Input/output cycles
A1	3.1	00	1
A2	4.1	00	1
A3	5.1	00	1
A4	6.1	00	1
A6	8.1	00	1
A7	9.1	00	1
A8	10.1	00	1
A9	11.1	00	1
C1	1.1	00	1
Gc	14.12	10	55
C1	14.9	11	49

The structure chart of the special processor of the Sobel operator is not shown because of its large size.

2.9.3. Implementation of a Threshold Noise Reduction Algorithm for Image Processing in the HCE

Let us analyze computing capabilities of the HCE, taking implementation of an image threshold noise reduction method as an example. The sense of the method is the following. An image is processed by a sliding square window of $N \times N$ elements. The window is moving from left to right and top-down with the step of one image element. An average brightness of a group of window elements, placed in neighborhood of the central element, is obtained at every step. New values of brightness of the central element X_c will be obtained according to the algorithm

$$X_c = \begin{cases} X_{av}, \text{ if } X_c < \varepsilon \\ X_C, \text{ if } X_c \geq \varepsilon \end{cases} \quad (2.1)$$

$$X_{env} = \frac{\sum_{i=1}^{N}\sum_{i=1}^{N} X_{ij} - X_{\left]\frac{N}{2}\right[,\left]\frac{N}{2}\right[}}{N \times N - 1} \quad (2.2)$$

where ε is a certain threshold value, $\left]\frac{N}{2}\right[$ is an integer value.

Let an input signal be a stream of the image elements, which are represented in sequential code and enter the computing structure starting from the less significant digit. Brightness of each element is set by 16 binary digits. The rate of data entry is equal to the HCE clock rate.

Implementation of window methods of processing provides for synchronization of a certain number of rows of the input image. The HCE is adjusted to data stream paralleling with a row shift of one stream about another. The program, which performs the data processing algorithm according to (2.1) and (2.2), may be placed into the HCE. The size of the sliding window may be 3×3, 5×5, 7×7 or 9×9 elements. The HCE also contains a program, that calculates the average brightness of the central element X_c within the window, and a program of definition of image edges. Figure 2.37 shows the functional diagram of the special processor of image processing in the HCE. In Figure 2.37 the microprogram module 1 sums up the values of brightness of the image elements within the window 3×3; the module 2 – within the window 5×5; the module 3 – within the window 7×7; the module 4 – within the window 9×9. The results of the modules 1÷4 enter the switch SW_1, which, depending on the window size, is adjusted to receiving of one of four sums of brightness of the window elements. The value $S_{N\times N}$ leaves the switch SW_1. Then the brightness of X_c is obtained. The result enters units of division.

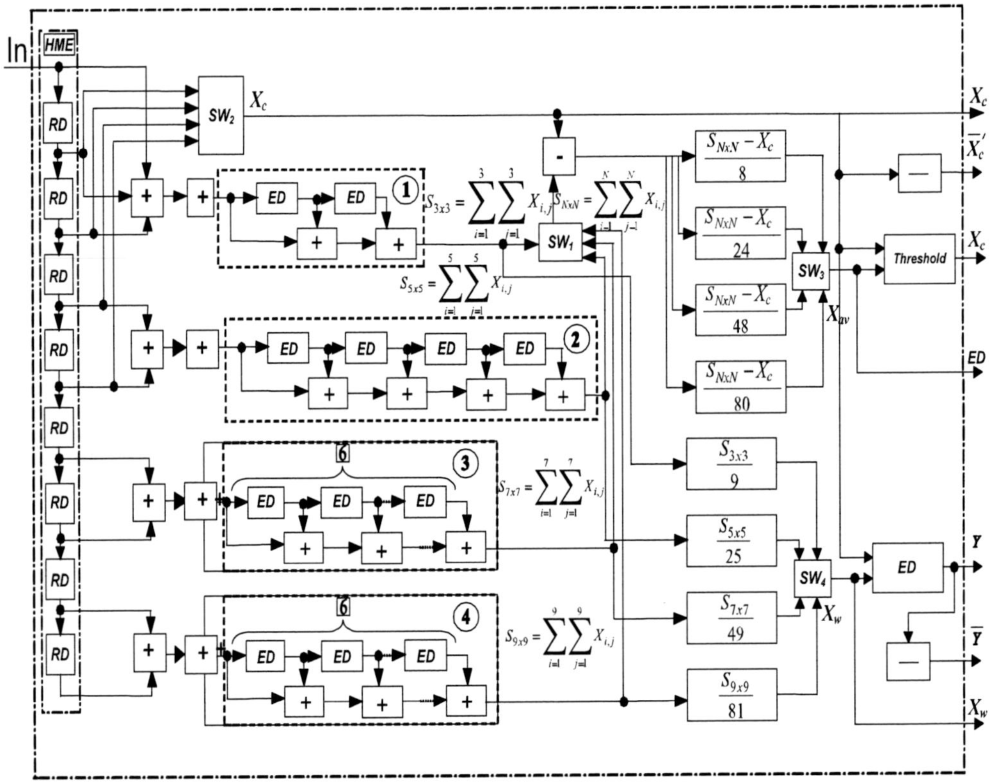

Figure 2.37. The functional diagram of the image processing algorithm.

The output streams enter the switch, which generates the averaged brightness X_{av} in the neighborhood of the central element in the $N \times N$ window. The value X_{av} enters the unit of threshold function, which also gets the brightness of X_c and the threshold values of the filter ε.

The output value of the switch SW_4 is X_w, which corresponds to an average brightness of an element within the window. Then an edge element is obtained according to the values X_w, X_c and ε_k. All units of division are simplified in comparison with that that is described in [54], because a divisor is a priori unknown.

The developed computing structure has 8 parallel outputs. The performance of the computing structure may be defined as the number of image elements, processed per unit of time, which depends on the clock rate F, the input data size r and the number of simultaneously performed functions. Figure 2.38 shows the structure chart of the special processor of image processing in the HCE.

The considered computing structure is the simplest example, which demonstrates wide computational capabilities of HCEs.

2.10. Development of Digital Filters in HCEs

Owing to maximally possible parallelization, the HCE special processors provide the potentially maximal performance and solve various real-world problems of digital signal processing. Different examples of HCE-based implementation of digital difference filtration devices [55, 74-76] are considered in this subsection. The filters are designed on the base of special processors, that carry out basic operations and process input signals at the clock rate of the HCE.

The algorithms, analyzed in [77], are the base of filtration devices. The re-adjusting filters, that provide the required performance at the specified clock rate of the HCE, are also considered in the subsection.

Hardware costs of the filter special processors are approximately proportional to the number of simultaneously used processors of basic operations, to the length of impulse response of the filter and to the size of processed data.

For the filters hardware costs and speed of processing depend on granularity of parallelism and on the length of impulse response. The special processor generates increment values, which then are summed with a constant value and are united to obtain the second difference of the output signal.

The values of the second differences, which are contained in two output accumulators, form a sequence of the first differences and the output signal of the filter by means of sequential code accumulators of larger capacity.

The following formats are used in the subsection: PCM (pulse-code modulation), LDM (linear delta-modulation), MDM (multilevel delta-modulation). For instance, the format LDM-PCM means that the input data consists of two signals: one signal is an LDM-signal, and the other is a PCM-signal.

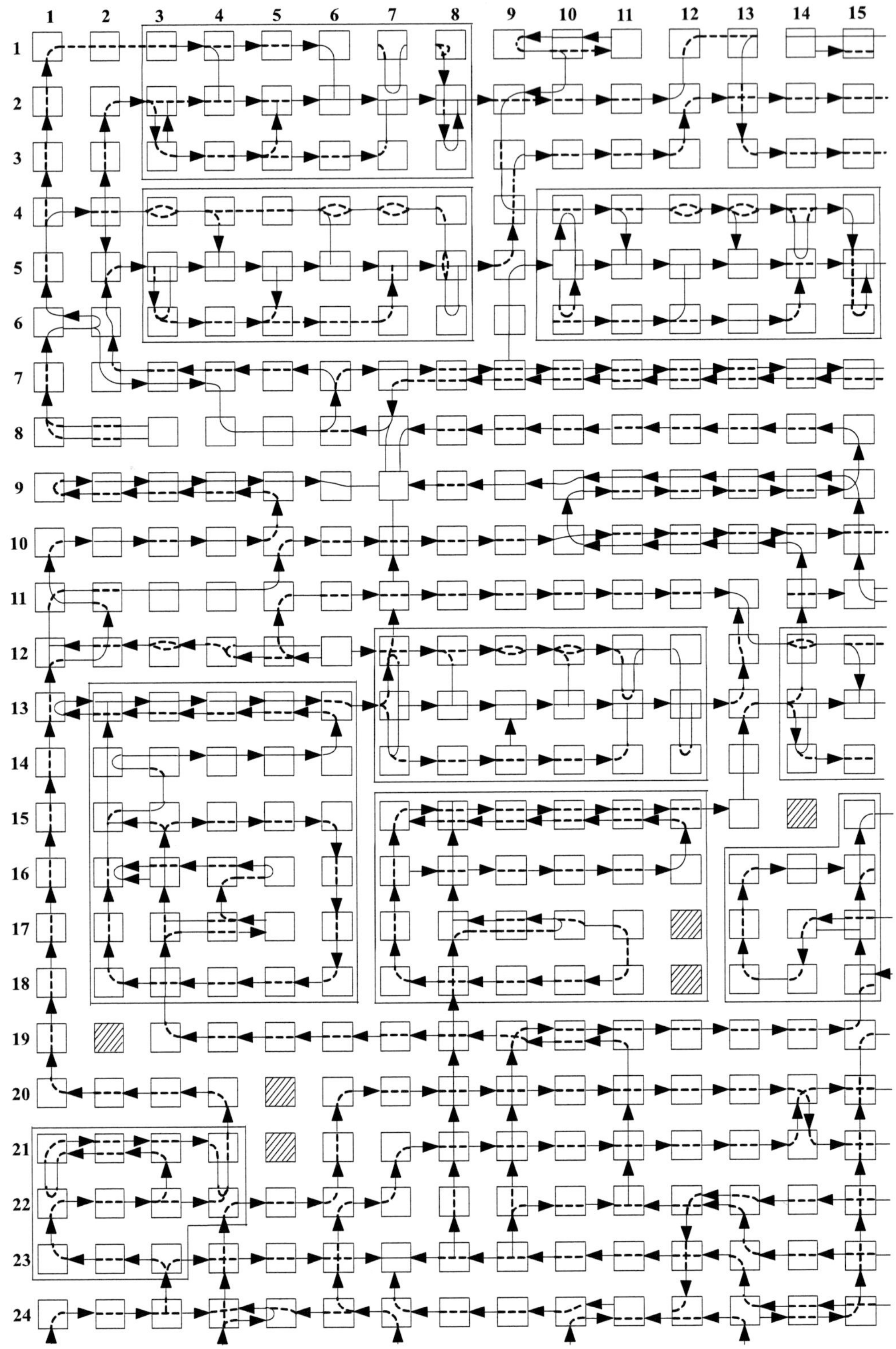

Figure 2.38. The special processor of image processing in the HCE (the beginning).

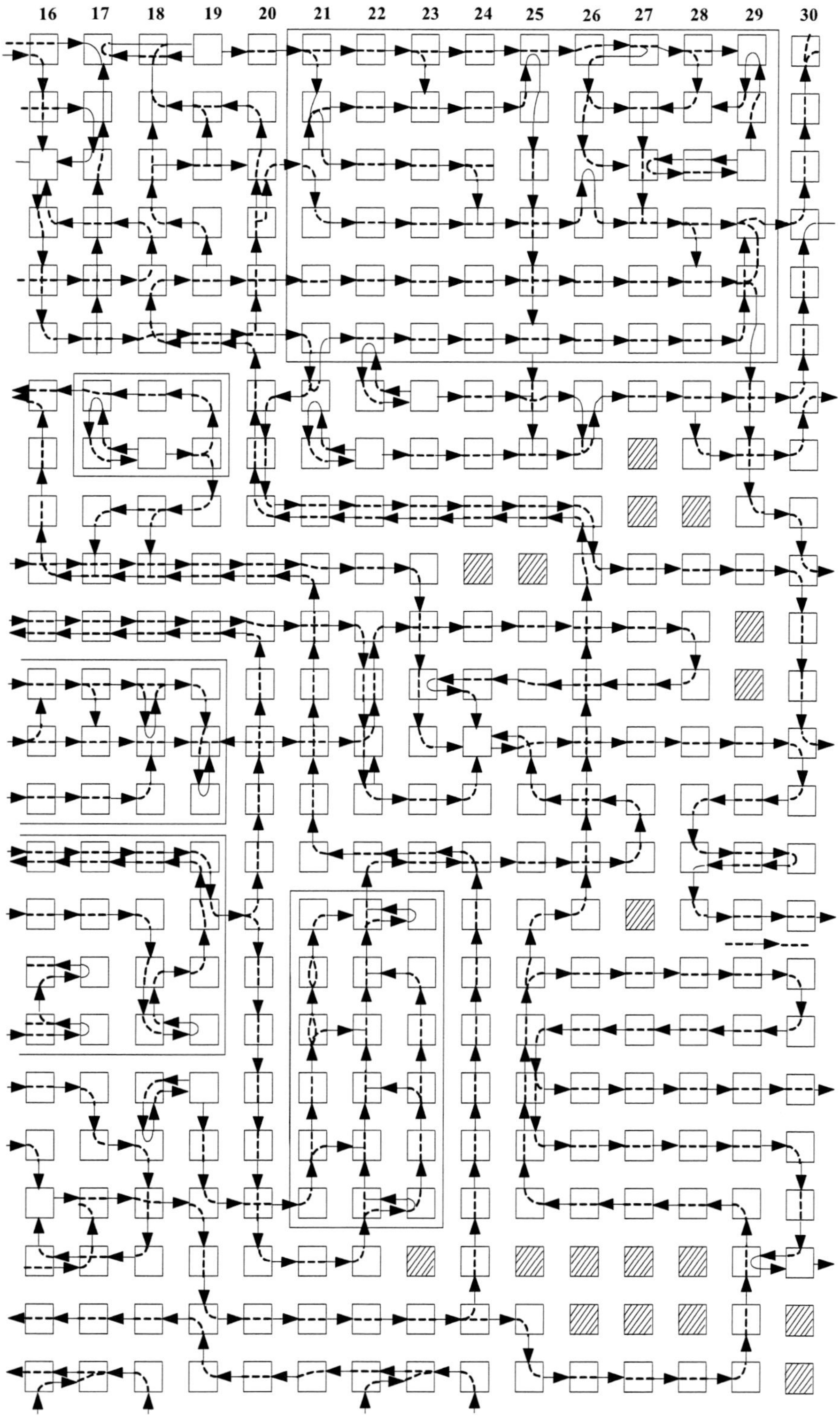

Figure 2.38. The special processor of image processing in the HCE (the end).

The Accumulator of a Parallel Input Stream

Accumulative addition is the most widely-used mathematical operation in algorithms of digital signal processing. In algorithms of digital filtration these operations are needed to obtain a convolution of an input signal with a filter impulse response.

Small resolution of delta-modulation simplifies operations of multiplication and reduces them to logical operations OR-NOT in the case of linear delta-modulation. Hence, accumulation with delta-modulation becomes one of the main operations. In fact, an accumulator that carries out the operation is a self-contained digital integrator.

When implementing the HCE accumulators, the main difficulty of using small-size delta-modulated signals consists in synchronization of the input data rate with the rate of generation of multi-digit output results. According to the classical algorithm, synchronization of a single-digit signal with an L-digit sum requires splitting of the input single-digit data stream into L decimated streams, each of which enters its own accumulator. Output data of all accumulators is summed to obtain the final result. Such method requires large computational resources and may only be applied, if the rate of the input signal is low. If the rate of the input stream of the single-digit accumulation signal is high, then the input stream is split into two streams: one stream consists of even samples and the other consists of odd samples.

The algorithm of the special processor of increments:

1. computation of the product $P_i = \overline{B_i^{(x)} \oplus B_m^{(h)}}$;
2. decimation of the streams $P_{1,i}$ and $P_{2,i}$ by the two-digit constant $K_1(2)$:
$P_{1,i}^{(pr)} = P_{1,i} \,\&\, K_1(2)$, $P_{2,i}^{(pr)} = P_{2,i} \,\&\, K_1(2)$;
3. nullification of partial sums of accumulators by the negate constant $\overline{K_2(4)}$:
$S_1^{(2)} = S_1^{(2)} \,\&\, \overline{K_2(4)}$, $S_2^{(2)} = S_2^{(2)} \,\&\, \overline{K_2(4)}$;
4. addition of the correspondent products:
$S_1^{(2)} = S_1^{(2)} + P_{1,i}^{(pr)}$, $S_2^{(2)} = S_2^{(2)} + P_{2,i}^{(pr)}$;
5. computation of result sums:
$S_1^{(4)} = S_1^{(2)} \,\&\, K_2(4)$, $S_2^{(4)} = S_2^{(2)} \,\&\, K_2(4)$.

Figure 2.39 shows the structure chart of the special processor of increments.

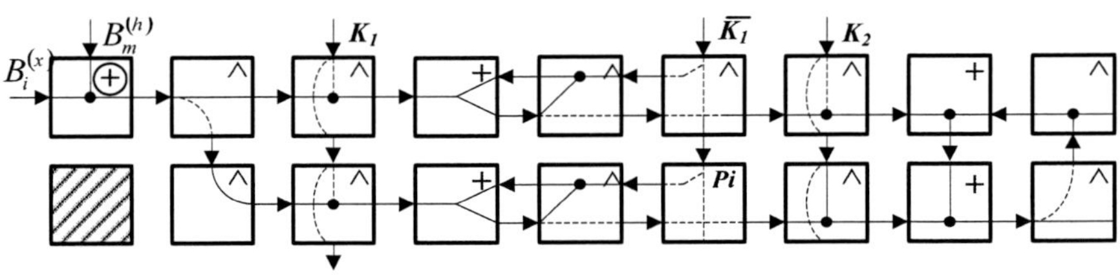

Figure 2.39. The structure chart of the special processor of increments.

The Special Processor of Stream Splitting with Size Doubling

The special processor of stream splitting with size doubling is used for implementation of digital filters of high resolution delta-modulation. The special processor algorithm is based on parallelization of a 16-digit input signal into two 32-digit ones and their addition with corresponding input signals. The signals are represented in radix complement representation.

The special processor splits the input 16-digit stream $\delta y_k^{(16,d)}$ into the streams P_1 and P_2, which are decimated by the constant $K_{16}(32)$. Before decimation, P_2 is delayed for 16 cycles. The results of decimation are 32-digit streams, which then are added to the 32-digit streams B and C, respectively. The constant $K_1(32)$ modifies the result buffer digit. Figure 2.40 shows the structure chart of the special processor of streams splitting.

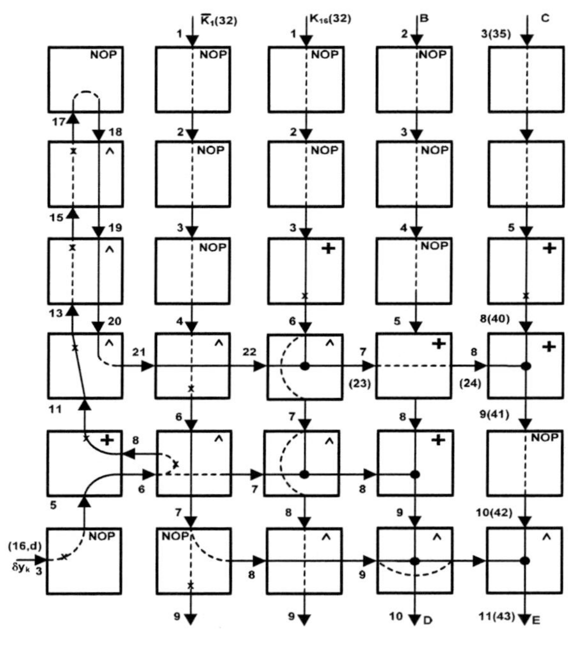

Figure 2.40. The structure chart of the special processor of stream splitting.

The Special Processor of LDM-LDM Signal Multiplication

The size of input signals is unity, the size of output signals is *r*. Multiplication is carried out at the clock rate of the HCE. Figure 2.41 shows the structure chart of the special processor.

The algorithm of multiplication of LDM-signals is based on modulo 2 addition of the signals $\left\{B_i^{(x)} \oplus B_i^{(y)}\right\}$. Then the result stream is split into *r* outputs.

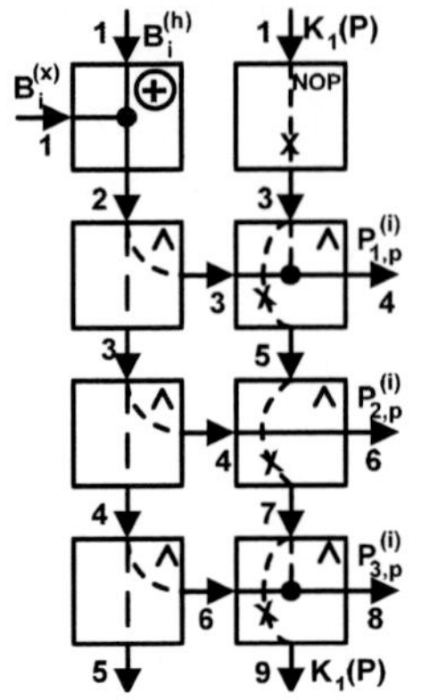

Figure 2.41. The structure chart of the special processor of LDM-LDM signal multiplication.

The Special Processor of LDM-PCM Signal Multiplication

The special processor of LDM-PCM signal multiplication processes two kinds of signals. The first signal is an LDM bit sequence; the second signal is a PCM or an MDM bit sequence. Both signals are represented in radix complement form. The size of the input operands and output results is similar and equal to r. The result size may be changed by changing of the PCM-signal size and size of the control constant $K_1(r)$. Figure 2.42 shows the structure chart of the special processor.

The special processor receives the PCM-signals $\{x_n\}$ on cycle 1, the LDM-signals $\{B_i^{(h)}\}$ on cycle 2, and the constant $K_1(r)$ on cycle 3. Starting from cycle 8, the multiplication result $\{P_{n,r}^{(i)}\}$ is obtained in radix complement form according to the equation

$$P_{n,r}^{(i)} = \sum_{k=0}^{r-1} B_n^{(p)k} 2^{r-k-1} + B_i^{(h)},$$

where

$$B_n^{(p)k} = B_n^{(x)k} \oplus \overline{B_i^{(h)}}.$$

The algorithm of the special processor is the following:

1. modulo 2 addition of the PCM-signals and LDM-signals $P_k^{(1)} = B_n^{(x)k} \oplus B_i^{(h)}$;
2. addition $P_k^{(1)} + K_1(r)$, i.e. obtaining the multiplication result $P_{n,r}^{(i)}$ in radix complement representation;
3. logical multiplication $P_k^{(2)} = B_n^{(x)k} \,\&\, B_i^{(h)}$, i.e. obtaining the multiplication result in radix complement representation;

4. logical multiplication of the result of the operation 2 with the negated LDM-signal $\overline{B_i^{(h)}}$;

5. logical addition of the result of the item 3 and item 4, i.e. obtaining $P_{n,r}^{(i)}$.

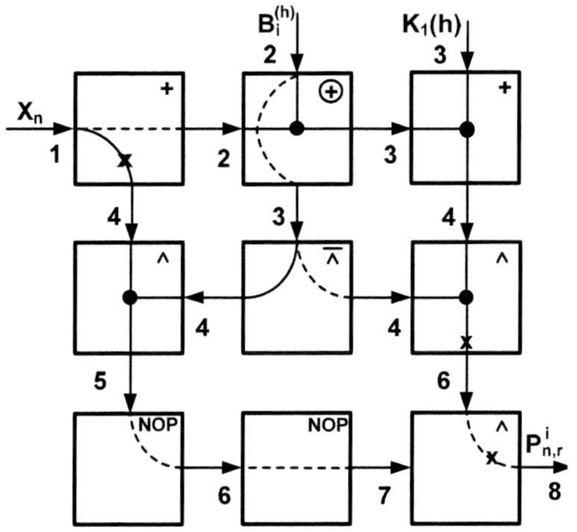

Figure 2.42. The structure chart of the special processor of LDM-PCM signal multiplication.

The Special Processor of PCM-MDM Signal Multiplication

The special processor level structure, consisting of $r/2$ levels, provides the required resolution of a PCM-MDM digital filter by a simple modification of the constant $K_1(r)$ and by increasing of the number of levels. The algorithm of PCM-MDM signal multiplication uses the operation of logical multiplication of $\{x_n\}$ by one of digits of $\{s_m^{(h)}\}$, and summing of the products with the required shift for all digits of $\{s_m^{(h)}\}$. The signals $\{x_n\}$ and $\{s_m^{(h)}\}$ enter the special processor, starting from cycles 1 and 4, respectively. The constant $K_1(r)$ enter the special processor, starting from cycle 2. The result is obtained, starting from cycle $(r+5)$. Figure 2.43 shows the structure chart of the special processor of PCM-MDM signal multiplication for r-digit data.

The algorithm of PCM-MDM signal multiplication is the following:

1. storage of a bit k, $k=\overline{1,r}$ of the signal $\{x_n\}$ after the unit digit of the constant $K_1(r)$: $B_n^{(x)k}$;

2. logical multiplication $B_n^{(x)k}$ & $s_m^{(h)}$;

3. the operations 1 and 2 are repeated for all r digits of x_n;

4. addition of the logical products, obtained as the results of the item 2; one cycle shift of the obtained result.

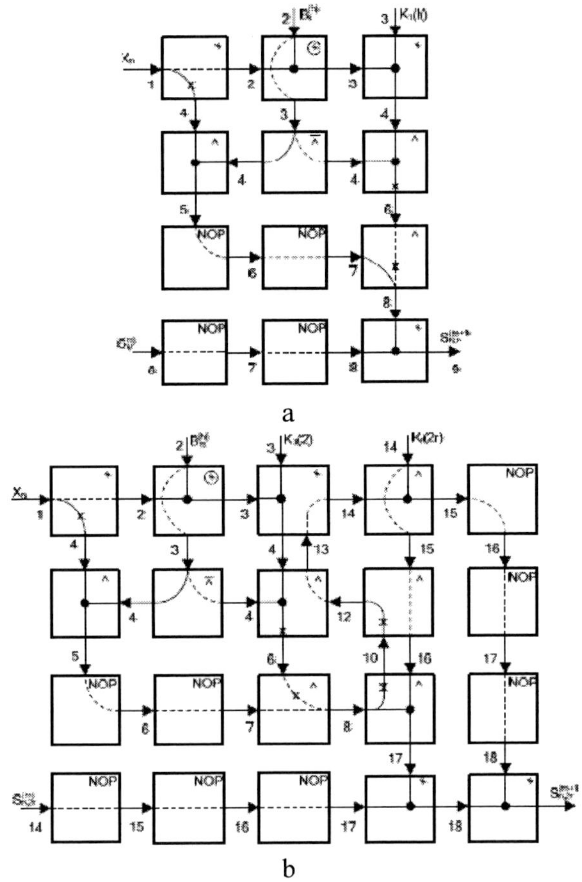

Figure 2.45. The structure charts of the special processors of LDM-PCM signal convolution.

The Special Processor of Convolution of MDM-MDM Signals in True Representation

The special processor of this operation may also process PCM-PCM signals. Figure 2.46 shows the structure chart of the special processor, which process signals in true representation.

The incomplete convolution values $S_{n,16}^{(m-1)}$, represented in radix complement form, enter the special processor, starting from cycle 5; the signal $\{s_m^{(h)}\}$ – from cycle 1; the signal $\{s_n^{(x)}\}$ and the constant $K_1(16)$ – from cycle 2. The constant $e_m^{(h)}$, which corresponds to the sign digit of $s_m^{(h)}$ - from cycle 3. The accumulation result $S_{n,16}^{(m)}$ is formed in radix complement representation, starting from cycle 13. The constant $e_m^{(h)}$ is equal to $K_0(16)=00...0$, if $s_m^{(h)}$ is positive, and is equal to $K_{16}(16)=11...1$, if $s_m^{(h)}$ is negative.

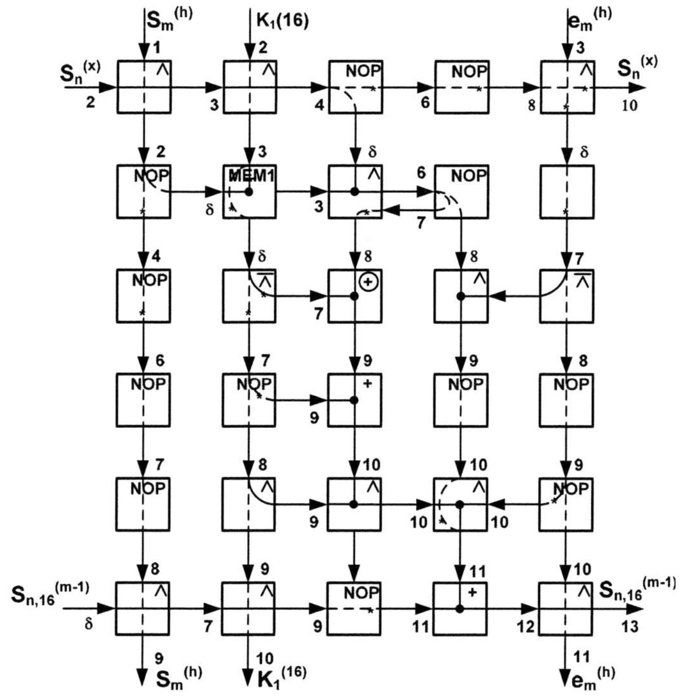

Figure 2.46. The structure chart of the special processor of convolution of MDM-MDM signals in true representation.

The special processor of signal convolution performs the following operations:

1. storage of a bit k of the signal according to the unit digit of the constant $K_1(16)$;
2. logical multiplication $B_m^{(h)k} \,\&\, s_n^{(x)}$. The result of the operation is equal to the input signal $s_n^{(x)}$, if $B_m^{(h)k} = 1$, and is equal to zero, if $P_n^{(d)} = B_m^{(h)k} \,\&\, B_{n-m}^{(x)}$;
3. negation of the constant $K_1(16)$: $\overline{K_1(16)}$;
4. modulo 2 addition $D_n^{(d)} = P_n^{(d)} \oplus \overline{K_1(16)}$;
5. addition $D_n^{(d)} + K_1(16)$. The result is equal to the input signal $s_n^{(x)}$, multiplied by $B_m^{(h)k}$, if $s_m^{(h)} < 0$;
6. logical multiplication $\left\{P_n^{(d)}\right\} \& \overline{B_m^{(h)k}}$. The result $s_n^{(x)}$ is multiplied by $B_m^{(h)k}$, if $s_m^{(h)} < 0$;
7. logical multiplication of the result of the item 5 by the constant $e_m^{(h)}$;
8. OR-uniting of the results of the items 6 and 7;
9. addition $S_{n,16}^{(m)} = S_{n,16}^{(m-1)} + P_{n,r}$.

The HCE programming system contains a great number of other microprogram modules, which adjust special processors of various algorithms of digital image and signal processing [50,52-54,57,74,78,79].

2.11. Hardware Implementation of Multimicropipeline Structures in the HCE

Homogeneity and simplicity of an HCE cell give wide prospects of its microelectronic implementation. Serial production of HCE VLSI chips "Raita-5" by the enterprise Alpha (Riga, Latvia, former Soviet Union) in 1980s was one of the first steps in this direction [5,69]. Figure 2.47 shows the structure chart of the "Raita-5" chip.

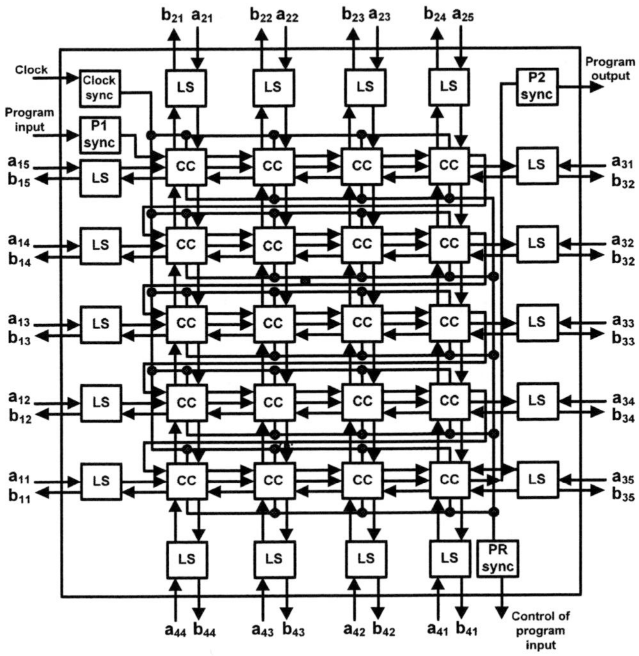

Figure 2.47. The structure chart of the "Raita-5" chip.

The HCE VLSI chip is the 5×4 matrix of computing cells. Inputs and outputs of peripheral computing cells of the matrix are connected with the chip inputs by connectors. The chip has 42 inputs/outputs, which are the following:

- a power supply bus;
- an input of program input control;
- a clock input;
- a programming input and output;
- 18 data inputs and 18 data outputs.

The "Raita-5" chips were used as basic components of circuit boards of homogeneous environments. The circuit board contained 132 chips, that implemented a computation field of 30×88 computing cells, 30 data inputs and outputs on its left and right sides and 88 data inputs and outputs on its top and bottom sides. In total, the circuit board contained 118 data inputs and 118 data outputs. For connection with other circuit boards and extension of the

HCE computation field, the circuit board contained a 135-pin connector, placed on its narrow side, and two 90-pin connectors on each of its wide sides. Owing to the design features, the sizes of the circuit board could be increased along two coordinate axis [7]. Figure 2.48 shows HCE boards, implemented on the base of the "Raita-5" chip.

Figure 2.48. Circuit boards of a homogeneous computing environment.

With the number of circuit boards $n = 16$, various configurations of a computer system are available (see Fig. 2.49). Optimal mapping of some application on the HCE may require configurations with an uneven width of the field.

a 1×16

1	2	3	4	5	6	7	8	9	10	11	12	13	14	15	16

b 2×8

1	2	3	4	5	6	7	8
9	10	11	12	13	14	15	16

c 4×4

1	2	3	4
5	6	7	8
9	10	11	12
13	14	15	16

Figure 2.49. Configurations of the HCE field.

For various possible configurations of the system, which consists of 16 circuit boards, a complicated and bulky electronic switch is required. However, the problem can be solved rather easily. It is sufficient to place two auxiliary connectors on each of the HCE circuit boards in addition to the main 135-pin connector, which is usually used for circuit board connection. One auxiliary connector is used as inputs and outputs of the left side of the field, and the other one is used as inputs and outputs of the right side of the field. Connecting the corresponding auxiliary connectors, and connecting the top and bottom connectors by cables, the circuit engineer may create arbitrary configurations. Figure 2.50 shows variants of the HCE field configurations.

The configuration, shown in Fig. 2.50 *a*, is a tape structure of the HCE field, which contains 16 boards, pairwise connected by cables and auxiliary connectors. The connector that corresponds to the left side of the most left board, shown in the Fig. 2.50, and the connector that corresponds to the right side of the most right board, are connected to an interface board for data exchange with the PC.

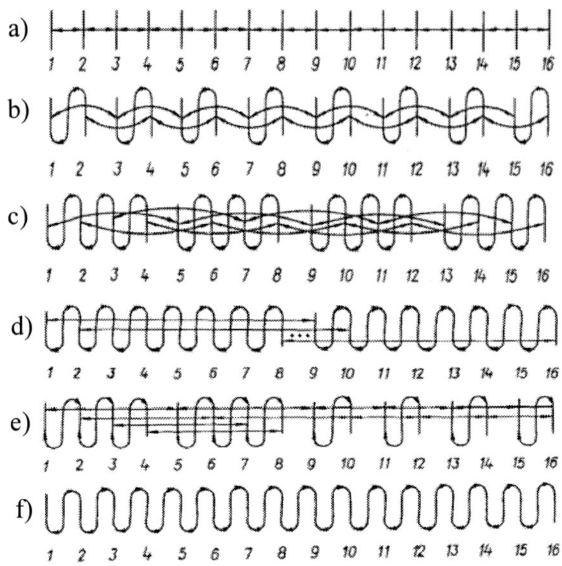

Figure 2.50. HCE fields of various configurations.

The configuration of 2×8 HCE boards, shown in Fig. 2.50 *b*, is created by pairwise connection of the bottom and top connectors of adjacent boards and by connection of the right side connector of the first board with the left side connector of the third board, by connection of the right side connector of the second board with the left side connector of the fourth board, etc. Making necessary connections, the circuit engineer may create any HCE field of any required configuration (see Fig. 2.50 *c, d, e, f*). The selected design of the circuit board and the computational block provides iterative data processing within the HCE field [7] (see Fig. 2.51). Organization of data processing in three-dimensional structures is practically similar to that in two-dimensional structures. Iterative processing may be organized in three-dimensional structures as well [7].

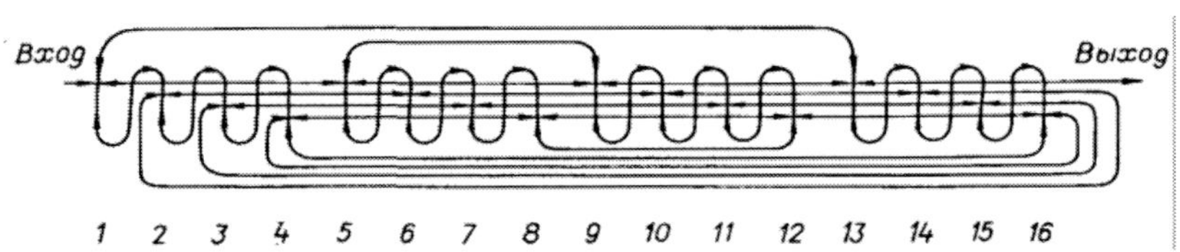

Figure 2.51. The structure of iterative processing.

When mapping a solving problem on an HCE, data transfer into computing cells of "remote" boards is often required. Long register channels, that may be adjusted for data transfer within the HCE, use a considerable number of computing cells ineffectively. Owing to auxiliary multiplexers, that are set on the each side of the circuit board, data may be transferred bypassing the computing cells. In addition, the multiplexers operate as amplifiers. Adjustment of each multiplexer requires a flip-flop in an auxiliary register, which is included into an interface unit.

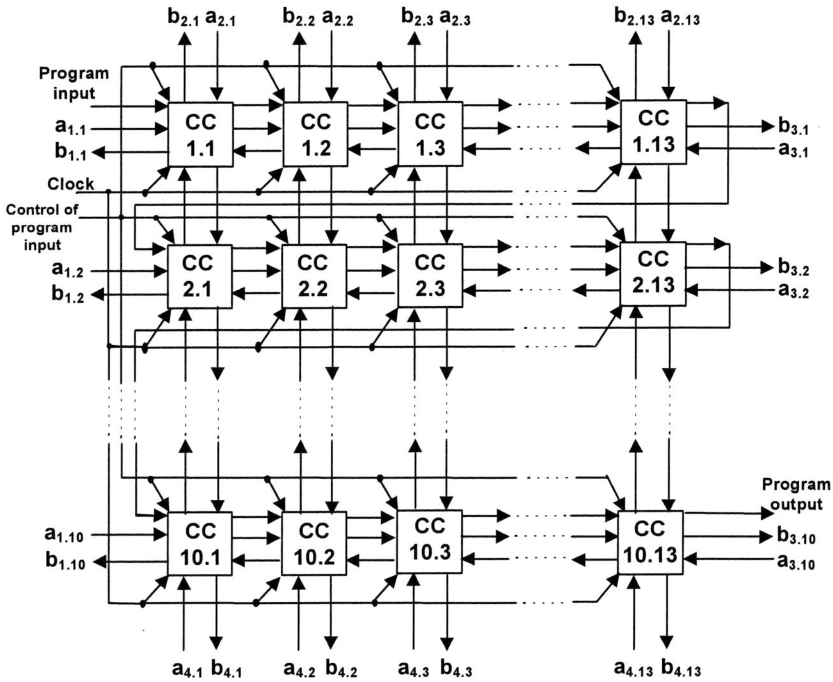

Figure 2.52. The structure chart of the HCE VLSI chip of the second generation.

The HCE VLSI chip of the second generation was designed in Scientific Research Institute of Precise Technology (SRI PT, Zelenograd, Russia) under the leadership of V.S. Sedov [33, 80] in 1990s. Reserve cells, included into a computing matrix, increase reliability of chips during manufacturing, and also their fault-tolerance during maintenance. The auxiliary logic cells provide testing of all cells, help to detect bad cells, and are used for connection of good cells and for organization of the matrix of the required dimension. The computing cell of the HCE VLSI chip of the second generation is a single-digit clocked computational unit with an instruction register. The instruction set of the cell corresponds to the instruction set of the computing cell of the "Raita-5" chip. The cells are united into a matrix as it is shown in Fig. 2.52.

The application developer works with the HCE, represented as the matrix of 10x10 computing cells. The VLSI chip contains 10 channels for horizontal loading of a program, inputs of program input control and the clock input. The matrix contains 13x10 processing elements and an auxiliary logic which provides reliability and fault-tolerance.

The HCE next-generation cells were considered in [65, 81-96].

Chapter 3

Structural-Procedural Organization of Computations in Multipipeline Structures

3.1. Principles of Multimacropipeline Computing Structure Functioning

As it was shown in Chapter 2, among advantages of HCE-based multimicropipeline computing structures there are the maximum rate of pipeline processing and simplicity of engineering implementation of the structures as VLSI chips and systems-on-chip. These advantages are achieved owing to single-bit operations and, hence, owing to the simplest structure of processing cells.

However, simplicity of a multimicropipeline structure cell causes the main disadvantages of these structures, such as complexity of programming and need of huge number of processing cells for solving complicated problems. At present, practically no examples of real application of HCEs remain because of disadvantages of HCE-based multimicropipeline computing structures.

As it was shown in Chapter 1, multimacropipeline computing structures designed according to the principles of MPCS with structural-procedural organization of computations are substantially devoid of disadvantages of HCE-based multimicropipeline structures. The design principles of MPCS-based multimacropipeline structures are suggested in [11, 22, 97-98]. At present these MPCSs are widely used, therefore all subsequent chapters of the book are devoted to theoretical development and practical implementation of multimacropipeline computations.

Unlike multimicropipeline computing structures, that are designed on the base of elementary single-bit processing cells, multimacropipeline structures are based on so-called *macroprocessors* (MAP) that are capable to perform large operations on multidigit operands [39]. A macroprocessor is a set of processing elements (PE), united into a single program-indivisible computational resource by a local spatial switch (SW_l) (see Fig. 3.1). As a rule, the internal switch of the macroprocessor connects all data inputs and outputs of the processing elements, as well as the external data inputs X and external data outputs Z of the macroprocessor according to crossbar switching. The processor is configured for execution of a large self-contained operation (macrooperation) with the help of a macrooperation unit.

Simultaneously the macroprocessor may perform only one macrooperation. The processing elements, that are not included into the macrooperation, remain idle.

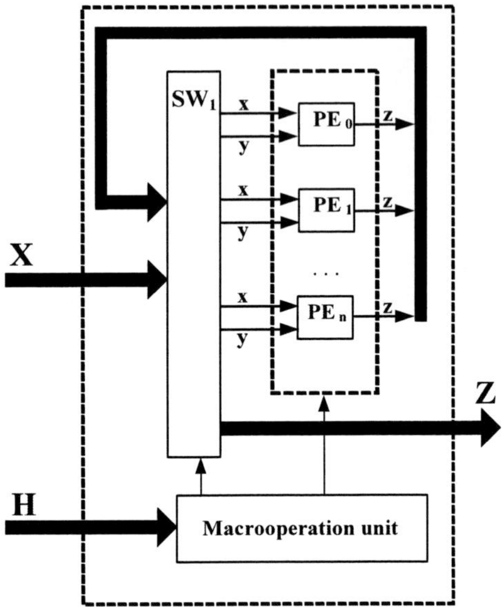

Figure 3.1. The structure chart of a macroprocessor.

At the same time, each *processing element*, in the general case, may be considered as an arithmetic-logical unit (ALU), completed with a unit of operand synchronization (UOS) and an instruction unit (IU) (see Fig. 3.2).

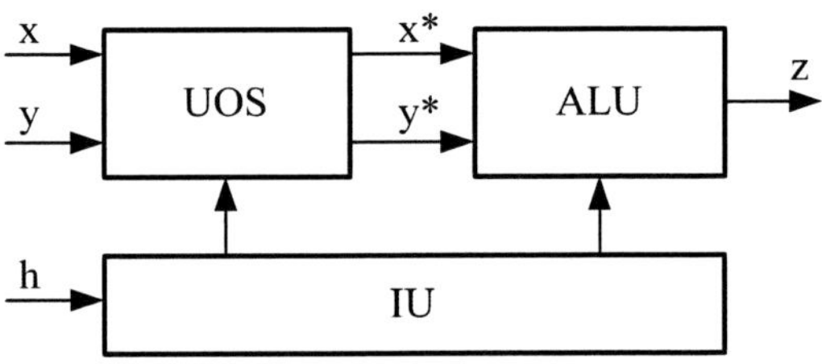

Figure 3.2. The structure chart of a processing element.

The macroprocessor provides structural implementation of a certain set of large self-contained macrooperations. Examples of the macrooperations are given in Table 3.1.

The processing elements, configured for execution of a certain arithmetic-logical operation, are designated on the structure charts as circles.

To implement an information graph of a solving problem, macroprocessors are to be capable of connecting into multipipeline chains. For this purpose, a multimacropipeline structure must contain *a second level switch* or *a system switch* which will provide various connections between macroprocessors.

Table 3.1. A basic set of operations of the macroprocessor

№	Name of macrooperation	A structure chart	Mnemonic	Algorithm of operation
1	Exponent	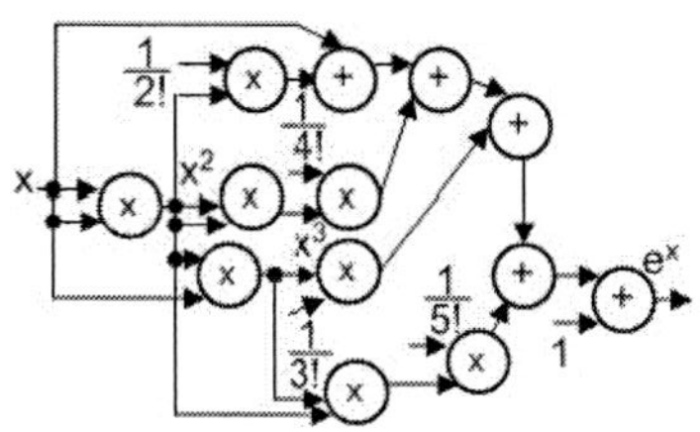	EXP	$e^X = 1 + X + \frac{X^2}{2!} + \frac{X^3}{3!} + \frac{X^4}{4!} + \frac{X^5}{5!}$
2	Sine	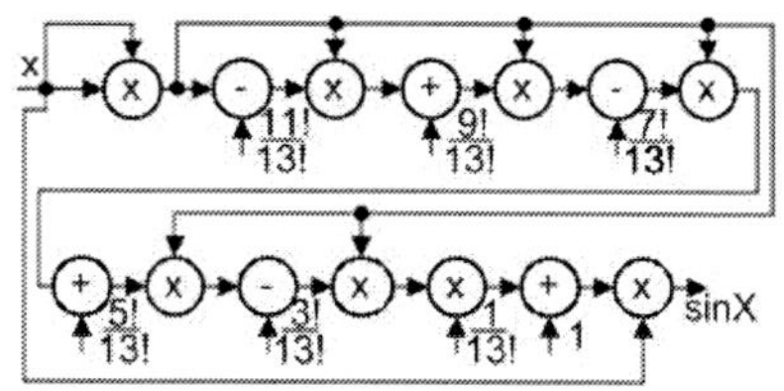	SIN	$\sin X = X - \frac{X^3}{3!} + ... + \frac{X^{11}}{11!} + \frac{X^{13}}{13!}$
3	Divergence	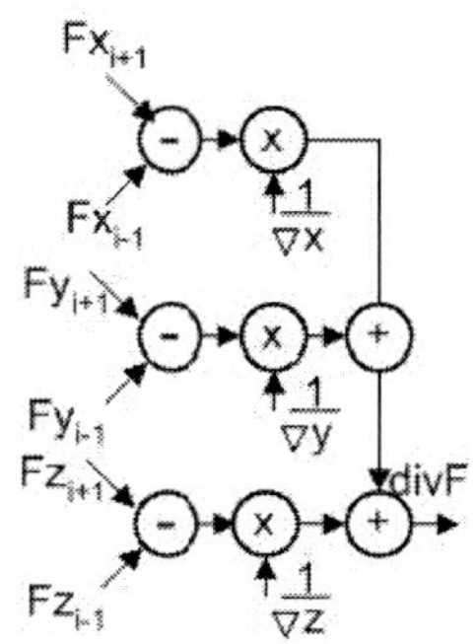	DIV	$divF = \frac{dFx}{dx} + \frac{dFy}{dy} + \frac{dFz}{dz}$

Table 3.1. (Continued)

№	Name of macrooperation	A structure chart	Mnemonic	Algorithm of operation
4	Gradient	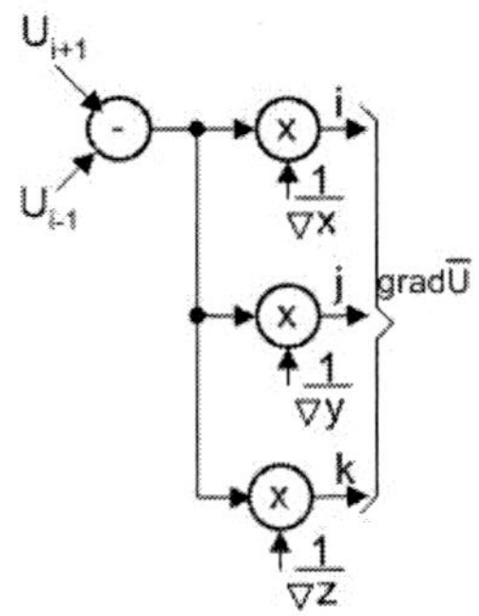	GRAD	$grad\overline{U}=\bar{i}\frac{dU}{dx}+\bar{j}\frac{dU}{dy}+\bar{k}\frac{dU}{dz}$
5	Rotor	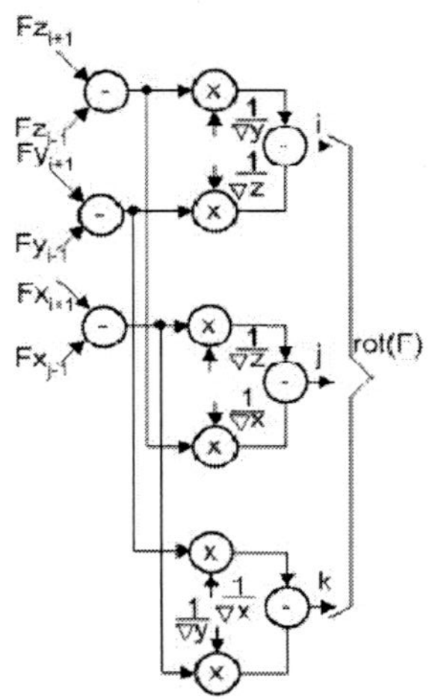	ROT	$rot(F)=(\frac{dFz}{dy}-\frac{dFy}{dz})i+(\frac{dFx}{dz}-\frac{dFz}{dx})j+(\frac{dFy}{dx}-\frac{dFx}{dy})k$

№	Name of macrooperation	A structure chart	Mnemonic	Algorithm of operation
6	Runge-Kutta integration	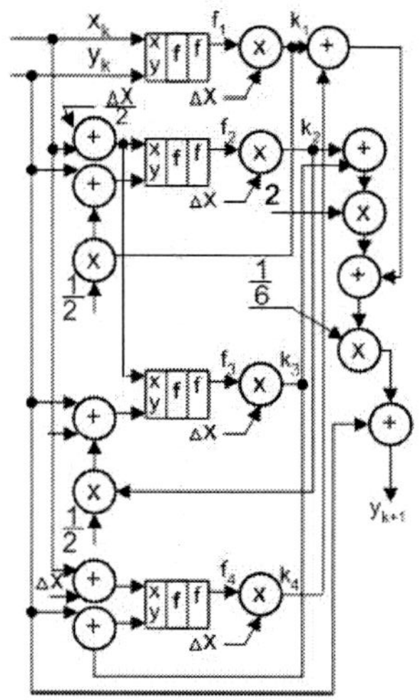	RKI	$y_{k+1} = y_k + \frac{1}{6}(k_1 + 2k_2 + 2k_3 + k_4)$ $k_1 = f(x_k, y_k)\Delta x$ $k_2 = f(x_k + \frac{\Delta x}{2}, y_k + \frac{k_1}{2})\Delta x$ $k_3 = f(x_k + \frac{\Delta x}{2}, y_k + \frac{k_2}{2})\Delta x$ $k_4 = f(x_k + \Delta x, y_k + k_3)\Delta x$
7	Decimation-in-frequency fast Fourier transform	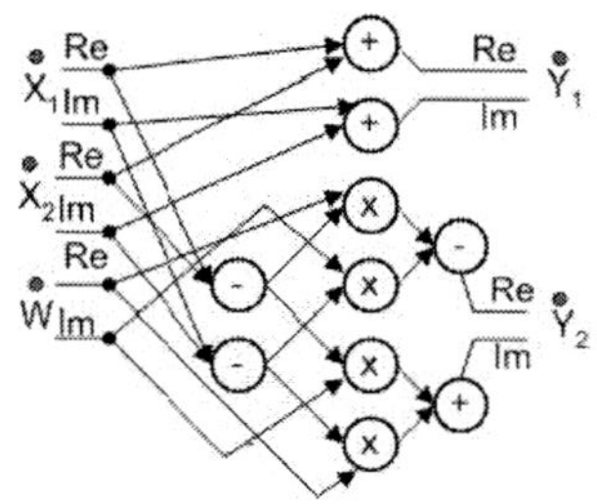	DF FFT	$\dot{Y}_1 = \dot{X}_1 + \dot{X}_2$ $\dot{Y}_2 = (\dot{X}_1 - \dot{X}_2)\dot{W}$

Figure 3.3 shows examples of multimacropipeline computing structures with various types of system switches.

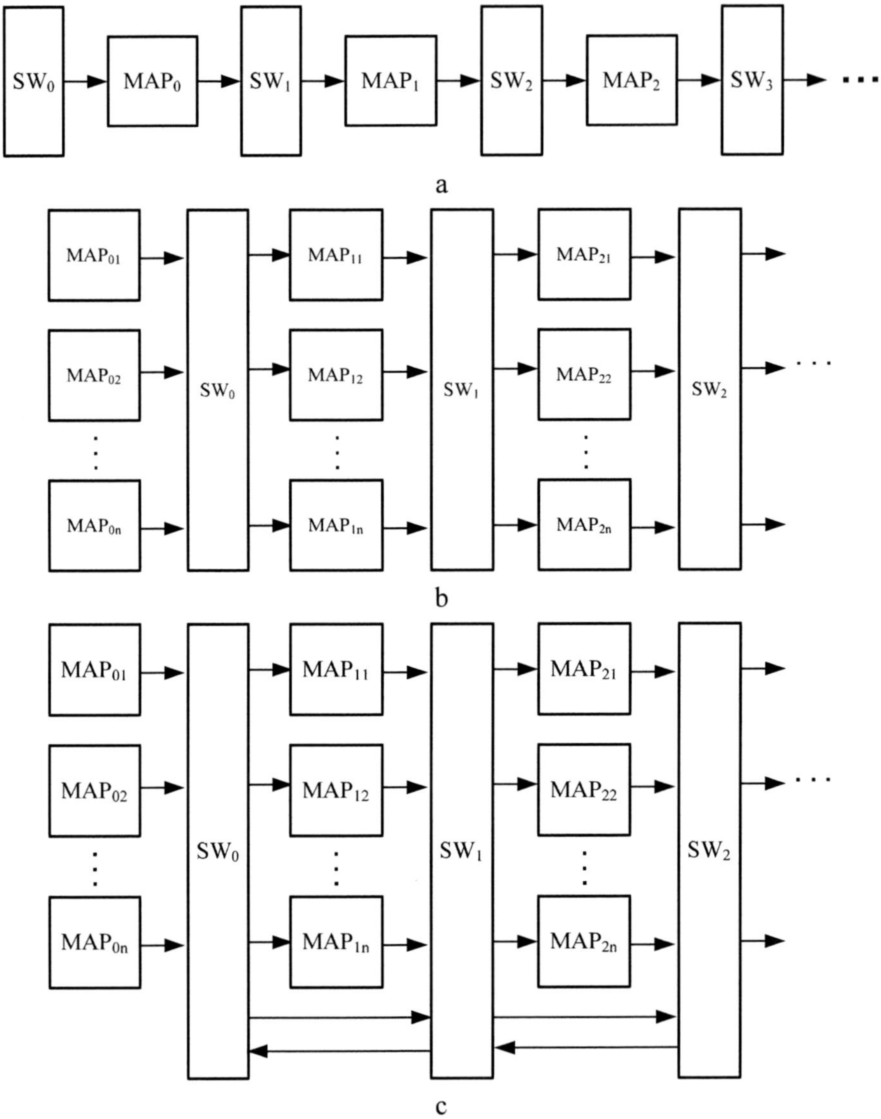

Figure 3.3. Examples of various types of system switches.

In addition, the multimacropipeline structure must contain distributed memory units that provide parallel reading and transfer of an input data array to inputs of macroprocessors, and writing of result data, that passes from outpurs of macroprocessors.

Figure 3.4 shows the generalized chart of the multimacropipeline structure. The structure contains several macroprocessors, each of which consists of processing elements connected by a local switch; distributed memory (DM), including a distributed memory controller (DMC) and a set of DM channels; a system switch SW_2 which connects the macroprocessors into pipeline chains and provides parallel access to the distributed memory channels.

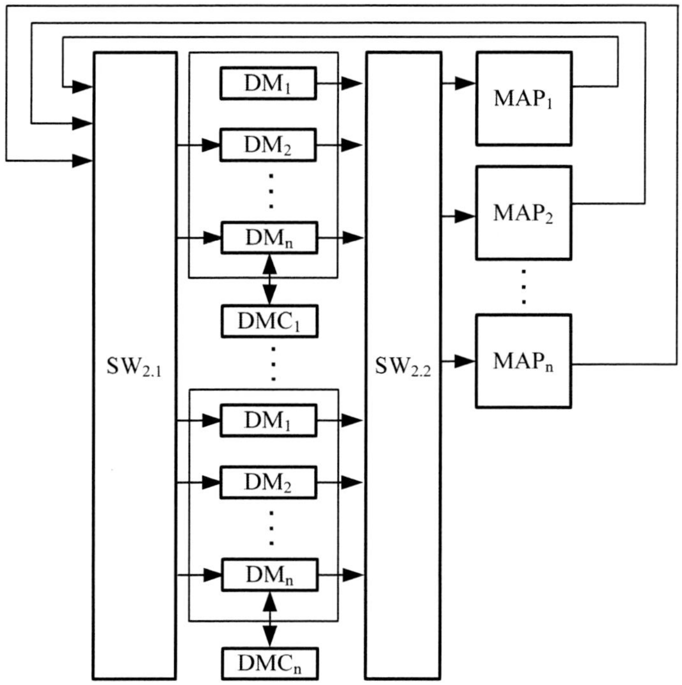

Figure 3.4. The generalized chart of the multimacropipeline structure.

The switch SW_2 may be split in two sections: the $SW_{2.1}$ which provides connections between the MAP and the DM, and the $SW_{2.2}$ which provides connections between the DM and the MAP.

Figure 3.5 shows mapping of the information graph of some solving problem on the multimacropipeline computing structure. The internal edges of the information graph are mapped on the set of communication channels, provided by the switch SW_1. Each operational vertex of the information graph of the problem is implemented with the help of the macroprocessor.

Besides, it is necessary to assign a macrooperation from a set of macrooperations, which can be structurally implemented by the processing elements of the macroprocessor, to the operational vertex of the graph.

It should be noted, that in contrast to a standard microprocessor, the processing element of the macroprocessor does not control data processing. It solely carries out a correspondent operation, receiving data from its own inputs. The set of input and output edges of the information graph is implemented by the switch $SW_{2.2}$, which connects the distributed memory channels to the inputs and outputs of the macroprocessors. Input data arrays are mapped on the correspondent distributed memory channels that may be accessed by the distributed memory controller.

While programming a multimacropipeline, each macroprocessor receives a macroinstruction that contains information about elementary operations, carried out by each processing element of the macroprocessor, and about connections of the inputs and outputs of the local switch SW_1.

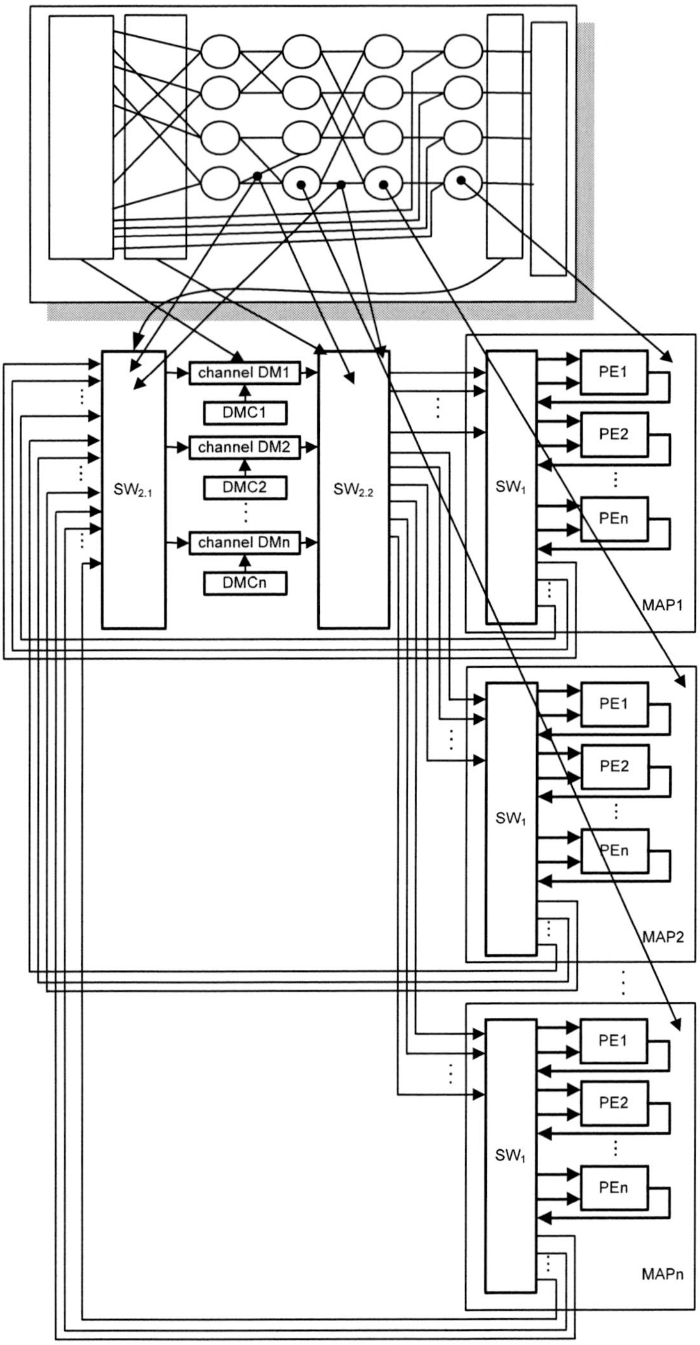

Figure 3.5. Mapping of the information graph of the problem on the multimacropipeline computing structure.

Adjustment of connections between macroprocessors, and connection of macroprocessor inputs and outputs with distributed memory channels is provided by the instruction which sets connections of the inputs and outputs of the switch SW_2. The operator of distributed memory access performs reading and writing procedures of distributed memory channels, and contains parameters of the reading and writing cycle.

In comparison with multimicropipeline structures, the considered multimacropipeline computing structure has some advantages. The first advantage is possibility of considerable simplification of computing structure programming owing to use of a set of large macrooperations. In this case, the application programmer uses large macrooperations, developed for a specialized problem domain, instead of single-bit microoperations as it is traditionally done for programming of multimicropipeline computing structures.

As an example, let us analyse a problem of a radar map creation [99]. Figure 3.6 shows the information graph of the problem.

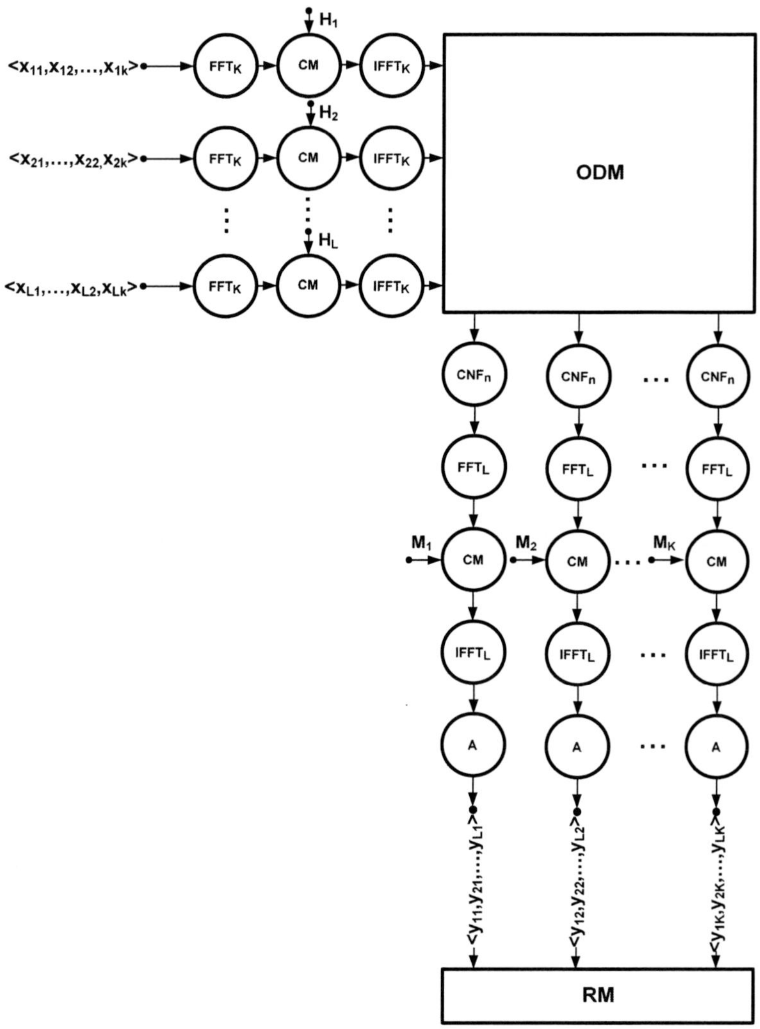

Figure 3.6. The information graph of the problem of radar map creation.

Information from a radar is simultaneously transferred through L channels that correspond to rows. The following sequence of operations is performed on information of each channel: the fast Fourier transform of the size K (FFT_K), complex multiplication by the coefficients H_i ($i = 1,2,\ldots,L$) and the inverse fast Fourier transform of the size K ($IFFT_K$). The results of processing enter a unit of orthogonal distributed memory (ODM) that performs data transposing.

Further processing is performed in parallel on the data of K channels, that correspond to columns. The following sequence of operations is performed on information of each channel: complex non-recursive filtration of the size n (CNF_n), the fast Fourier transform of the size K (FFT_K), complex multiplication (CM) by the coefficients M_i ($i = 1,2,\ldots,K$), the inverse fast Fourier transform of the size L ($IFFT_L$) and determination of signal amplitudes (A). The set of amplitude values forms the radar map (RM).

The information graph may be implemented in the multimacropipeline computing structure by means of a library of macrooperations, intended for solving problems of digital signal processing. Each vertex of the graph is put in correspondence with a certain macrooperation, a communication structure or macroaccess to the set of distributed memory channels. This correspondence eliminates a semantic gap in mapping of the information graph on the architecture of the multipipeline structure.

Nevertheless, a very important circumstance should be taken into account. The larger are the macrooperations, corresponding to the vertices of the information graph of the problem, the more complicated is optimal mapping of the macrooperations on the macroprocessors, i.e. the larger are the macrooperations, the less effective is use of computational resources of the structures, because a part of the macroprocessor processing elements may remain idle.

It is obvious, that this disadvantage may be avoided by simplification of the macrooperations, assigned to the vertices of the information graph. In extreme case, each vertex of the information graph may receive some elementary operation, carried out by some processing element. As a result, it is possible to provide the most intensive use of the whole computational resources in a computational process. Thus, when developing some multipipeline structure, the application programmer may choose between two alternative approaches. The programmer may enlarge macrooperations, simplifying programming of the structure and reducing effectiveness of computational resource use. The other approach is quite opposite.

Another advantage, which distinguishes multimacropipeline computing structures from multimicropipeline ones, is the two-cascade communication system which consists of the macroprocessor local switches and the system switch. Such organization of the communication system provides wide communication capabilities and does not require resources of the main computational units of the processing elements. At the same time, the number of processing cells, used for data transit in multimicropipeline structures, comes to 30%.

3.2. Transformation of Information Graphs into Structural-Procedural Programs

In the previous subsection, while presenting the main idea of organization of reconfigurable structures, we considered the special case, when the information graph of the problem may be entirely mapped on hardware resources of the system. However, in reality hardware resources are not sufficient for mapping of complex information graphs of application problems. As it is shown in subsection 1.5, structural-procedural organization of computations is to be used in this case. Let us make a detailed analysis of special features of structural-procedural organization of computations.

In general form an information graph may be represented as follows

$$G = \langle (X, A_X), (Q, A_Q), (Y, A_Y) \rangle,$$

where X is a set of input vertices;

Y is a set of output vertices;

Q is a set of operational vertices.

A union of an input edge set A_X, an operational edge set A_Q and an output edge set A_Y describes information dependences of the problem.

For any edge $<u,v>$, contained in A_X, $u \in X$ and $v \in Q$. For any edge $<u,v>$, contained in Q, $u \in Q$ and $v \in Q$. For any edge $<u,v>$, contained in Y, $u \in Q$ and $v \in Q$. Each operational vertex of Q is to receive some macrooperation. As it was shown in the previous subsection, optimality of mapping of an information graph of any problem on a reconfigurable computing structure depends on interpretation of a macrooperation. The best mapping results may be achieved, when each vertex of Q receives an elementary operation contained in the operation set of the processing element. Therefore, for the sake of generality it is assumed, that the vertex $q_j \in Q$ of the information graph receives the elementary operation $o_j \in O$, where O is the set of elementary operations that are carried out by the processing element.

If the number of processing elements of the reconfigurable structure is not less than the number of operational vertices of Q, and all edges of A_X, A_Q and A_Y may be implemented by the local switches *SW* and the system switch SW_1, then the information graph of the problem is trivially implemented in the reconfigurable structure (see Fig. 3.7).

In this case the operational vertex of the graph corresponds to the processing element, adjusted to execution of a certain arithmetic-logical operation. The input information vertex is to correspond to a distributed memory channel that stores initial data, and the output information vertex is to correspond to a distributed memory channel that will contain results of computations, obtained during execution of the whole task.

Such organization of computations, when the entire information graph of the problem is mapped on resources of the multimacropipeline structure, is called *structural organization of computations*. It is obvious that structural organization of computations provides the minimum time of input data flow processing, because the special-purpose calculator, oriented on the given problem, is formed within the multimacropipeline structure. However, in practice it is impossible to provide structural organization of computations, because, as a rule, the number of operational vertices of the information graph of the problem considerably exceeds the number of processing elements of the reconfigurable structure.

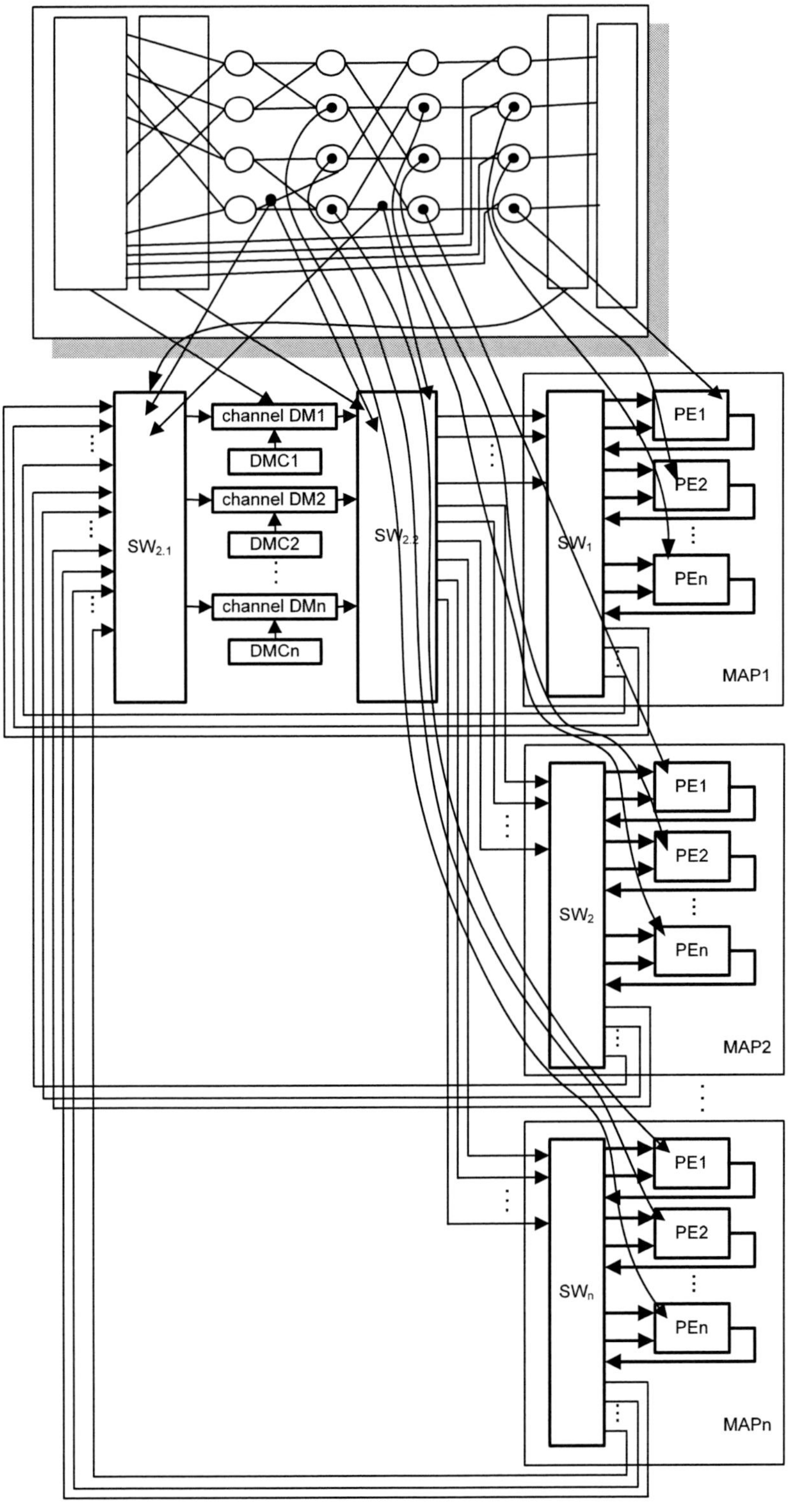

Figure 3.7. Implementation of the information graph of the problem in the reconfigurable structure.

The evident solution to this difficulty is partitioning of the information graph of the problem into disjoint subgraphs that may be structurally implemented one-by-one. The problem, represented as the information graph G, may also be represented as a set of large self-contained computational fragments P_i,

$$G = \bigcup_{i=1}^{M} P_i .$$

If there are vertices a and b, such that $a \in P_i$ and $b \in P_j$, and these vertices are connected by at least one edge $< a,b >$, then $i \leq j$. The edges, that connect the subgraphs, are called *external*; the edges, that are contained in the subgraph P_i, are called *internal*.

The informational graph subgraphs, which are subsequently implemented in the reconfigurable structure, are called *cadrs* [11]. Representation of the information graph as a sequence of disjoint subgraphs is called a *structural-procedural program*.

Definition 3.1. *Functioning of a reconfigurable multipipeline computing structure according to a single control program is called structural-procedural organization of computations. The cadr is the minimum and program-indivisible structurally implemented object.*

It is possible to say that reconfigurable structures with structural-procedural organization of computations are a hybrid of von Neumann architecture and data flow architecture. Such organization of computations provides a high performance of parallel computations for various problems, and determinism of program execution, which, in general case, cannot be provided by multiprocessor systems with a traditional multiprocedural architecture.

The cadr is defined as a set of processing elements and distributed memory units, united by a reconfigurable communication system into a single computational resource, that performs functional transformation of input data flows into output ones [11]. Essentially, the cadr is executed in the operand flow mode [1, 100]. It may be claimed that the cadr is a specialized computing structure, developed by the application programmer for hardware solution of some problem fragment.

The cadr is often described as an object, defined as a triplet $< Q,I,O >$, where Q is a cadr graph, which consists of operational vertices, corresponding to processing elements adjusted to operations, that remain unchanged during execution of the cadr; I is a set (or a cortege) of data flows, connected one-to-one with input vertices; O is a set of data flows, connected one-to-one with output vertices. Figure 3.8 shows the graphic chart of the cadr.

The input data flow is the vector $I =< I_1, I_2, ... I_n >$ of independent flows with one-to-one correspondence between the flow of the input vertices of the functional graph of the cadr and each element I_j. Similarly, the output flow is the vector $O =< O_1, O_2, ..., O_n >$ of independent flows with one-to-one correspondence between the flow of the output vertices of the functional graph of the cadr and each element O_i. The elements of the input data flow sequentially enter the input vertex. The elements of the output data flow are sequentially read from the output vertex.

independent isomorphic subgraphs into a cortege, the subgraph cortege can be transformed into the cadr with the help of the only rule.

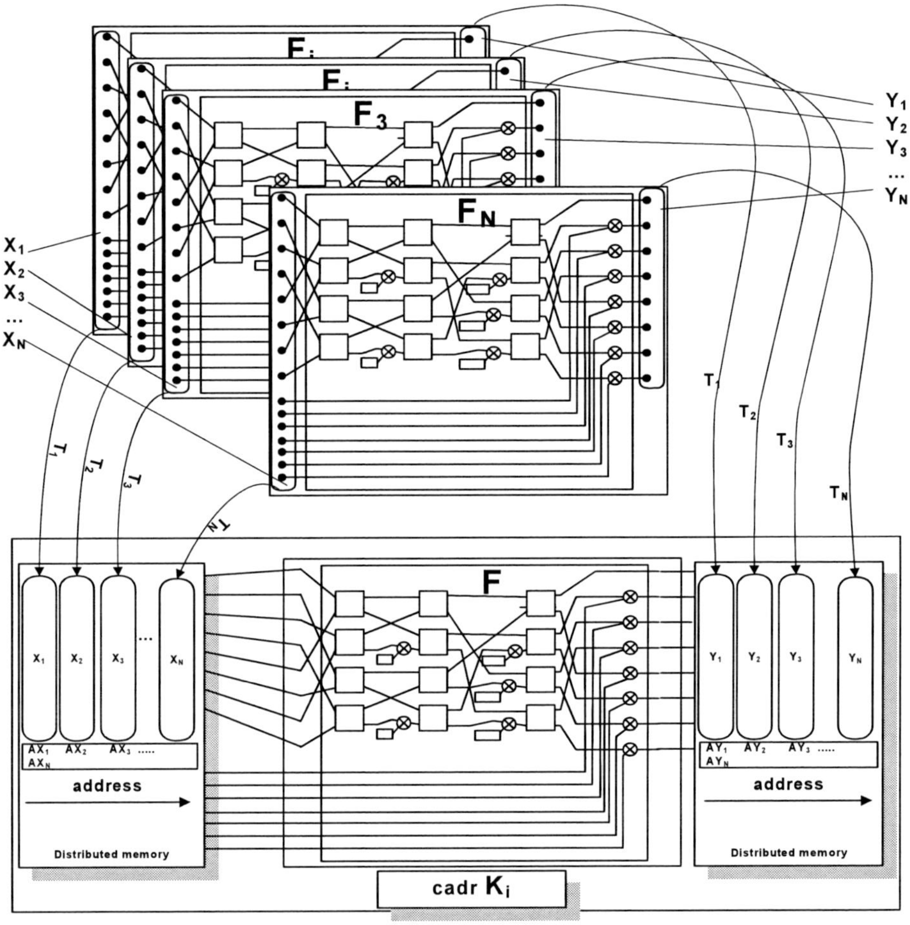

Figure 3.10. Transformation of the cortege of the isomorphic basic subgraphs into the cadr form.

Existence of recurrent rules of mapping of edges of the subgraphs g is the supplementary condition for representation of the subgraph set in the cadr form:

$$F_R : (x, a_X)_i \to (x, a_X)_{i+1},\ i = 1,2,...,N-1, \quad (3.1)$$

$$F_W : (y, a_Y)_i \to (y, a_Y)_{i+1},\ i = 1,2,...,N-1, \quad (3.2)$$

where x_i is the set of input vertices of the graph i;

a_{Xi} is the set of input edges of the graph i;

y_i is the set of output vertices of the graph i;

a_{Yi} is the set of output edges of the graph i.

It should be noted, that isomorphic graphs, which are transformed into a cadr, are to be executable, i.e. when the cadr is initialized, all information elements, corresponding to input vertices of all information graphs, are to be calculated and are to be stored in distributed memory channels according to a reading function.

Let us assume that there are two isomorphic subgraphs G_1 and G_2, which are not information-independent. Let us also assume that there is the only edge which connects the start vertex $a_1 \in G_1$ and the end vertex $b_2 \in G_2$, and there is no other path, that could connect any vertex of G_1 with any vertex of G_2. Under such conditions G_1 and G_2 may be ordered and represented as a cadr according to the only ordering rule $< G_1, G_2 >$, defined by the edge $< a_1, b_2 >$.

In this case a supplementary transformation of the information graph is performed. The edge, which connects a and b, is created. The edge accepts an initial element that provides feedback. The initial value is defined by the vertex of the edge, contained in G_1. The vertex b_1 corresponds to the vertex b_2. The graphs G_1 and G_2 are isomorphic. The edges $< x_{b_1}, b_1 >$ and $< a_1, b_2 >$ are deleted from the cadr graph. Figure 3.11 shows an example of such transformation.

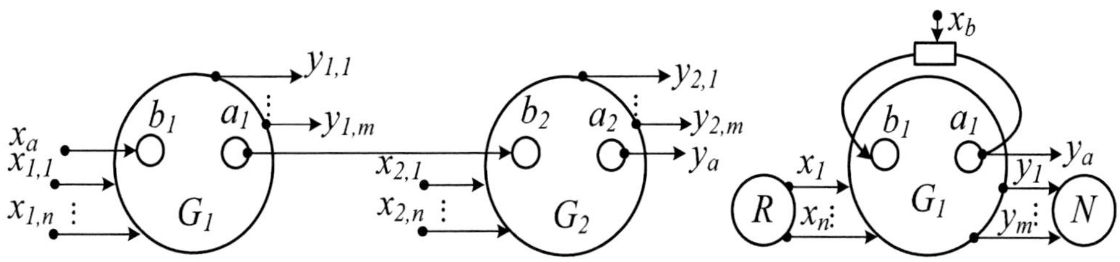

Figure 3.11. An example of transformation of joint information graphs into a cadr.

If the isomorphic subgraphs $G_1, G_2, ..., G_n$ are information-dependent solely by the adjacent edges a_i, which connect G_i and G_{i+1}, then $G_1, G_2, ..., G_n$ may be transformed into a cadr.

If a_i is connected with the input vertex X_a, then such subgraphs may be transformed into a cadr. The rule of subgraph ordering is defined by the path which connects the subgraphs $< G_1, G_2, ..., G_n >$. An example of such graph is shown in Fig. 3.12.

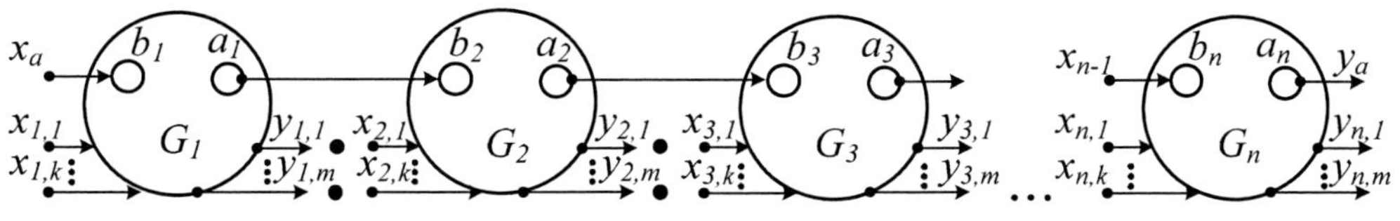

Figure 3.12. The graph containing isomorphic joint subgraphs.

It should be noted, that subgraphs may be connected by several information edges. The edges may connect any fragments, not only adjacent ones, but every connection is to be described by the functional dependence, which exists between the subgraphs, according to the equations (3.1) and (3.2). In more general case, it is necessary to define the functional dependence between the start vertex $a_i \in G_i$ and the end vertex $b_{f(i)} \in G_{f(i)}$. Connections may exist not only between single subgraphs, but between a certain set of subgraphs as well.

3.3. Transformation of a Problem into the Structural-Procedural Form

Sequentially implemented cadrs may be connected solely through RAM, which is to be distributed and multichannel for multiprocessor computer systems with structural-procedural organization of computations. Each input edge a_j of the cadr is put in correspondence with the reading function r_j, which provides input of data flows into the cadr from distributed memory channels. Similarly, each output edge b_j of the cadr is to be put in correspondence with the writing function w_j, which provides storage of result data flows, that pass from the cadr outputs, in the distributed memory channels.

The informational cadr is the triplet

$$K = < R, Q, W >,$$

where R is the reading node;

Q is the structurally implemented subgraph of the problem;

W is the writing node.

Here

$$R = \bigcup_{j=1}^{N(A)} < a_j, r_j >,$$

where $N(A)$ is cardinality of the input edge set A of the cadr.

The writing node is a union of ordered pairs:

$$W = \bigcup_{j=1}^{N(B)} < b_j, w_j >,$$

where $N(B)$ is cardinality of the output edge set of the cadr.

When partitioning the information graph into a set of subgraphs, which correspond to cadrs, a certain condition is to be fulfilled. Otherwise, if the condition is not fulfilled, effectiveness of parallel computations may be considerably reduced, or algorithmic conflicts may occur, when data required for calculations is not ready. According to the supplementary condition of partition of the information graph of the problem into structurally implementable subgraphs, each cadr is to be represented as a set of subgraphs:

$$P_i = \bigcup_{j=1}^{N} S_{i,j} .$$

The subgraphs S_{ij} are called *layers*. Edges, that connect the layers, are called *interlayer edges*, and edges, that connect the operational vertices of one and the same layer, are called *intralayer edges*.

Each subgraph S_{ij} may be represented as a set of subgraphs:

$$S_{i,j} = \bigcup_{k=1}^{K} g_{i,j,k} ,$$

and all the subgraphs $g_{i,j,k}$ are to be isomorphic. The subgraphs $g_{i,j,k}$ are called *basic subgraphs*. In addition, the intralayer edges are to be contained in the basic subgraph. If there is the edge $< a,b > \in S_{i,j}$, then $a \in g_{i,j,k}$ and $b \in g_{i,j,k}$. Hence, there are no edges, connecting the vertices of the basic subgraphs of one and the same layer.

The basic subgraph $g_{i,j,k} \subset S_{i,j}$ if and only if there is at least one edge $< a,b >$ for which $a \in q_{i,j-1,p}$, $b \in q_{i,j,k}$. Such representation of the graph as a set of subgraphs is called *a graph sequencing*. The sequenced graph of computational process of the problem may be represented as follows

$$P_i = \Phi_{i,j}(F_{i,j,k}(g_{i,j,k})) ,$$

where $F_{i,j,k}$ is the function, which maps the edges of $g_{i,j,k}$ on $g_{i,j,k+1}$;

$\Phi_{i,j}$ is the function, which maps the set of the edges of $S_{i,j}$ on the set of the edges of $S_{i,j+1}$.

The set of functions F describes the parallel implementation of the basic subgraphs. The set of functions Φ defines the informational dependence between the layers.

If after sequencing of P_m the subgraphs $g_{i,n,k} \subset S_{i,n}$ and $g_{i,m,k} \subset S_{i,m}$, $k = 1,2,...,K$, are isomorphic for all n and m, $n = 1,2,...,N$, $m = 1,2,...,N$, and there are generalizing functions of basic subgraph intralayer mapping and functions of layer-to-layer mapping, then such partition of the graph G of computational process of the problem into the subgraphs P_i will be called *a functional sequencing*. In this case each subgraph P_i may be represented as follows

$$P_i = \Phi_i(F_i(g_i)) .$$

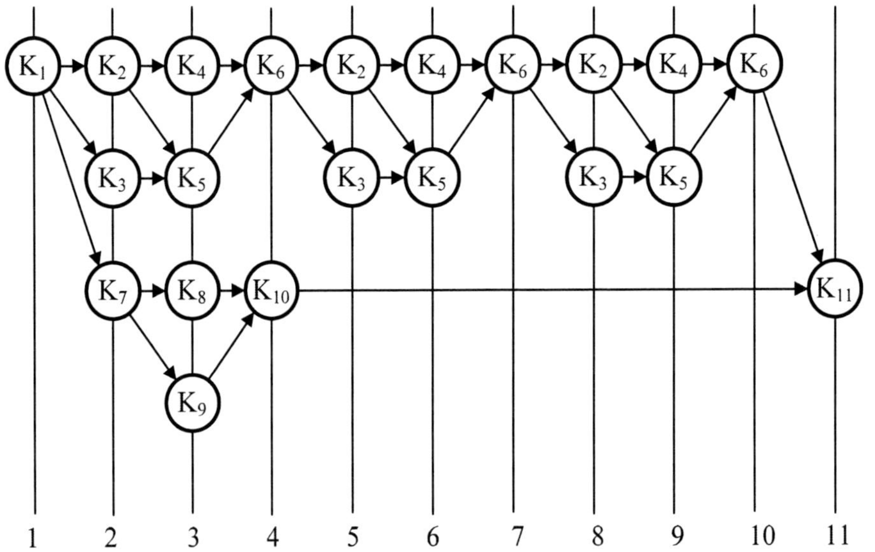

Figure 3.13. A problem graph, represented as a cadr multilevel structure.

The two conditions are to be fulfilled for placing cadrs a and b into one and the same level. Firstly, a and b are to be information-independent, i.e. there must be no path which can connect a and b by the edges of informational dependence. Secondly, the path, which contains the control edges and connects a and b, cannot pass through any decision-vertex. Hence, the cadrs will be placed in one and the same level if and only if they can be executed simultaneously.

Figure 3.14 shows the example of the structural-procedural form of the problem. Here the control edges are marked by dotted lines, and the information edges are marked by solid ones. The vertex P_1 is a decision-vertex. In contrast to the multilevel structure of the information graph, in which sets of vertices, placed in one and the same level, may be implemented simultaneously, the cadrs that are contained in one and the same level of the structural-procedural form are implemented solely sequentially.

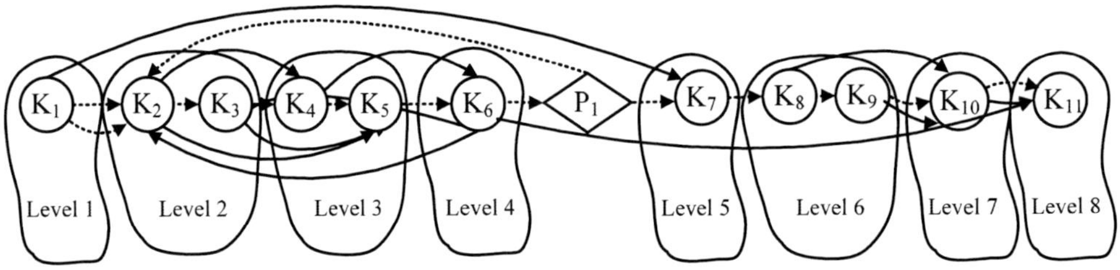

Figure 3.14. The example of the structural-procedural form of the problem.

During analysis of cadr structures of problems from various problem domains it is possible to select standard procedures of access to distributed memory channels, called macroaccess procedures, and standard procedures of parallel access, called macroswitch procedures, and standard components of structural implementation of problem fragments, called macrooperations.

When developing parallel programs, the application programmer may use pre-developed macroaccesses and macroswitches along with macrooperations. Such approach considerably simplifies programming and allows development of effective parallel programs for solving problems from various domains.

The cadr components of the solving problem are naturally mapped on the computation field of the MPCS (see Fig. 3.15).

The operational edges of the basic subgraph are mapped on the communication channels, implemented in the macroprocessor switch. The operational vertices of the basic subgraph are mapped on the processing elements. The input and output edges of the basic subgraph are implemented in the communication channels of the switches. Data arrays are mapped on the addresses of distributed memory. The control program that provides cadr changing and performs the procedure of access to the distributed memory channels is executed in the distributed memory controller. Thus, when the application programmer develops cadrs, he creates a set of special-purpose calculators. Owing to mathematical formulation of algorithms, some problems may be directly transformed into the cadr form.

3.4. Transformation of Functionally-Regular Information Graphs

A functionally-regular graph may be implemented as a single cadr, represented as a structural basic subgraph or as a set of joint basic subgraphs with a data flow passing through them. An element of the data flow corresponds to a set of information vertices at the inputs of the structurally implemented subgraph. An element of the result data flow corresponds to a set of information vertices at the outputs of the structurally implemented subgraph. Therefore, transformation of the functionally-regular graph is rather simple. Figure 3.16 shows an example of some functionally-regular subgraph.

When information-dependent basic subgraphs are united into a cadr, feedbacks are generated if and only if system resources are sufficient for structural implementation of the subgraph set, and the time, needed for operands of the same level to pass through the structure, is less than the pipeline filling time.

If the resources, allocated for the problem, are sufficient solely for structural implementation of one basic subgraph, then the subgraph cortege, transformed into the cadr, may be represented as $< G_{1,1}; G_{1,2}; G_{2,1}; G_{2,2}; G_{3,1}; G_{3,2} ... >$ or $< G_{1,2}; G_{1,1}; G_{2,2}; G_{2,1}; G_{3,2}; G_{3,1} ... >$. The basic subgraphs, contained in one and the same level, are implemented in the first place. Generation of recursive connections is required if and only if

$$t_{pipeline}(a,b,G) + t_{setsw} > 2\tau ,$$

where $t_{pipeline}(a,b,G)$ is the depth of pipeline filling for structural implementation of G ;

t_{setsw} is the time of switching of data flow sources;

τ is the time of an operand passing.

If the time of passing of two operands is above or equal to the time of data appearing, then generation of feedbacks is optional. Certainly, the feedback is implemented, but through memory, and the time of the problem execution is not increasing.

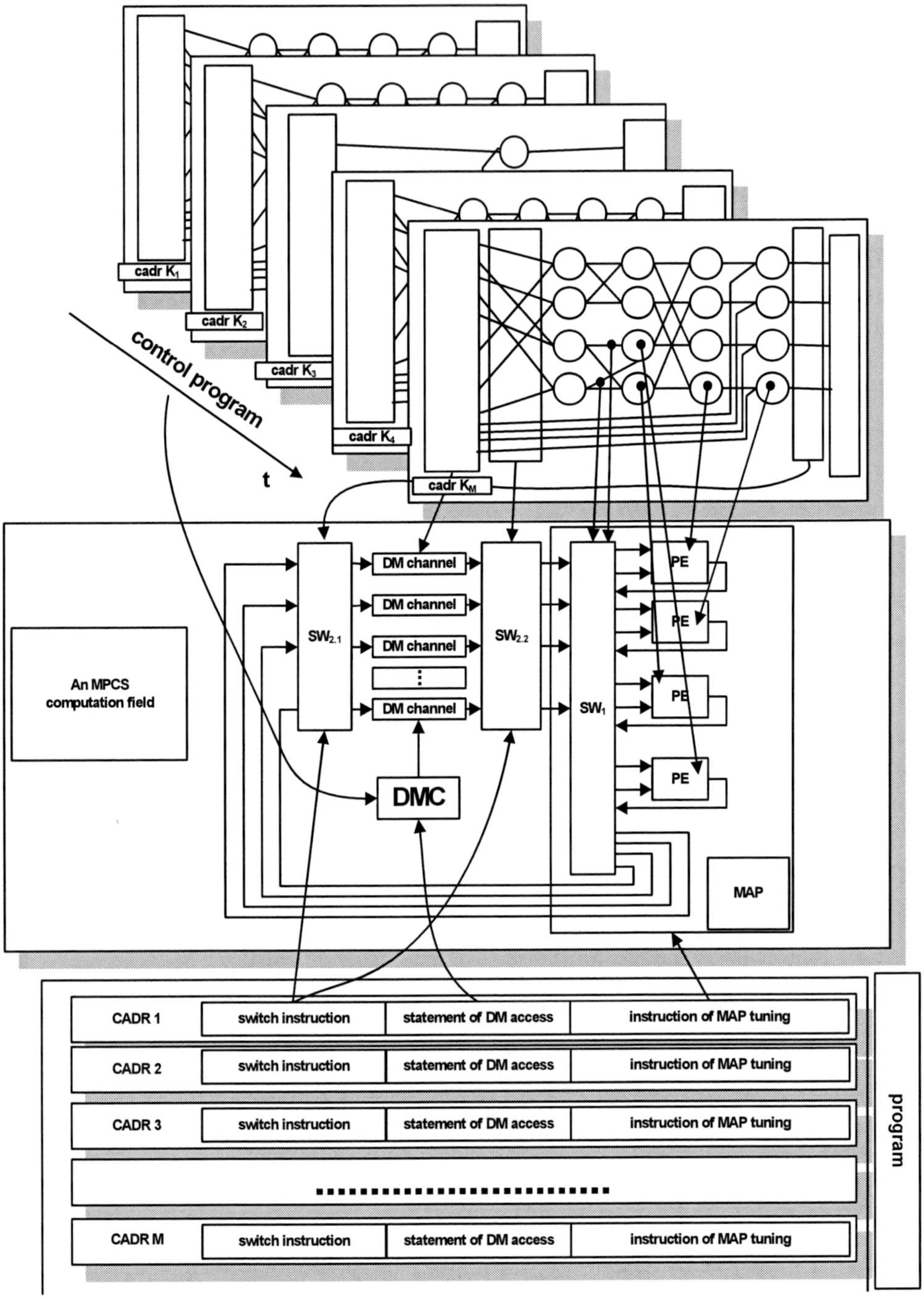

Figure 3.15. Solving the application problem on the MPCS.

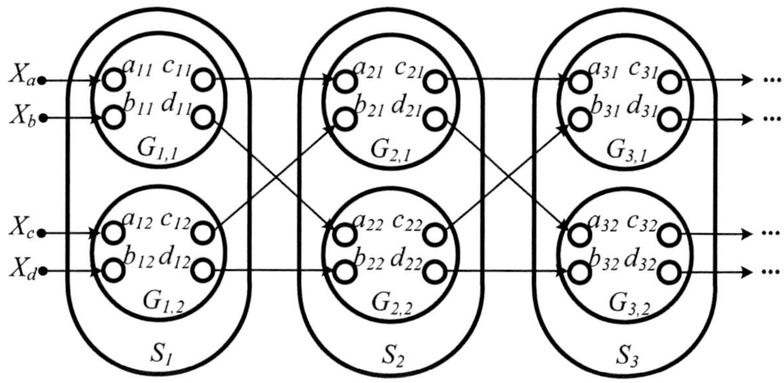

Figure 3.16. A fragment of a functionally-regular subgraph.

The time of cadr execution is obtained according to the equation

$$T_k = t_{pipeline}(Q) + t_{setup} + \tau N ,$$

where $t_{pipeline}(Q)$ is the time of cadr pipeline filling;

t_{setup} is the time of cadr adjustment;

N is the size of the data flow, passing through the cadr;

τ is the time of operand passing.

The time of operand passing is estimated according to the equation

$$\tau = \max(t_l(Q), t_0),$$

where t_0 is the time of operand passing in the flow;

t_l is the maximum feedback delay in Q.

If N is large, then the time of pipeline filling and the time of cadr adjustment can be neglected. The time of cadr execution may be estimated as

$$T_k \approx t_0 N .$$

However, if a recursive connection exists, then the time of cadr execution increases to the value

$$T_k \approx t_l(Q) N .$$

This time may significantly exceed the minimum time of cadr execution for rather large structural fragments. In the case, when the layer of the information graph fragment, which is transformed into the cadr, contains two basic subgraphs, non-recursive implementation requires fulfillment of the inequality

$$2t_0 \geq T_p .$$

The value T_p is defined as the time, required for writing of an operand and switching data flows by a switch:

$$T_p = t_{setup} + t_{setsw} + \tau .$$

In more general case, when the layer of the information graph fragment, that is being transformed into the cadr, contains n basic subgraphs, the requirement of recursive connection may be written as follows

$$t_0 \geq \frac{T_p}{]n/p[}, \qquad (3.3)$$

where p is the number of simultaneously executed basic subgraphs.

Therefore the granularity of parallelism p is to be increasing while the inequality (3.3) is being fulfilled. Otherwise, the time of problem execution will be considerably increased: if the layer of the fragment contains n basic subgraphs, then, with execution of all basic subgraphs of the layer, the time of execution of the whole problem is obtained according to the equation

$$T_k = t_{pipeline}(Q) + t_{setup} + t_l L ,$$

where L is the number of layers.

If the number of executed subgraphs is m , then

$$T_k \approx t_Q + t_0 L]n/p[, t_l = mt_0 .$$

Hence, effective organization of computations requires the following inequalities to be fulfilled:

$$Lmt_0 \leq t_0 L]n/p[$$

or

$$m \leq]n/p[.$$

Usually $\sqrt{N} \le]n/p[$, and hence it is possible to organize structural calculations with no feedbacks in the cadr graph.

The functionally-regular graph may be implemented as the only cadr, represented as the data flow passing through the hardwarily implemented subgraph of the problem. Since the subgraphs g in the cortege are information-independent and may be executed simultaneously, then, during transformation of the functionally-regular graph into the cadr form, it is possible to implement cadr structures for various granularity of parallelism.

3.5. Transformation of the Poisson Problem into the Cadr Form

As an example let us consider the cadr form transformation of a problem of numerical solution of partial differential equation in a rectangular domain D. Assume that the Poisson differential equation

$$\frac{\partial^2 U}{\partial^2 x} + \frac{\partial^2 U}{\partial^2 y} = f(x, y) \tag{3.4}$$

is to be solved with the boundary conditions of the first kind.

Let us analyze a numerical solution of the differential equation in grid points of a uniform grid. After transformation of the equation (3.4) into a finite-difference form [103], the continuous functions $U(x, y)$ and $f(x, y)$ are replaced by a set of values of grid nodes $U_{i,j}$ and $f_{i,j}$, and a system of linear algebraic equations is generated. To solve the system an iterative method of simultaneous under-relaxation [104] is used. According to the method the k-th approximation of the grid function $U_{i,j}^k$ in the grid points (x_i, y_i) is obtained according to the recurrent formula

$$U_{i,j}^k = c_1(U_{i,j-1}^{k-1} + U_{i,j+1}^{k-1}) + c_2(U_{i-1,j}^{k-1} + U_{i+1,j}^{k-1}) + c_3 U_{i,j}^{k-1} - f_{i,j},$$

where c_1, c_2 and c_3 are the coefficients of the differential equation.

The equation is solved for indices $i = 1, 2, ..., N-1$, $j = 1, 2, ..., M-1$. The values M and N are defined by horizontal and vertical dimensions of D and by the grid step. Iterative process is to be continued, until the following inequality is fulfilled for all values of i and j

$$\left| U_{i,j}^r - U_{i,j}^{r-1} \right| \le \varepsilon,$$

where ε is a constant value which defines accuracy of the solution;

r is the iteration index.

Thus, computational process of the problem solving may be represented as an ordered sequence of self-contained computational fragments $<P^1, P^2, ..., P^k>$, where k is the number of the last iteration. The result of execution of P^k is the set $U^k = \{U^k_{i,j}, i = 1,2,...,N-1; j = 1,2,...,M-1\}$. The elements of U^k are input data for P^{k+1}, during execution of which the set $U^{k+1} = \{U^{k+1}_{i,j}, i = 1,2,...,N-1; j = 1,2,...,M-1\}$ is obtained.

Should computational process of the solving problem be represented as a computational graph G, an information subgraph S^k will correspond to each fragment P^k. Informational dependence between P^k and P^{k+1} is defined by the set A^k of edges, that connect subgraphs S^k and S^{k+1}. For any edge, contained in A^k, its operational vertex-source is included into S^k and its operational vertex-target is included into S^{k+1}. Hence, G may be represented as the cortege of information-dependent subgraphs $<S^1, S^2, ..., S^{k\_end}>$.

If the number of processing elements of the reconfigurable structure is limited, then the problem, represented in the parallel form, is to be transformed into the structural-procedural form, i.e. in the cadr form. In this case the initial information graph is to be partitioned into a set of isomorphic successively executed subgraphs. The number of operational vertices in each subgraph cannot exceed the number of processing elements, allocated for the problem. The communication system must provide connections between the processing elements according to the edges of the subgraph.

For transformation of the subgraph cortege into the cadr form, all subgraphs of the cortege are to be isomorphic, and a rule Φ of mapping of one subgraph on another is required. In addition, it is desirable to represent the subgraph of the cortege as the set of information-independent basic subgraphs, and to define a rule of mapping of basic subgraphs. Such rules provide effective transformation of information graphs into the cadr form. On the base of the obtained cadr form a scalable program, executable at various types of structural-procedural parallelization of computations, can be created.

For the Poisson problem both of conditions are fulfilled. Since all the computational fragments P^k are identical, then all the information subgraphs S^k are isomorphic. The mapping rule Φ is uniquely defined by A^k. Each S^k may be represented as a set of isomorphic basic subgraphs $\{g^k_{i,j}, i = 1,2,...,N-1; j = 1,2,...,M-1\}$. The basic information subgraph $g^k_{i,j}$ corresponds to calculation of $U^k_{i,j}$ in a grid node. If a problem of mathematical physics is solved by the method of simultaneous displacements, then the basic subgraphs, included in one and the same subgraph S^k, are information-independent, because S^k does not contain any edge, that connects an operational vertex-source of one basic subgraph and an operational vertex-target of another basic subgraph. Owing to informational independence of $g^k_{i,j}$, which are contained in one and the same S^k, it is possible to develop a scalable parallel program for solving mathematical physic problems.

For structural implementation of the basic subgraph in the reconfigurable structure, the number of processing elements, which corresponds to the number of the operational vertices of the basic subgraph, is allocated. The number of vertices of the basic subgraph is equal to the number of arithmetic and logical operations. The edges of the basic subgraph are implemented in the communication system of the reconfigurable structure. According to programming methods of mathematical physics problems, the structurally implemented basic subgraph, which corresponds to calculations performed in a grid node, will hereinafter be called *a nodal processor*. Figure 3.17 shows the basic subgraph of the Poisson problem.

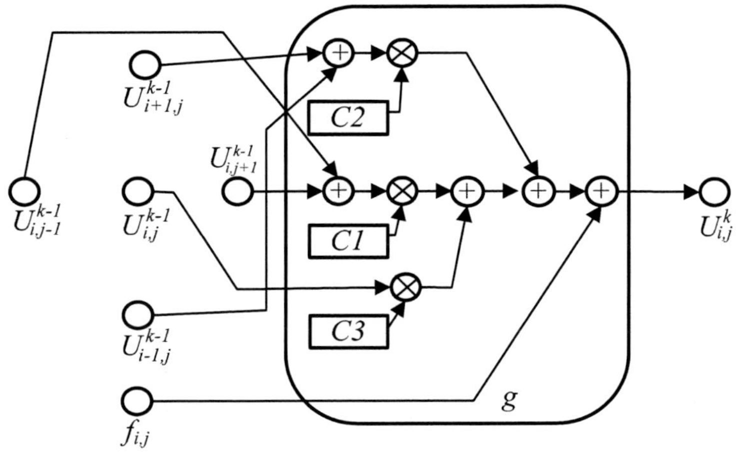

Figure 3.17. The basic subgraph of the Poisson problem.

Each $g^k_{i,j}$ contains six input information vertices

$$< f_{i,j}, U^{k-1}_{i,j}, U^{k-1}_{i,j-1}, U^{k-1}_{i,j+1}, U^{k-1}_{i-1,j}, U^{k-1}_{i+1,j} >$$

and one output information vertex $U^k_{i,j}$.

Intermediate results of calculations of $U^k_{i,j}$ on the k-th iteration are written into the distributed memory of the reconfigurable structure. There is a one-to-one correspondence between each approximated value of the grid function $U^k_{i,j}$ and the vertex of the same name, contained in the information graph.

When the information vertex $U^k_{i,j}$ is added into the graph, then every edge of A^k, is replaced by a pair of edges, which are incident to $U^k_{i,j}$. For the first edge the information vertex $U^k_{i,j}$ is the target vertex. For the second edge it is the source vertex. Figure 3.18 shows the information graph of the Poisson problem, partitioned into a number of the basic subgraphs $g^k_{i,j}$.

For solving the problem of mathematical physics on L nodal processors, each information subgraph S^k is represented as a set of isomorphic subgraphs $\{F_m, m = 0,1,...,q-1\}$. To obtain q the following equation is used:

$$q = (M-1)\left[\frac{N-1}{L}\right].$$

Each subgraph F_m is the set of L basic subgraphs (or nodal processors).

Hereinafter such group of basic subgraphs will be called *an L-node processor*. Let us denote the sets of input and output information vertices of F_m as X_m and Y_m, respectively. The set X_m is the union of subsets, which contain the input information vertices of the basic subgraphs, included into F_m.

Similarly, Y_m contains the output information vertices of the basic subgraphs, included into F_m.

For cadr synthesis, $\{F_m\}$ is to be transformed into a cortege which uniquely defines the sequence of operations of the L-node processor. Here the relation, which sets correspondence between X_m and the time t_m of carrying out the operation of the L-node processor, is defined.

The inverse transformation is a reading function. A writing function can be defined similarly. Hereinafter, while synthesizing access procedures to data, located in the distributed memory channels, the reading (writing) function is transformed into a function of "addressing of reading (writing)" and into a function of "commutation of reading (writing)". The addressing function specifies successive access to data located in one and the same memory channel, and the communication function specifies concurrent access to a group of data located in different memory channels.

When the problem is transformed into the cadr form, the data structure, which imposes constraints on implementation of one or another variant of problem parallelization, is generated.

For instance, if the problem is solved by the method of simultaneous displacements, it is necessary to allocate different areas M_1 and M_2 of distributed memory for the data U^k and results U^{k+1} of the grid function. In this case, when the odd fragments P^{2k+1}, $k=0,1,2,...$ are executed, $U_{i,j}^{2k}$ is transferred to the inputs of the L-node processor from M_1. The results $U_{i,j}^{2k+1}$ are written into M_2. When the even fragments P^{2k}, $k=0,1,2,...$ are executed, $U_{i,j}^{2k-1}$ is transferred to the inputs of the L-node processor from M_2, and the results $U_{i,j}^{2k}$ are written into M_1. Each iteration is put in correspondence with a cadr C. The whole iterative process may be represented as a loop, in which the cadr C is executed.

Another constraint is a conflict-free data location in distributed memory, which guarantees, that concurrent access to data, placed in one and the same channel, but in memory cells with different addresses, will never be required. Data location in memory of reconfigurable structures is specified by generation of the subgraphs F and by their order in the cortege.

If S^k is arbitrarily partitioned into the set $\{F_m\}$, then X_m may contain up to $6L$ information vertices.

Concurrent data transfer to the inputs of the L-node processor will require up to $6L$ distributed memory channels, and the addressing and switching functions may be too complicated for implementation in reconfigurable structures.

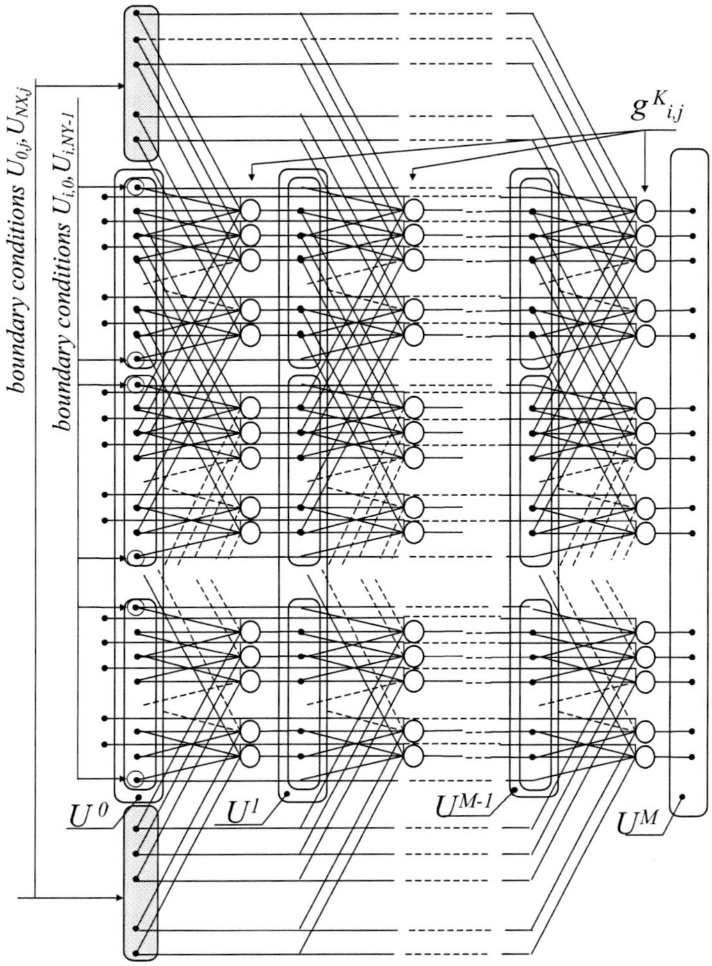

Figure 3.18. The information graph of the Poisson problem.

At the same time, there are several rational variants of partition of S^k into $\{F_m\}$, that simplify addressing and switching procedures. The most interesting partition is the one, in which F_m corresponds to calculations, carried out in adjacent grid nodes of one and the same column, and contains the basic subgraphs

$$F_m = \varphi\{g^k_{r+p,s}\}, \tag{3.5}$$

where indices r and s are functionally related to the index m.

Hereinafter, the index p possesses the values 0, 1, 2, ..., L-1. The subgraphs of the set (3.5) correspond to calculation of $\{U^k_{r+p,s}\}$ of the grid function in the s-th column, starting with the r-th row of the grid.

Let us analyze the generalized formula, which provides transformation of one-dimensional indexing of a sequence of object groups, specified by an index x, into a two-dimensional indexing of columns and rows $< y_1, y_2 >$, when the groups are located in a rectangular region. The first group, which contains K objects, is located in the initial column y_2^0, starting with the row y_1^0. The next group is located in the column $y_2^0 + 1$, etc. After placement of one of such groups in the last column, the next group will be located in the initial column, starting with the row $y_1^0 + K$. For such placement of the object groups, the row number y_1 and the column number y_2 relate to the one-dimensional index x according to the equations:

$$y_1 = y_1^0 + K\left[\frac{x + x_0}{Z}\right] \tag{3.6}$$

$$y_2 = y_2^0 + \text{mod}_Z(x + x_0) \tag{3.7}$$

where [*] is the operation of rounding down;

$\text{mod}_a(b)$ is a residue of division a/b;

y_1^0 and y_2^0 are the initial values of the indices;

Z is the number of elements in the row;

x_0 is the basic displacement.

Owing to use of the basic displacement, the column index y_2 can be transformed into y_2^0 and the row index y_1 can be transformed into y_1^0, if the initial value x differs from zero.

The indices r and s may be obtained according to the equations (3.6) and (3.7), respectively. It is obvious, that x corresponds to the index m of F_m, and y_1 and y_2 correspond to r and s, respectively. When calculating the grid values, solely the internal grid nodes are processed, therefore the number of processed nodes, which are contained in one row of the grid, is $M - 1$. The coefficient K defines the number of the basic subgraphs of the set (3.5) and is equal to L. Since the initial row and column indices of the grid nodes, that are being processed, are equal to unity, and the indices of the basic subgraphs correspond to the indices of the grid nodes, then the initial displacements y_1^0 and y_2^0 are equal to unity. The basic displacement $x_0 = 0$, because indexing of F_m starts with $m = 0$. Hence, functional dependences between the one-dimensional index m of F_m and the indices r and s are

$$r = 1 + L\left[\frac{m}{M - 1}\right];$$

$$s = 1 + \text{mod}_{M-1}(m). \tag{3.8}$$

For solving the Poisson equation a 5-point pattern is used. Let us denote the pattern point (x_i, y_j) as the central point, and the points (x_{i-1}, y_j) and (x_{i+1}, y_j) as upper and lower, and the points (x_i, y_{j-1}) and (x_i, y_{j+1}) as left and right. Similar denotations may be used for the input information vertices of the basic subgraph $g_{i,j}^k$. The input information vertex $U_{i,j}^{k-1}$ is the central one, and the vertices with indices $(i-1, j)$, $(i+1, j)$, $(i, j-1)$, and $(i, j+1)$ are upper, lower, left and right, respectively.

When the subsets of the input information vertices, which are contained in the subgraphs of the set (3.8), are united, then information vertices, that have identical names, are "absorbed". The upper vertex of $g_{r,s}^k$, the lower vertex of $g_{r+L-1,s}^k$, and the left and right vertices of all $\{g_{r+p,s}^k\}$ are unique. At the same time, the lower vertex of $g_{r+p,s}^k$ and the central vertex of $g_{r+p+1,s}^k$ have identical names. Similarly, the upper vertex of $g_{r+p,s}^k$ and the central vertex of $g_{r+p-1,s}^k$ have identical names. Hence, when the information vertices are united, the central vertices $\{U_{r+p,s}^{k-1}\}$, the left vertices $\{U_{r+p,s-1}^{k-1}\}$, the right vertices $\{U_{r+p,s+1}^{k-1}\}$, and the unique vertices $U_{r-1,s}^{k-1}$ and $U_{r+L,s}^{k-1}$ are included into X_m.

The set Y_m of the output information vertices of F_m contains $\{U_{r+p,s}^k\}$. These output vertices are the central input vertices of F_m, which are contained in S^{k+1}.

When the subgraph cortege $K_F =< F_0, F_1, ..., F_{q-1} >$ is transformed into the cadr $C =< R, Q, W >$, the sets of the input information vertices X_m are replaced by the reading function $r(t)$, and the sets of the output information vertices Y_m are replaced by the writing function $w(t)$. The reading function $r(t)$ puts the time moment t_m in correspondence with X_m. Similarly, the writing function $w(t)$ puts the time moment t_m in correspondence with Y_m. The cadr component Q is the graph, whose vertices and edges correspond to the operational vertices and operational edges of F_m.

According to the accepted order of F_m in K_F, the equation for $r(t_m)$ may be written as

$$r(t_m) = \left\{ \left\{ f_{r+p,s} \right\}, U_{r-1,s}^{k-1}, \left\{ U_{r+p,s-1}^{k-1} \right\}, \left\{ U_{r+p,s}^{k-1} \right\}, \left\{ U_{r+p,s+1}^{k-1} \right\}, U_{r+L,s}^{k-1} \right\}.$$

So, on each cycle t_m, $4L+2$ values are to concurrently enter the L-node processor for parallel calculation of $\{U_{r+p,s}^k\}$. If on the cycle t_m all data is to be read directly from memory channels, then the minimum number of memory channels will be $4L+2$.

If the number of distributed memory channels is limited, it may be insufficient for use of all processing elements of the reconfigurable structure. In this case effectiveness of the problem solution will decrease. However, it is possible to reduce the number of distributed memory channels, that are necessary for the problem.

Let us partition the computational fragment P^k into a number of single-type fragments p_i^k. The fragment p_i^k corresponds to processing of the grid nodes, which are located in the columns, starting from the first to the $(M-1)$-th one. In this case $C<R,Q,W>$ is partitioned into $(N-1)/L$ cadrs C_i. The reading function r_i of C_i is the reading function r of the initial cadr C, determined in the range $T_i=[t_a,t_b]$, where $a=(r-1)(M-1)$, $b=r(m-1)-1$. Similarly, the writing function $w_i(t)$ of C_r is the writing function $w(t)$ of the initial cadr C, determined in the range T_r.

For a certain set $\{H_i\}$, $i=1,2,...,N_H$ a function $Z(H_i)$, which puts the element H_i in correspondence with the element H_{i-1} except H_1, is defined as follows:

$$Z(H_i)=\begin{cases}H_{i-1}, & \text{if } i>1\\ G, & \text{if } i=1\end{cases} \tag{3.9}$$

Taking into account Z, the composition $Z(Z(H_i))=Z^2(H_i)$ is defined:

$$Z^2(H_i)=\begin{cases}H_{i-2}, & \text{if } i>2,\\ G_1, & \text{if } i=2,\\ G_2, & \text{if } i=1.\end{cases} \tag{3.10}$$

The functions $Z(H_i)$ and $Z^2(H_i)$, which accept the vector H_i as an argument, can be similarly defined.

Let us denote $U^{k-1}_{r,s+1}=\{U^{k-1}_{r+p,s+1}\}$ and $f_{r,s}=\{f_{r+p,s}\}$. Then the values of $r(t_m)$ may be written as follows

$$r(t_m)=\left\{f_{r,s},Z(U^{k-1}_{r-1,s+1}),Z^2(U^{k-1}_{r,s+1}),Z(U^{k-1}_{r,s+1}),U^{k-1}_{r,s+1},Z(U^{k-1}_{r+L,s+1})\right\}.$$

Hence, it is obvious that $4L+2$ memory channels are not needed for implementation of the reading function. It is sufficient to concurrently read $f_{r,s}$ and $U^{k-1}_{r,s+1}$, and $U^{k-1}_{r-1,s+1}$ and $U^{k-1}_{r+L,s+1}$ from the memory channels on the cycle t_m. Entry of $U^{k-1}_{r,s}$ and $U^{k-1}_{r,s-1}$ to the L-node processor is provided by means of the buffer memory, which performs the functions Z and Z^2. Thus, for obtaining $U^{k-1}_{r,s}$ on the cycle t_m, the values $U^{k-1}_{r,s+1}$ enter the L-node processor directly from the memory channels. The values $U^{k-1}_{r,s}$ and $U^{k-1}_{r,s-1}$ are read from the buffer memory, in which they were written on the cycles t_{m-1} and t_{m-2}, respectively.

Use of one operand delay elements is a variant of buffer memory implementation. A two operand delay may be implemented by sequential connection of two elements of one operand delay. In this case $U^{k-1}_{r,s}$ enter the L-node processor with one operand delay, and $U^{k-1}_{r,s-1}$ – with two operand delay. Since $U^{k-1}_{r-1,s}$ and $U^{k-1}_{r+L,s}$ enter the L-node processor with one operand delay, then they are to enter simultaneously with $\{U^{k}_{r+p,s}\}$. Such data entry requires two vector delay operations $z1$ and $z2$, and two delay operations $z3$ and $z4$.

Calculation of the grid function values $U_{r,l}^{k-1}$ of the first column is a special case. Here, $z1$ and $z2$ are to be preloaded with $U_{r,0}^{k-1}$ and $U_{r,l}^{k-1}$, and $z3$ and $z4$ are to be preloaded with $U_{r-1,1}^{k-1}$ and $U_{r+L,1}^{k-1}$. Then, when calculating $U_{r,1}^{k}$, the values $U_{r,0}^{k-1}$ and $U_{r,l}^{k-1}$ will enter the L-node processor from the delay elements $z1$ and $z2$; the values $U_{r-1,1}^{k-1}$ and $U_{r+L,1}^{k-1}$ - from the delay elements $z3$ and $z4$; the values $U_{r,2}^{k-1}$ – from the memory channels. So, use of the delay elements modifies the reading function of the cadrs. Figure 3.19 shows the structure chart of the L-node processor with the delay operations z.

The minimum number of memory channels, required for data concurrent entry to the L-node processor with the elements of one operand delay, is to be $2L+2$.

In $L+2$ memory channels the values of the grid function U are stored; the other L channels store the values of the function f. Results of calculations are simultaneously written into L memory channels. However, for implementation of the reading function one and the same input of the L-node processor is to be connected to different memory channels. A number of variants of connections is determined as the least common divisor of the number of memory channels, allocated for storage of U, and of the number of nodal processors.

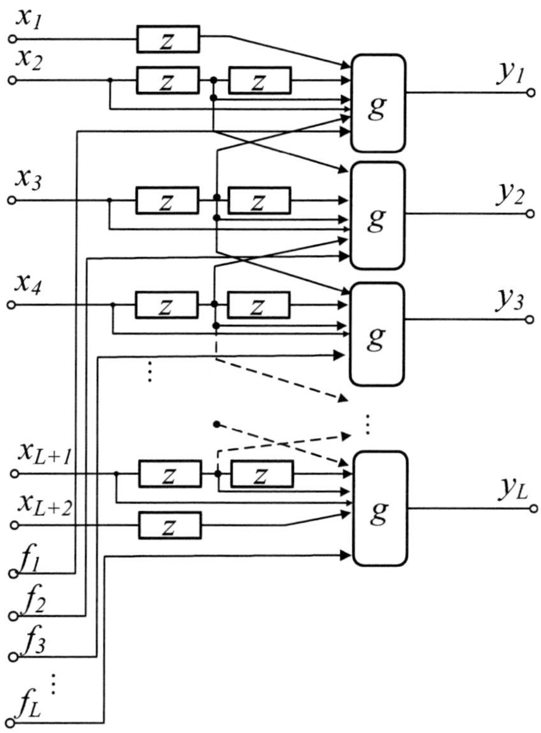

Figure 3.19. The structure chart of the L-node processor.

So, owing to increasing of the number of the channels, that store $U_{i,j}^{k-1}$, up to $2L$, only two kinds of connections may be used. The current variant of parallelization requires $3L$

distributed memory channels in total. When creating the data structure, $U_{i,j}^{k-1}$ with equal j are to be located in one and the same memory channel. In this case, there is no need to change connections with memory, when reading elements of one and the same row from the memory channels.

According to the reading function, data is read from memory in ascending order of i. Therefore, in order to simplify the addressing procedure, the elements of the grid row are to be located in the memory channel in a natural order. Owing to the natural order of data location, it is possible to obtain simple recurrent formulas for calculation of sequences of data reading and writing addresses.

Owing to the equations (3.9) and (3.10), two-dimensional data addressing $< N_S, A_C >$ may be used in the distributed memory channels. In this case, an object, located in the distributed memory channels, is a grid row, that contains $(M+1)$ elements. The variable x corresponds to the row number i, and y_1 and y_2 correspond to N_S and A_C, respectively. The rows of the grid are located in $2L$ distributed memory channels.

The choice of the basic displacement x_0 is specified by the channel number, from which location of the grid rows starts. If the initial channel of the grid row is N_S^0, then the basic displacement is equal to zero. Hence, $< N_S, A_C >$, corresponding to $U_{i,j}^{k-1}$, are obtained according to the equations

$$A_C = (A_C^0 + j) + (M+1)\left[\frac{i}{2L}\right]; \tag{3.11}$$

$$N_C = N_C^0 + \mathrm{mod}_{2L}(i).$$

Owing to the data structure, the order of the connections between the memory channels and the inputs of the L-node processor in C_r, implemented according to location of U-values in the distributed memory channels, may be easily specified. When executing C_i with $i = 2[(r-1)/2L]$, the set of data

$$< U_{r-1,s+1}^{k-1}, U_{r,s+1}^{k-1}, U_{r+L,s+1}^{k-1} >$$

is to simultaneously enter the first $L+2$ inputs of the L-node processor on the cycle t_m from the range T_i.

It should be noted, that $r = (i-1)L + 1$. Here and then, the index of the element of the cortege corresponds to the number of the input of the L-node processor. The numbers of reading channels, connected to the inputs of the L-node processor in these cadrs, and numbers of writing channels, connected to the outputs of the L-node processor, may be obtained according to the equation (3.11).

So, the first input of the L-node processor is connected to the channel N_C^0, from which the values $U_{r-1,s+1}^{k-1}$ are read. The next L inputs of the L-node processor are connected to the channels $N_C^0 + j$ with $j = 0,1,2,...,L-1$. The last input is connected to the memory channel $N_C^0 + L$. The calculated values $U_{r,s}^k$ are written into the channels $N_C^0 + j + 1$. Numbers of distributed memory channels, connected to the inputs of the L-node processor in C_i for odd values i with $r = (i-1)L + L + 1$, are calculated in the similar way. Figure 3.20 shows C_2 and C_4, which use the abovementioned scheme of structural-procedural computations.

Owing to rather a simple functional dependence between parameters of reading (writing) cycles and the number L of nodal processors and the grid dimensions M and N, the parameters may be easily modified. Program implementation of the procedure, which calculates the grid function, is a loop with a loop variable r and four cadrs $K1-K4$, contained in the loop body.

In $K1$ and $K3$ the delay elements z are initialized by initial values, required for calculation of the grid function. In $K2$ and $K4$ values of the grid function for $s = 2,3,...,M-1$ are obtained. The predicate P_{out} is obtained according to the equation (3.10) and is written into a separate memory channel concurrently with calculation of the grid function values. After execution of the loop, the program executes a conditional jump depending on P_{out}.

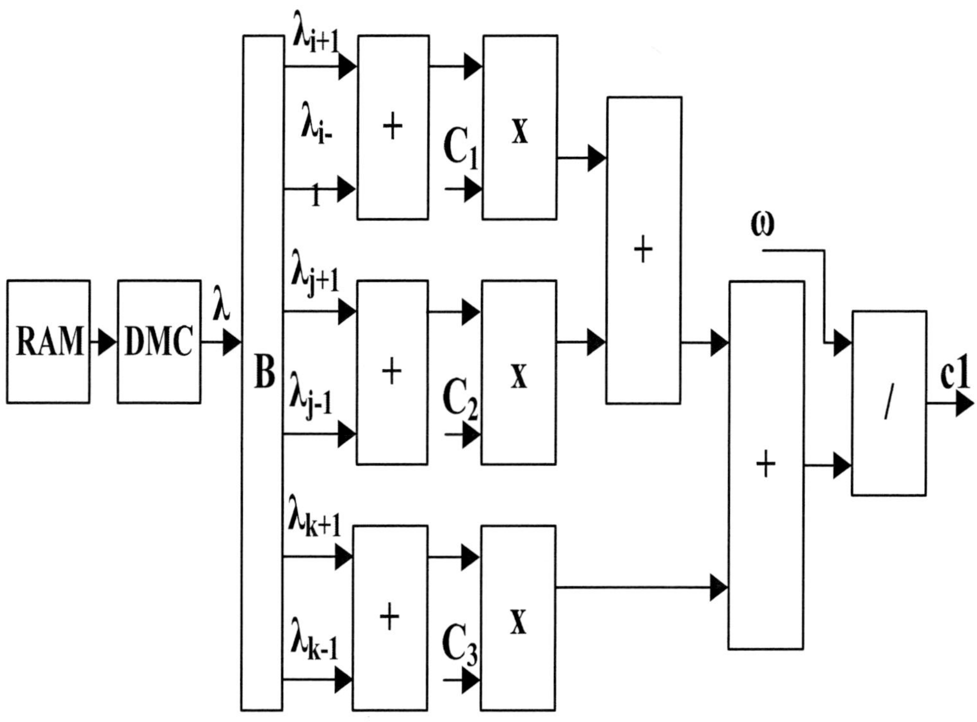

Figure 3.20. Structural-procedural computations in C_2 and C_4.

Other variants of organization of computations, that are defined by partition of the information graph, and by creation of the subgraphs, are also acceptable for the problem. The cadr form of the problem will be defined according to specific architectural features of reconfigurable structures, including the number of memory channels, the communication system topology, the number of processing elements, and the methods of organization of address calculations.

Structural-procedural organization of computations provides highly-effective problem solving on MPCSs, because data transfer between system components is combined with data processing in nodal processors. In addition, the cadr form of the problem provides a computational pipeline with a depth, which is determined by amount of processed data.

The time of problem solving on the MPCS may be estimated according to the following equation:

$$Tp \approx N_{it}\left\{M\left[(N-2)/L\right]t_{op} + 2(t_{pipeline} + t_{setup})\right\},$$

where N_{it} is the number of iterations;

t_{op} is the time of one operand processing;

$t_{pipeline}$ is the pipeline delay;

t_{setup} is the time of adjustment of computer system components.

When increasing dimension of the grid domain, the time values $t_{pipeline}$ and t_{setup} are reducing, and effectiveness is approaching to unity. As a result, almost linear growth of the real performance of the MPCS is guaranteed at increasing of the number of concurrently functioning nodal processors.

The analyzed example of implementation of the problem on the MPCS is rather simple. Nevertheless, it completely demonstrates abilities of the technology of problem transformation into the structural-procedural form. Owing to the technology of structural-procedural programming, the problem may be transformed from its initial representation into the form of a parallel program with no semantic gap. In addition, effective parallel programs for solving problems from various problem domains may be developed on-the-fly.

It is possible to extend the obtained results, when the information graph of the problem consists of several functionally-regular non-isomorphic subgraphs $G_1, G_2, ..., G_n$. In this case each subgraph $G_1, G_2, ..., G_n$ will be transformed into a separate cadr.

It should be noted, that reading/writing functions may be more complicated, than in the previous case. For instance, connections between the distributed memory channels and the elements of the structurally-distributed graph may vary. Detailed descriptions of structural-procedural algorithms of problems of mathematical physics are given in [16, 98, 105-109].

Below there is given analysis of transformation into the cadr form of a more complicated problem: a procedure of convolution in a frequency domain with the help of the fast Fourier transform (FFT) algorithm [110, 111].

3.6. TRANSFORMATION OF IRREGULAR INFORMATION GRAPHS INTO THE STRUCTURAL-PROCEDURAL FORM

It is possible to select several basic types of information graphs, that require special informational-equivalent transformations for their conversion to the structural-procedural form. Though such information graphs have a deterministic structure and may be represented as a cortege of information-dependent subgraphs, the functional dependence between the basic subgraphs is implicit. The distinctive feature of the information graphs is a pre-defined number of layers in each subgraph, but each layer contains a different number of basic subgraphs. In addition, there is a rule of mapping of the basic subgraphs within each layer. The rule of layer mapping is either too general, or specified for each particular variant of the problem. Another type of the information graph is typical to the problems, whose layers contain equal numbers of basic subgraphs, but the mapping rules of the basic subgraphs within each layer differ from each other, and the layer mapping rule is specified by each particular variant of the problem implementation. At last, information graphs of problems may contain layers with different and a priori unknown number of basic subgraphs. Cardinality of a layer, that contains the basic subgraphs, depends on computational results, obtained on previous layers of the graph.

A trivial solution to the problem of structural-procedural organization of computations for the above-listed types of graphs is creation of a parallel information graph, which has a rectangle multilevel structure, and transformation of this graph into the cadr form. However, the proposed method is ineffective on account of functional redundancy of such graphs: the majority of its basic subgraphs correspond to "void" (empty) operations, when the results of the operations are identically equal to input data [112].

If the procedure of graph sequencing does not directly give functionally-regular graphs, that provide an effective parallel program, then it is possible to make special informational-equivalent transformations of such graphs and to obtain the structural-procedural form as a sequence of cadrs. Besides, a computational pipeline can be organized and problems can be solved rather effectively on reconfigurable structures.

An algorithm of a non-regular problem is represented in the absolutely-parallel form, i.e. as a graph of computational process. Then functional redundancy of the graph of computational process is eliminated, and an intermediate functionally-irregular graph is formed. Finally, a special transformation, called vectorization of the information vertices of the graph [113], is performed. This transformation considerably differs from vectorization of loops and procedures [40, 114-116]. According to vectorization of the information vertices of the graph, a set of disjoint subsets of the information vertices, which are connected by a specified acyclic path, or ordered by any other deterministic rule within each subset, is replaced by a special object called *an array vertex*. The vertex corresponds to the position of the data structure in the distributed memory channels of the computing structure. Each subset of the information vertices, connected by some acyclic path in the information graph, or selected according to any other ordering rule, will be located in a separate distributed memory channel. The address of each vertex, placed in the channel, is put into one-to-one correspondence with the index of the vertex in the path, which connects the vertices in the graph, or with the index in the ordering cortege. As a result of vectorization, the information edges with the vectorized information endvertices are removed and replaced by the edges with array endvertices.

The vectorized graph is completed with address edges, each of which is formed according to the following rule: a uniquely named information address vertex, which corresponds to the address of the information vertex, is the start vertex of the address edge, and a label of the information edge, which outgoes (incomes) from (in) the array vertex, is its target vertex. The address corresponds to the number of the distributed memory channel, i.e. to the number of the information vertex subset, whose elements are ordered into a cortege.

Each address edge has the attribute t, which corresponds to the time mark of access to the cell in the distributed memory channel.

Let V_i be a set of information vertices of a graph P_i, represented as a cortege of layers $< S_{i,1}, S_{i,2}, ..., S_{i,K} >$ of information-dependent subgraphs

$$S_{i,j} =< (Vx_{i,j}, Ax_{i,j}), (Vq_{i,j}, Aq_{i,j}), (Vy_{i,j}, Ay_{i,j}) >,$$

where $Vx_{i,j}$ is a set of input vertices;

$Ax_{i,j}$ is a set of input edges;

$Vq_{i,j}$ is a set of operational vertices;

$Aq_{i,j}$ is a set of edges, which connect the operational vertices of the basic subgraph $q_{i,j}$;

$Vy_{i,j}$ is a set of output vertices;

$Ay_{i,j}$ is a set of output edges.

The set V_i may be represented as the union of two disjoint subsets VX_i and VY_i:

$$VX_i = Vx_{i,1} \cup Vx_{i,2} \cup Vx_{i,3}, ..., \cup Vx_{i,K},$$
$$VY_i = Vy_{i,1} \cup Vy_{i,2} \cup Vy_{i,3}, ..., \cup Vy_{i,K}.$$

When performing the procedure of vectorization, the set of information vertices VX_i is replaced by an array vertex W_X, and the set of information vertices VY_i is replaced by an array vertex W_Y. The edges, contained in AX, outgo from W_X. Each $a \in AX$ is described as follows:

$$a =< W_X, < F(x), t >, b, p >,$$

where W_X is the name of the array vertex;

$< F(x), t >$ is a pair of attributes;

b is a name of an operational vertex;

p is a number of an output.

The attribute $F(x)$ is the address attribute, which uniquely corresponds to the vertex name $x \in VX_i$, and t is the time mark, which indicates that x is contained in $S_{i,t}$. Similarly, the edges, which are contained in AY and income to W_Y, may be defined as

$$a =< b, p, W_F, < F(y), t >> .$$

In several cases it is sufficient to perform single vectorization of information vertices to obtain a functionally-regular graph and, hence, to formulate the mapping rules F and Φ.

Figure 3.21 shows the W-graph, obtained as a result of primary vectorization of information vertices of a functionally-irregular graph, that was formed for solution of a problem of histogram creation.

The procedure of histogram creation is widely used in various problems of signal and image processing. Here an operational vertex S corresponds to an increment operation. After transformation of a redundant information graph, the operation is performed on the corresponding element of the histogram array according to the obtained address. The address is obtained by $F(x_i)$, which determines what element of the histogram the accepted x_i refers to. The obtained results are written into the intermediate array vertex WX.

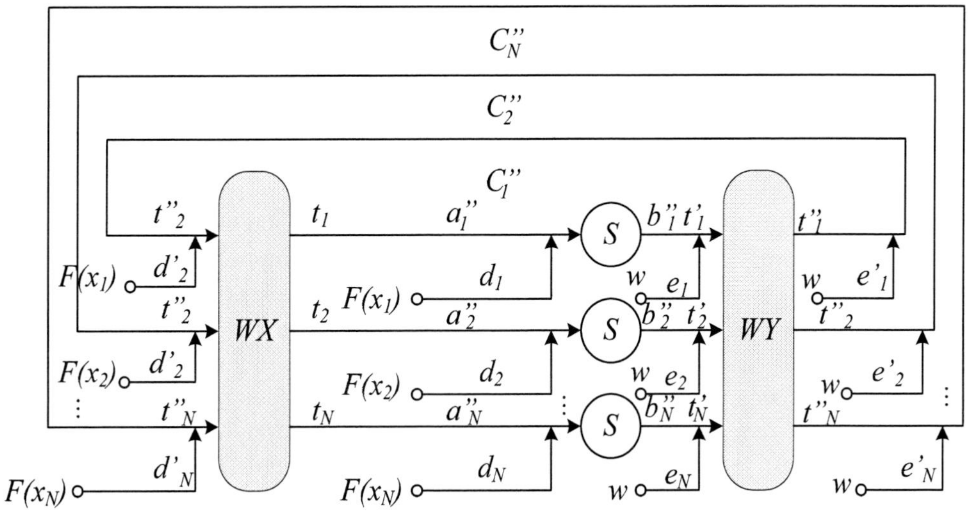

Figure 3.21. The W-graph, formed as a result of vectorization of vertices.

While solving a number of problems, for instance, a routing problem [128], single vectorization of information vertices is not sufficient and, therefore, secondary vectorization is required. In contrast to the primary vectorization, the array vertex contains the address attributes $F(x)$ and $F(y)$, which may be considered as information address vertices.

Figure 3.22 shows the W-graph, formed as a result of the secondary vectorization.

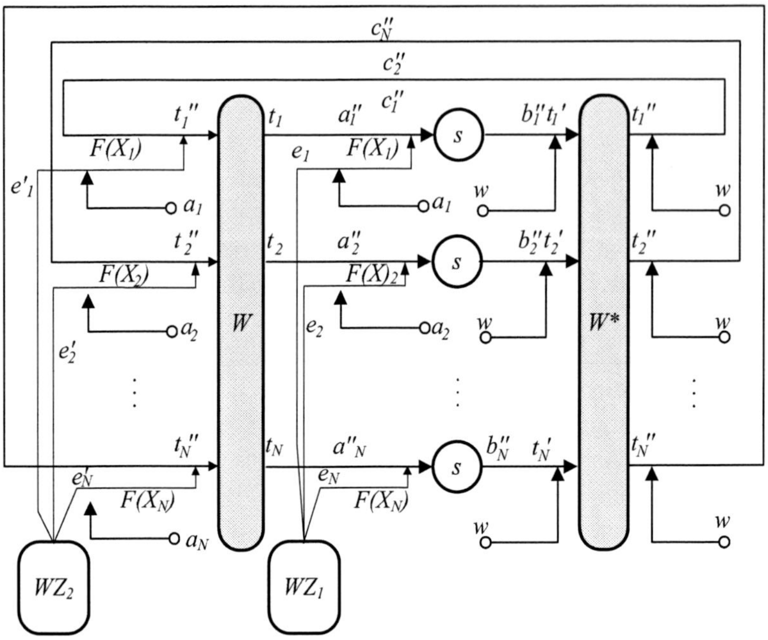

Figure 3.22. The secondary vectorization of address vertices in the W -graph.

Here WZ_1 and WZ_2 are the address array vertices, obtained by vectorization of the primary address vertices. As a result of the secondary vectorization of the address vertices, a functionally-regular graph is formed. The mapping rules of the edges of the subgraphs are determined by the mapping rules of the address vertices $a_i \rightarrow a_{i+1}$.

The operation of vectorization of information vertices allows the problem to be represented in the form of functionally-regular information subgraphs. The rules of mapping of basic subgraphs on each other are determined by the mapping rules of address vertices. The information subgraphs are transformed into the cadrs, that can be executed for arbitrary granularity of parallelism. Creation of a conflict-free data structure, and definition of addressing and switching functions for such problems have a particular feature, which, except certain special cases, requires an indirect addressing for vectorization of information vertices.

As a result, owing to the procedure of vectorization of information and address vertices, effective structural-procedural implementations can be obtained for several complicated problems with non-regular structure of information graphs, such as a problem of wave routing problem, analyzed in [117, 118].

3.7. Structural-Procedural Implementation of a Routing Problem

VLSI routing is one of the most important steps of microelectronic product design. Characteristics of a developed device considerably depend on quality of routing. Owing to modern microelectronic technologies, it is possible to design VLSI circuits with integration over 10^8 elements on chip. Rivalry in the market of microelectronic technologies requires reducing of the lead time. Besides, the time for development of VLSI topology becomes practically equal to the preproduction time. Therefore, introduction of high-performance computer aids, that will reduce the time of development of VLSI topology, becomes particularly urgent.

During the last years developers of VLSI circuits show a great interest in parallel computer systems, whose potential performance enormously exceeds performance of monoprocessor workstations. A wide range of research has been carried out in development of parallel algorithms for solving problems of computer-aided design of VLSI chips on multiprocessor systems with various architectures.

At the same time, the structure of the routing problem requires intensive data exchange between processors, when it is implemented on multiprocessor systems with distributed memory. It turned out that the time, needed for organization of parallel computations, considerably exceeds the time of computations. Hence, the ratio "real system performance/peak system performance" is about 0.05-0.1 [109,119] during execution of the routing problem. Owing to structural-procedural organization of computations, which is an alternative to well-known parallel algorithms, effectiveness of parallelization of the routing problem may be considerably increased.

At present, various routing algorithms exist, but the most universal one is *the Lee routing algorithm* [120]. In comparison with other algorithms, such as ray tracing and channel routing algorithms, which are faster and more cost-effective, the Lee routing algorithm has two significant advantages. Firstly, it provides selection of the trace with the minimum possible cost between two elements, if such trace exists. Secondly, it creates the maximum number of traces that may be routed. Taking into account a performance reserve of multiprocessor systems, implementation of the Lee routing algorithm is justified.

The solution to the routing problem, based on the Lee routing algorithm, consists of two steps. On the first step all possible decisions are determined. On the second step the decision, that fulfills all conditions of the problem, is selected. The most laborious and complicated for parallelization is wave expansion, therefore this step of the algorithm is considered in details.

A model of a VLSI chip for the routing problem implementation is a cell set D. Each cell $d_{i,j}$ of D, has a mass coefficient $W_{i,j}$, which defines the cell position on the field of the model. If the cell is free, then $W_{i,j} = 0$. If the cell is occupied, then $W_{i,j} \neq 0$. At the initial time all $W_{i,j}$ are equal to zero.

The source of a trace is the cell with coordinates $< i_b, j_b >$. If the cells, adjacent to the source of the trace, are free, then the wave is formed, and the mass coefficients W, corresponding to the adjacent cells, possess the value of unity. Then, the cells, adjacent to the ones with the modified W, are analyzed. If the cells are free, then their mass coefficients possess the current wavefront value. The wave is expanding until the trace target point with the coordinates $< i_e, j_e >$ is achieved. In the general case, on the k-th iteration the wavefront is formed, and it consists of coordinate pairs of the cells of the field. Here and then, it is assumed that a step of coordinate increment is constant for the whole field. Hence, a scaled unit increment of coordinates can be used.

Let us denote a set of coordinates of the cells, contained in the k-th wavefront, as B_k. Let us define a relation $Z(a,b)$, which indicates that a is adjacent to b. In the simplest case, the cells with the coordinates (X_i,Y_i) and (X_j,Y_j) are called *adjacent*, if the condition $|X_i - X_j| + |Y_i - Y_j| = 1$ is fulfilled.

The elements of the set A_k, which contains coordinates of the cells, adjacent to the elements of B_k, are determined according to the following relational formula:

$$A_k = \{(a,b) \mid Z(a,b) \wedge B_k(a,b)\}.$$

Here and then, $B(a)$ means that a, represented as $< x, y >$, is contained in B. The set B_k, which contains elements of the next wavefront, is determined according to the equation

$$B_k = \{f \mid (W_f = 0) \wedge A_k(f)\}.$$

The elements of W, whose indices are contained in B_{k+1}, possess $k+1$. The main idea of organization of parallel computations for the routing problem is simultaneous and independent calculation of all elements of A_k on each step of the algorithm. All coordinates, contained in B_k, may also be obtained simultaneously.

The information graph of numerical wave expansion is represented as a cortege of information layers $S =< S_0, S_1, ..., S_M >$. The layer S_k corresponds to the procedure, that obtains the elements of the k-th wavefront. The layer may be represented as a set of isomorphic basic subgraphs g. Each subgraph corresponds to generation of $W_{i,j}^k$ according to the equation

$$W_{i,j}^k = (k+1) \,\&\, P\{W_{i,j}^{k-1} = 0\} \vee W_{i,j}^{k-1} \,\&\, P\{W_{i,j}^{k-1} \neq 0\},$$

where $W_{i,j}^k$ is the element of W of the k-th iteration;

$P\{...\}$ is a predicate.

Basic subgraphs are implemented not for the entire array W, but for the elements $< i, j > \in A_k$, therefore the number of the basic subgraphs of S_k, and the topology of connections between the input and output information vertices of the layer are defined by the elements of W of the previous iteration. The non-modified elements of W are rewritten from the previous iteration. It corresponds to the edge, which connects the information vertices $W_{i,j}^{k-1}$ and $W_{i,j}^k$. Hence, the information graph of the algorithm is functionally-irregular, because, in the general case, there is no single rule $\Phi : S_{k-1} \to S_k$ of layer mapping.

At the same time, for effective implementation of calculations the number of cadrs should be reduced for providing a greater data flow through the pipeline of the computing structure.

Thereto, the graph of the problem is to be transformed in the functionally-regular form by means of several informational-equivalent transformations.

Figure 3.23 shows a fragment of the information graph of the problem solution. The fragment is simplified; it demonstrates the vertices that are adjacent according to one coordinate.

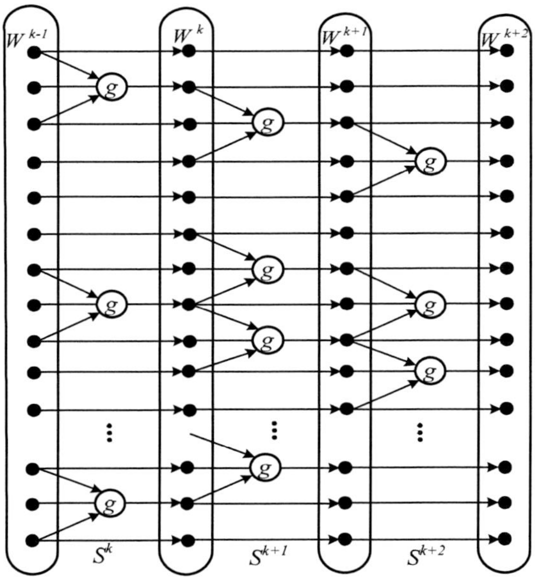

Figure 3.23. A fragment of the information graph of the problem solution.

One of the main methods of transformation of the problem graph into the functionally-regular form is vectorization. The path, that connects the cells $W_{i,j}^{k}$, which have the similar coordinates $<i,j>$, and are contained in different layers k, is used for ordering of the information vertices in the subset.

Figure 3.24 shows the vectorized subgraph of the problem, which corresponds to creation of the k-th wavefront. Each subgraph g_1 contains the number of the subgraphs f, that corresponds to the displacement of the wavefront element according to the adjunct relation Z in several directions. In the simplest case, the number of directions is 4. The set A^k is created on the base of B^{k-1} and contains elements, that are considered as an address of an array-vertex of the coefficients W of the field. The subgraphs g_2 form the elements of the next wavefront B^k. The similar graph is generated for each new wavefront.

After the primary vectorization, the graph of the wave routing problem remains functionally-irregular, because the ordering rule of address vertices of different wavefronts, that are specified by the parameters of the problem, is not determined. The secondary vectorization transforms the graph of the problem into the functionally-regular form.

When performing auxiliary vectorization of the vectorized graph, B_k is replaced by the array vertex B. Each information edge is put in correspondence with an address edge, which has an information address vertex M_B as a source vertex. The elements of the address vertex subsets are ordered according to the time marks. One and the same distributed memory channel may contain only those address vertices, which have different time marks.

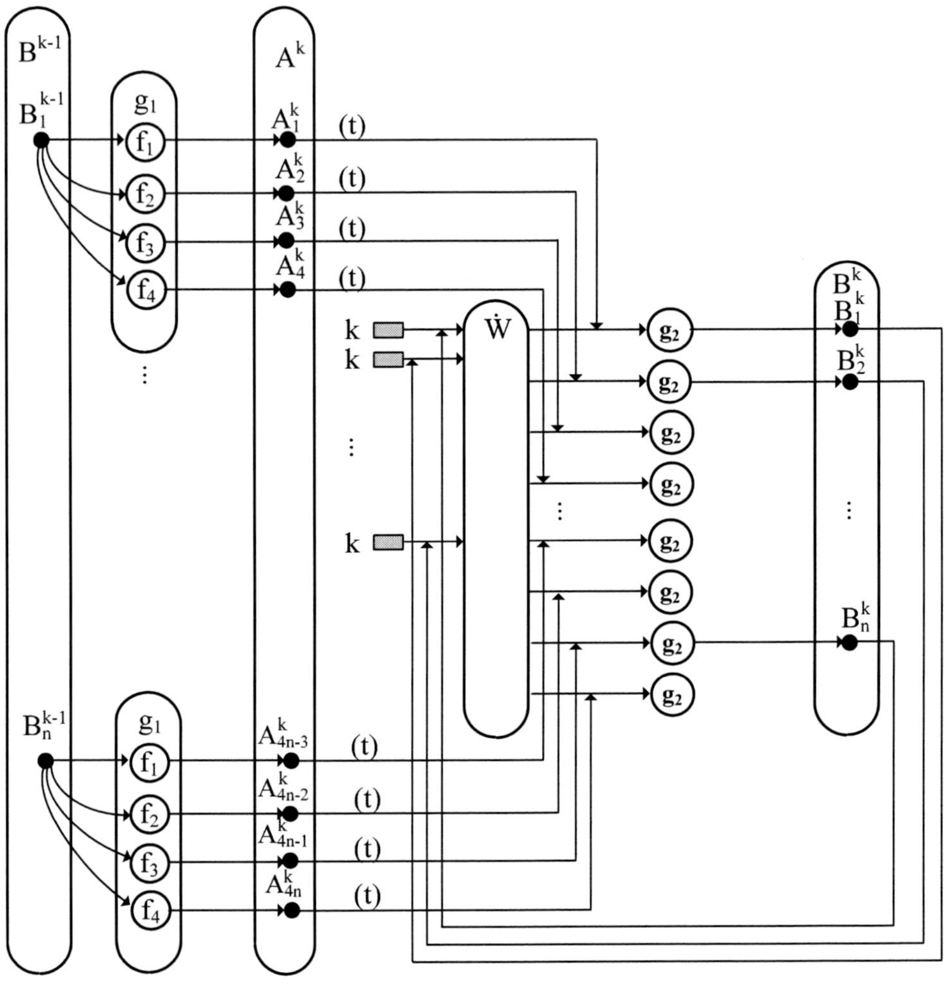

Figure 3.24. The vectorized graph of the problem.

The vectorized information graph, obtained as a result of vectorizations of the address vertices, is functionally-regular, because the single rule, which does not depend on results of data transformation, and which is applicable for mapping of the vertices of all layers of the graph, is defined.

A direct structural-procedural implementation of the vectorized graph as a single cadr is impossible, because it requires simultaneous access to different read and write addresses in the same distributed memory channels.

The reading function and the writing function put one input value and one result of calculations in correspondence with a time moment. Hence, the procedure of wavefront creation is implemented in two cadrs.

In the first cadr the following operations are performed: the coordinates of the adjacent elements A_k are obtained according to the coordinates of the wavefront elements B_k; the elements of W are read; the coordinates of B_{k+1} are obtained. In the second cadr the elements of W, whose addresses are contained in B_{k+1}, possess $k+1$. At the initial time moment, the array vertex contain the only one element, corresponding to the starting element of the trace Z_1^1. The starting element is placed in a cell with address $P1_1$.

With the help of the starting element of Z_1^1 and structurally implemented subgraph g_1, which contains 4 subgraphs f_1, f_2, f_3, f_4, each of which corresponds to a specific direction of adjacency from the starting element, the coordinates of four elements $q_1^2, q_2^2, q_3^2, q_4^2$ of the next wavefront are obtained.

These information vertices are the source vertices of the address edges, which are used to access to W. The elements, read from W, enter g_2. If the element, read from the address q_1^2, is equal to zero, then a_1^2, which corresponds to the coordinates of the next wavefront, is obtained; otherwise, no element of the wavefront is obtained.

The rest elements of the wavefront are processed similarly. The obtained coordinates $a_1^2, a_2^2, ..., a_{m2}^2$ of the wavefront are written in the array Mb according to the addresses $d_1^2, d_2^2, ..., d_{m2}^2$.

In addition, the elements of the wavefront are addresses themselves, according to which the constant, corresponding to the second wavefront, is written to the array vertex W. It should be noted, that the moment of time, when data is written in Mb and W, does not concur with the moment of data reading. For the k-th wavefront the coordinates are read from the array vertex $d_1^k, d_2^k, ..., d_{mk}^k$ of the wavefront, which are transformed into the coordinates of the next wavefront $q_1^{k+1}, q_2^{k+1}, ..., q_{4mk}^{k+1}$ by means of the array of the subgraphs f_1, f_2, f_3, f_4.

With the help of structurally implemented subgraphs g_2, the wavefront elements are transformed into the wavefront elements $a_1^{k+1}, a_2^{k+1}, ..., a_{n_{k+1}}^{k+1}$. According to the addresses of these elements, data is written into the array vertex W. The elements are written in Mb. Processing of the wavefront elements continues, until one of them reaches the end point of the trace.

Development of a parallel application for the routing problem is not the only important issue. The parallel application is to be executed for different variants of parallelization and on different configurations of reconfigurable structures. If the granularity of parallelism of the problem is L, then the matrix W may be placed in L distributed memory channels. A pair of Cartesian coordinates $<x, y>$ corresponds to each element of the matrix. Placement in the distributed memory channels requires transformation from Cartesian coordinates into so-called communication-memorial coordinates (A, K), where A is the address in the memory channel, and K is the number of the memory channel.

The Cartesian coordinates $<x, y>$ are transformed into the communication-memorial coordinates according to the equation

$$K = \text{mod}_{Ly} A = x + N\left]y / L\right[,$$

where N is the number of elements in the row of the matrix of the field;

$]...[$ is rounding down.

Storage of wavefront coordinates requires an L-channel memory array B. To access the matrix element with the communication-memorial coordinates $< K, A >$, storage of Cartesian coordinates and structural implementation of their transformation into the communication-memorial coordinates are not needed. It is sufficient to write the address A and the channel number K in B.

In parallel implementation for L wavefront elements, specific transformations of communication-memorial coordinates correspond to operations of increment of Cartesian coordinates.

Increment of x: $x' = x \pm 1$ corresponds to similar increment of the address $A' = A \pm 1$. Positive increment of y corresponds to the following modifications of the address and the channel number:

$$K' = K \oplus_L 1,$$

$$A' = \begin{cases} A + N, & \text{if } K = L - 1, \\ A, & \text{if } K \neq L - 1. \end{cases}$$

Decrement of y corresponds to the following modifications of the address and the channel number:

$$K' = K \ominus_L 1,$$

$$A' = \begin{cases} A - N, & \text{if } K = 0, \\ A, & \text{if } K \neq 0. \end{cases}$$

As long as the granularity of parallelism may be less, than the number of subsets of the vectorized information vertices, then, in general case, conflicts of access to the distributed memory channels may occur. To prevent simultaneous operations with data, located in the cells with different addresses of one and the same memory channel, a special “monitor” can be developed. However, development of the hardware monitor unit requires more processing elements, than it is used for calculation of wavefront elements, and leads to considerable reduction of effectiveness of parallelization.

The problem of conflict-free access to the distributed memory can be solved by splitting each cadr of the procedure of wavefront expansion into 4 fragments (according to the number of directions). Figure 3.25 shows the structure chart of the cadr, where the subgraphs f_i correspond to the directions of wave expansion.

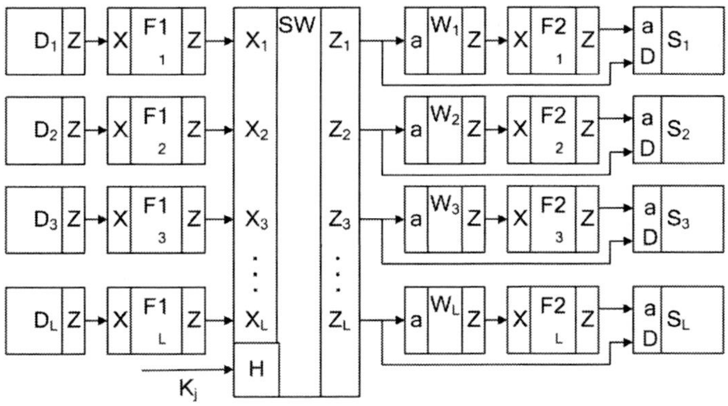

Figure 3.25. The structure chart of K_{2i}.

Let us analyse the structure of the cadr, which calculates coordinates of a new wavefront.

Here D_i is the set of distributed memory channels for storage of the wavefront coordinates, S_i is the set of distributed memory channels for storage of the obtained coordinates of the wavefront $k+1$. For D_i, $i=0,1,...,L$ reading of the address vectors is performed.

The structurally implemented fragments (macrooperations) $F1$ and the switch SW perform transformation of addresses and channel numbers, which corresponds to a unit increment of Cartesian coordinates.

The transformed addresses enter the address inputs of the distributed memory channels W_i. From the outputs of the memory channels that store W, data passes to the macrooperation $F2$, which forms the write addresses.

The write address $AW_{m,i}$ of the channel S_i is obtained according to the equation

$$AW_{m,i} = A_m \ \& \ P\{W_{m,i} = 0\};$$
$$A_{m,i} = A_{m-1,i} + P\{W_{m,i} = 0\}.$$

The wavefront coordinates enter the inputs D of S_i, and then are written into the memory cells. When the wavefront is formed, the buffer cell of each $F2$ contains the address of the last occupied memory cell. The difference between the address of the last occupied cell and the initial address is equal to a number v_i of occupied cells. When performing the cadr of calculation of the k-th wavefront, the read cycle parameter, which specifies amount of data, read from D_i, is obtained as $C_{k+1} = \max_i(v_i)$.

A detailed analysis of structural-procedural organization of computations for the routing problem is given in [118], which also contains a complex analysis of effectiveness of implementation of the wave routing problem on reconfigurable computing structures.

Hence, owing to structural-procedural organization of computations, implementations of complicated and computationally laborious problems from various domains may be effectively performed on MPCSs.

Chapter 4

FPGA-BASED RECONFIGURABLE MULTIPIPELINE COMPUTING STRUCTURES

As it was shown in Chapter 1, a multipipeline computing structure (MPCS) is a computation field that consists of a number of single-type computing elements such as cells, processing elements, macroprocessors, etc., and has a flexible system of interconnections. Multipipeline computing structures, similar to the information graph (or a subgraph) of a solving task, are created by programming of computing elements and adjustment of the communication system of the computation field.

Technical implementation of the concept of MPCS computation field adjustment to the information graph mapping requires appropriate components that provide any capabilities of hardware adjustment. Such components are to match the following basic criteria:

- large-scale integration for implementation of large computational fragments;
- relatively low price of small-scale production of computer systems;
- capability of design of special-purpose computing structures;
- capability of creation of various schemes of parallelization of computations;
- capability of reconfiguration of computing structures when algorithms and schemes of parallelization are changed;
- use of CAD-systems for design engineering solutions.

Let us analyze how modern element base matches these criteria.

4.1. COMPONENTS OF RECONFIGURABLE MULTIPIPELINE COMPUTING STRUCTURES

4.1.1. Custom VLSI Chips

Custom VLSI chips may be conditionally placed into two groups: full-custom VLSI chips and custom VLSI chips based on IP-cores (IP semi-custom VLSI chips). The full-custom VLSI chips have the most optimal structure, performance, die size and power consumption, and are used for mass production of various devices. It is assumed, that standard elementary units (registers, counters, decoders) and more complex units (bus transceivers, interfaces, etc.), included into libraries of standard units of CAD-systems, can be used for design of the IP semi-custom VLSI chips. Such approach provides reduction of the price of VLSI structure design with keeping optimal characteristics of performance and die size. But the main advantage of the approach is considerable reduction of the design period. Such VLSIs are called ASICs – Application Specific Integrated Circuits.

Disadvantages of the custom VLSI chips are high price of small-scale production, long time of design, modernization and verification. But the main disadvantage is a fixed structure, inadaptable to modifications of any task algorithm. The circuit engineer has to modify connections between transistors after any slight modification of the task algorithm. The results of such modifications are labour costs and financial expenses that are almost equal to the costs of a new project design. Hence, it is unreasonable to build multipipeline computing structures on the custom VLSI chips.

4.1.2. Uncommitted Logic Arrays

For further reduction of the design time and cost, special semi-custom VLSI chips were created. They were called uncommitted logic arrays (ULA). The abbreviation "ULA" reveals the main sense of the chips: standard elementary units are placed in the points of a rectangular lattice (matrix). Besides the standard elementary units, the ULA contains connection facilities (traces and standard interface units with customizable properties) for interconnection of the elements. To design a complete digital circuit in the ULA, it is enough to connect the elements by prefabricated traces or connections, created by the circuit engineer. As a rule, a photo mask, designed by the customer with a help of some CAD-system, is quite sufficient for creation of the traces. Besides, full-scale tests are not necessary for ULA-based circuit solutions, because, in most cases, they are operable. However, placement of the standard elementary units in the points of the rectangular lattice and some limitations of ULA tracing lead to some non-optimality and redundancy of the circuit solution in comparison with custom VLSI chips. The ULAs provide the performance, comparable with the performance of custom VLSI chips. Modification and revision of algorithms are much easier and cheaper in the ULAs than in custom VLSI chips, because it is necessary to retrace the connections between transistor structures, but not to modify the transistor structures themselves. Effective CAD-systems support design of photo masks that are used for interconnection of the logic units. At the same time, ULA modifications and revisions are practically equal to a new project design, though they are not so laborious and expensive as that of custom VLSI chips. Owing to these advantages, it was possible to build MPCSs of early generations on the ULAs. However, even slight modifications that may be made by MPCS developers, require design of new photo masks and reissue of chips. As a result, use of the ULAs in modern reconfigurable multipipeline computing structures is unprofitable because of the expense.

4.1.3. Systems-on-Chip

Owing to progress in the field of integrated circuit production, it has become possible to create chips that contain tens and hundreds of millions of transistors. However, such results lead to a contradiction between universality and speciality of the chips. On the one hand, a general-purpose VLSI chip, i.e. a microprocessor, has been quantity-produced and therefore has a low price. At the same time, the general-purpose VLSI chip is redundant for the most part of applications. This leads to non-optimal performance and power consumption of the designed circuit solutions, especially if they are used in quantity-produced devices. On the other hand, modern VLSI chips enable to place an electronic circuit of a designed device on a single chip. Such approach is quite effective in the terms of power saving, increasing of performance and reliability, reduction of production costs, etc. But it also leads to design of special-purpose VLSI chips, and as a result, to rising of production costs, because such chips cannot be quantity-produced. Creation of a new class of VLSI chips, called a System-on-Chip (SoC), has become a compromise solution.

The System-on-Chip may consist of a great variety of complex self-contained units, such as general-purpose processors, memory units, various controllers and co-processors, interface units, etc. Besides the self-contained units, the SoC usually contains programmable logic like FPGAs (Field Programmable Gate Array) or ULA-like blocks. Using the programmable logic, the circuit engineer can design various missing units that are necessary for his circuit solution.

Methodology of SoC design is based on IP-cores. The IP-cores are designed by the circuit engineer or specialized vendors of IP-cores. Multiple use of the IP-cores significantly reduces the time of design and production costs of the special-purpose VLSI chip – SoC, even for small-scale production.

The SoC components may be designed as custom special-purpose units or units implemented in the FPGA-part of the chip. In the first case, the IP-cores are presented as a layout configuration, and are implemented as custom units that are analogues of complex circuit components. In the second case, the IP-cores are presented as an HDL-description, as a fragment of an FPGA.

In both cases, the Systems-on-Chip are the custom VLSI chips, because the structure of the most IP-cores is fixed and its modification during execution of a task is impossible. This is the main disadvantage that reduces effectiveness of the SoCs, when they are used for building of reconfigurable computing structures.

4.1.4. Programmable Logic Devices

The main advantages of programmable logic devices (PLD) are:

- possibility of implementation of complicated parallel algorithms;
- availability of CAD-tools for complete system simulation;
- possibility of programming or modification of in-system configuration;
- compatibility of various design projects when they are converted in a VHDL, AHDL, Verilog descriptions or in any other hardware description language.

The history of the PLD architectures started in the end of the 1970s, when the first PLDs with programmable-AND and programmable-OR arrays appeared. Such architectures were called FPLAs (Field Programmable Logic Array) and FPLSs (Field Programmable Logic Sequencers). Their main disadvantage is weak use of programmable-OR array.

In the end of the 1980s PLD developers suggested new architectures that were simpler and cheaper: PALs (Programmable Array Logic) and GALs (Gate Array Logic). Such architectures were used in the PLDs of Intel, Altera, AMD, Lattice, etc. The PLDs had rather low integration density, a programmable-AND array and a fixed-OR array. Another approach to reduction of programmable-OR array redundancy was a programmable macro-logic. Macro-logic-based chips contained either a programmable-NAND array or a programmable-NOR array, but it was possible to form complicated logic functions owing to numerous feedbacks.

In the beginning of the 1980s there were three leading PLD vendors on the programmable logic device world market. Altera Corporation was founded in June 1983, Xilinx Inc. – in February 1984, Actel Corporation – in 1985. At present these three companies are holding about 80% of PLD world market and determine the ideology of PLD applications. In the past the PLD chips were only one product among various products of such enterprises as Intel, AMD, etc. But in the middle of the 1980s leading positions on the PLD market were taken by enterprises specializing solely in design and production of the PLDs.

New vendors offered new PLD architectures, such as CPLDs (Complex Programmable Logic Devices). The CPLD has rather high integration density and contains several functional blocks connected by a switch matrix. Each functional block contains a programmable-AND array and a fixed-OR array. This PLD class includes Altera MAX7000, Xilinx XC9500 and various PLDs of other vendors, such as Atmel, Vantis, Lucent, etc.

Introduction of FPGAs (Field Programmable Gate Array) ignited revolution of devices with programmable logic. The FPGA class includes Xilinx XC2000, XC3000, XC4000 and Spartan, Actel ACT1, ACT2, Altera FLEX8000 family and some Atmel and Vantis PLDs.

The FPGA configurable logic blocks (CLBs) are connected by a programmable switch matrix. The logic blocks consist of one or several rather simple logic cells based on a 4-input look-up table (LUT), a program-controlled multiplexer, a D-flip-flop. Input/output blocks (IOB) that provide bidirectional input/output, tri-state, etc., are typical for the FPGA-architectures. The FPGA chips have the following advanced features:

- a JTAG port that supports all mandatory boundary-scan instructions specified according to the IEEE 1149.1 standard;
- a master configuration mode (that required a build-in oscillator).

The FPGAs with a dedicated block RAM were the result of further development of the FPGA architecture, owing to which the FPGAs can be used with no external memory devices. The FPGAs have a high logic capacity, an easy-to-use architecture, a quite high reliability and an optimal ratio "price/logic capacity", therefore they match various requirements, claimed by circuit engineers.

During the last years, the FPGAs, along with custom VLSIs, have become a basis of systems-on-chip. IP-cores of such systems have been designed separately and can be used in various projects. The final structure of an FPGA-based SoC is implemented by means of CAD tools.

The ideology of SoC design spurred the leading FPGA vendors on, and in the end of 1998 – in the beginning of 1999 they introduced products with an equivalent logic capacity about a million equivalent gates and more. The Altera ApEX20K family is an example of new FPGA families for SoC design.

The Xilinx Virtex family has a similar architecture with a great variety of high speed routing resources, a dedicated block RAM, an optimal high-speed carry logic. The Virtex family provides a high speed of data exchange between the chips – up to 200 MHz (HSTL IV standard). The chips of the Virtex family have a relatively low price (not more than 40% from the equivalent price of XC4000XL series) owing to the advanced production technique and perfected verification methods.

In 1998-1999 augmentation of FPGA equivalent logic capacity had changed attitude to CAD-tools of both software developers and users. Till the end of the 1990s the main tool of project description and schematic entry was a graphic editor and libraries of standard primitives and macros such as logic elements, elementary combinational and sequential functional units, analogues of standard integrated circuits of small-scale and medium-scale integration. At present, circuit engineers widely use hardware description languages for FPGA-based implementation of algorithms. Besides, up-to-date CAD-tools support both standard hardware description languages (such as VHDL, Verilog) and specialized hardware description languages developed by FPGA vendors specially for their own needs, CAD-tools and FPGA families with special architecture features. Such example is AHDL (Altera Hardware Description Language), which is supported by the Altera CAD-tools MAX PLUS II and Quartus.

Xilinx corporation designs IP-cores of frequently used functional blocks, including blocks of digital processing, bus interfaces, processors and etc. Owing to design of the IP-cores, based on use of the Xilinx LogiCORE™ development package and on use of some software analogues developed by second-party vendors, it is possible to reduce the time of project design, to minimize risks, and to achieve the highest performance. In addition, with the help of the CORE Generator™ the circuit engineer can design his own IP-cores with predictable and repeatable timing characteristics. The CORE Generator™ has a quite simple user interface for generation of parameterized IP-cores specially perfected for use in the Xilinx FPGAs.

The Virtex-4 Xilinx family has an advanced architecture, designed on basis of modular blocks (Advanced Silicon Modular Block, ASMBL). The family consists of three subfamilies (LX, FX and SX), which, in the aggregate, match requirements of various areas of FPGA applications. The FPGAs contain several hard-IP cores capable to perform a number of standard tasks. The set of hard-IP cores contains processors PowerPC with a new APU interface, a Tri-Mode Gigabit Ethernet MAC (TEMAC), sequential transceivers with the transfer rates from 622 Mbit/sec to 11.1 Gbit/sec, a voltage and temperature system monitor, dedicated DSP slices, a high frequency clock distribution network and numerous interface blocks. The Virtex-4 basic blocks are improved versions of the similar blocks of the previous Virtex families: Virtex, Virtex-E, Virtex-II, Virtex-II Pro and Virtex-II Pro X, and are compatible with previously designed projects. The Virtex-4 FPGAs are produced on a 90-nm copper process using a 300-mm wafer technology.

The Virtex-4 FPGA is a regular structure that contains the following basic elements:

- input/output blocks (IOB);
- configurable logic blocks (CLB);
- block RAMs;
- built-in digital signal processing (DSP) slices;

- a digital clock manager (DCM);
- routing resources.

The configurable logic blocks CLB are connected to a switch matrix to access the general routing resources, and contain four slices, grouped into SLICEM and SLICEL pairs. Each pair contains an independent carry chain; besides, the SLICEM LUT can be configured as a distributed memory, or as a shift register.

Each CLB slice includes two function generators, two storage elements (D-flip-flops), MuxF5 and MuxFX multiplexers, fast carry logic with look-ahead, and arithmetic logic gates.

The function generators can be configured as 4-input look-up tables (LUT) for implementation of any 4-input Boolean functions. The delay of a signal, when it passes through the LUT, does not depend on the implemented function. Each function generator of the SLICEMs can be configured as a 16x1-bit distributed RAM, or as a 16-bit shift register (SRL) with no D-flip-flops, i.e. each LUT can delay the data sequence from 1 to 16 cycles. To produce long delays, the SRL can be cascaded using ShiftIn and ShiftOut pins. Any single CLB can produce a delay up to 64 cycles. Owing to such programming of delays, it is possible to create balanced pipeline structures.

The multiplexers effectively combine the look-up tables within any single CLB and between adjacent CLBs, providing implementation of various logic functions even with a relatively big number of variables. If a function has less than nine variables, it can be implemented in a single slice. An arithmetic logic is a XOR-primitive for 2-bit addition within a single slice. An additional AND-primitive improves routing of multiplication.

In 2006 Xilinx announced a new FPGA family – Virtex-5. The Virtex-5 FPGAs were built on a 65-nm copper process and with the core supply voltage 1 V. At present, the FPGAs with a capacity up to 20–25 million equivalent gates are available.

The FPGA architecture cardinally changed. Now the function generators are 6-input LUTs that can be configured as 32-bit shift registers, or as 64x1 RAMs.

Each function generator has 2 outputs, and can be configured as an arbitrary 6-input Boolean function, and as two arbitrary 5-input Boolean functions with shared inputs. Owing to such feature, it is possible to considerably reduce the number of sequentially connected logic gates that carry out a multivariable function in large complex designs. As a result, signal delays are reduced, and the clock rate of the design is increased.

The block memory also changed. The basic block RAM resources are 36Kb, and each block can be used as two independent 18 Kb RAM blocks. Multiplication of 25-digit and 18-digit numbers, and addition of three operands within a single unit is available in the DSP slices. Also, an independent 48-pin input was added into the DSP slices. Besides, there were modified routing resources. In the Virtex-5 FPGAs diagonal interconnections between their logic cells are available in addition to vertical and horizontal routing lines. There was designed a new version of the Clock Manager Technology (CMT), based on the two Digital Clock Managers (DCM), used in the Virtex-4 FPGA family, and on one Phase-Locked Loop (PLL).

The maximum clock rate of the Virtex-5 FPGAs amounted to 550 MHz, and power consumption reduced by about 40% in comparison with the Virtex-4 family.

In February 2009 Xilinx Inc. presented a new FPGA family – Virtex-6, produced on a 40-nm process and with core supply voltages 1.0 V and 0.9 V. There are three subfamilies: LXT, SXT and HXT. The LXT subfamily is optimized for carrying out high performance arithmetic-logical operations, and for implementation of high speed serial interfaces. The SXT subfamily is optimized for the highest DSP performance. The HXT subfamily is optimized for telecommunication. In comparison with the Virtex-5, the FPGAs of the new family have the following characteristics: power consumption is reduced by 50%, the price is reduced by 20%, the capacity of logic resources is increased in 2.3 times, the number of DSP units is increased in 2 times. Besides, the number and variety of hard-IP cores are considerably extended, the number of supported I/O standards is increased.

The Xilinx FPGAs have a super-large scale integration (up to $(20–25)\times10^6$ of equivalent gates), various built-in logic blocks, that can be used for design of compact memory blocks, multipliers, adders, a clock manager, various interfaces, built-in RISC-processor cores, etc. Every 1.5–2 years Xilinx launches new FPGA generations that exceed the previous ones by power consumption, clock rates and functionality (see Fig. 4.1).

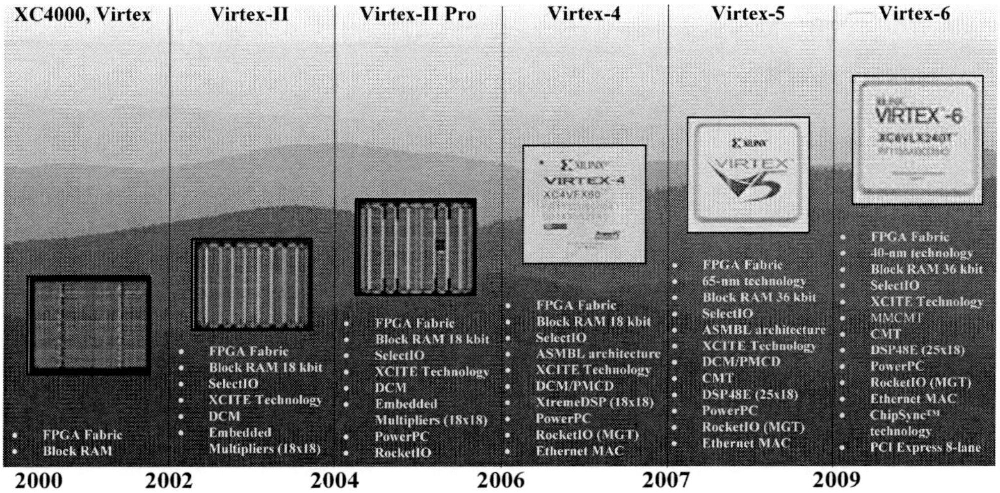

Figure 4.1. Extension of functionality of the Xilinx high performance FPGAs.

Figure 4.2 shows the graph of the general trend of Xilinx high performance FPGA development.

According to the graph of Fig. 4.2, the FPGA price (see Line 1) is reducing. The performance (see Line 3) was estimated by placing the maximum possible number of 32-digit IEEE-754 floating-point units into an FPGA, running at the FPGA clock rate.

The Line 2 shows the dependence of growth of the performance/price ratio. The value grows exponentially, and indicates growth of the FPGA economic efficiency.

It is obvious, that functionality of the FPGAs is extending, and their performance is increasing while the price is steadily decreasing. Besides, availability of effective CAD-systems considerably simplifies reconfiguration of the FPGA internal structure.

So, circuit engineers have been given unique opportunities owing to rapid development of the FPGA technology during the last decades. The youngest products of the leading FPGA vendors enable implementation of various circuit solutions in a single chip. The variety of solutions is rather wide: from general-purpose microprocessors, DSP blocks, neural networks and various interfaces to special-purpose computers. The circuit engineer can modify any circuit solution at any time, if it is necessary for implementation of the task. Summarizing the aforesaid, it becomes quite evident that the FPGA-based technology of MPCS design is the most promising.

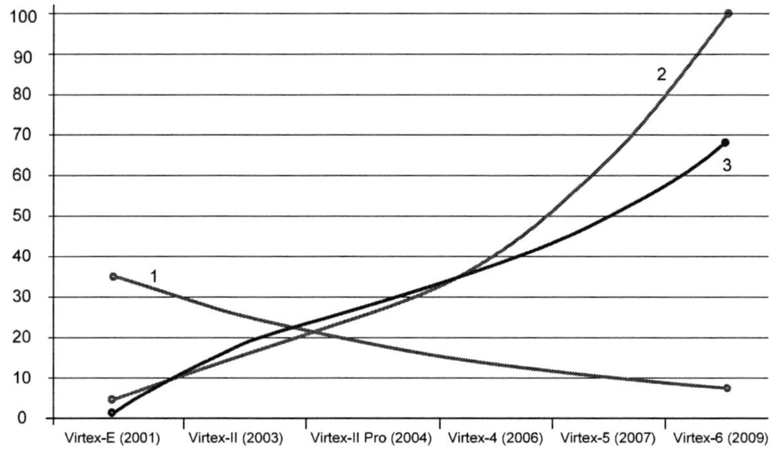

Figure 4.2. General trends of development of Xilinx high performance FPGAs: 1 – price, thousands of rubles, 2 – performance/price, 3 – performance, 10e+08.

4.2. Modular-Scalable Implementation of Reconfigurable Multipipeline Computing Structures

The principles of organization of reconfigurable multipipeline computing structures, given in Chapter 1, lead to conclusion that a computation field of an MPCS may be effectively implemented by interconnection of a number of FPGAs into a single computational resource (see Fig. 4.3).

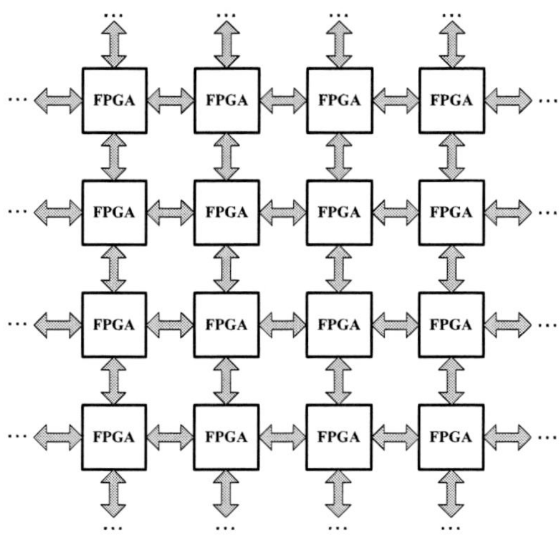

Figure 4.3. The FPGA-based computation field of an MPCS with an orthogonal communication system.

The larger is the computation field, the bigger information graph (subgraph) of the task can be mapped. In certain extreme cases, it is possible to create the FPGA computation field large enough for entire mapping of the information graph of any task. However, technology and business factors restrict hardware implementation of MPCSs, and, as a result, impose constraints on creation of large computation fields. Engineering feasibility of the large computation fields is rather a complicated problem, because it is impossible to place a large number of FPGAs on a single board.

An evident approach to the problem of creation of the large FPGA computation fields is a principle of modular scalability. The computation field is assembled of single-type basic modules (BM), implemented as single boards according to technological limitations.

For example, the basic modules can be connected as it is shown in Fig. 4.4 for further expansion of the FPGA computation field up to the required size. Owing to such modular-scalable organization, design and production of FPGA-based reconfigurable multipipeline computing structures can be considerably simplified.

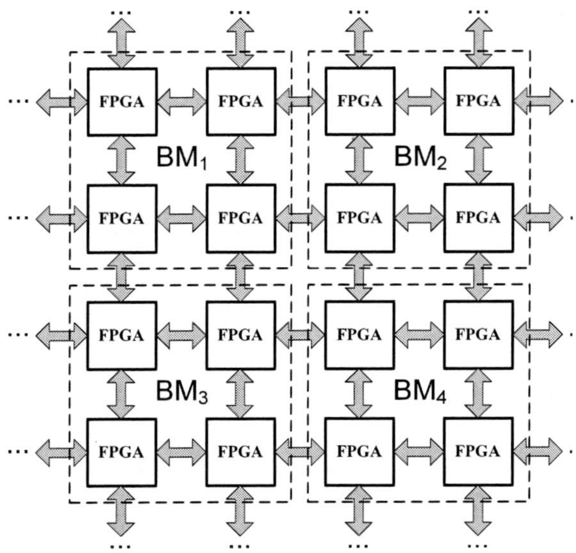

Figure 4.4. Modular-scalable organization of the MPCS computation field.

It is clear, that besides the elements of the computation field the basic module must contain a number of subsystems, that provide design of the MPCS with the required performance and purpose specialization. On the one hand, the basic modules have technological limitations, and on the other hand, they must completely preserve all features of the MPCS architecture. Therefore, all requirements and limitations that affect the characteristics of the MPCS basic module may be divided into architectural and technological types.

The architectural requirements are system characteristics that must be reflected in the main characteristics of the basic modules for further design of MPCSs with all their distinctive features. Any fragment of the computation field, organized within the basic module, must provide creation of a sufficient number of functional units and memory units, connected by a full-accessible communication system of high speed connections, that are used for mapping (complete mapping, if possible) of basic subgraphs of various tasks. According to the operating system requirements, the basic module is to be a program-indivisible hardware resource available for computations in the MPCS.

The basic module design depends on and is determined by the current level of technology. The characteristics of the basic module depend on many factors: FPGA family features that define computational capabilities of the basic module; the dimensions and the technology process of the printed circuit board; characteristics of cables and connectors for external connections; the power supply subsystem, etc. All these factors limit the architecture capabilities of the basic modules. Nevertheless (it will be shown below), modern MPCSs have computational resources of the basic modules that are sufficient for mapping of basic subgraphs of various tasks from different problem domains [11, 18].

The abovementioned requirements and limitations of architecture and design are taken into account during creation of general-purpose MPCSs, intended for implementation of a wide range of problems. Capability of complete mapping of basic subgraphs from some particular problem domain on a single basic module may become a crucial requirement for special-purpose MPCSs. In this case, the basic module may be implemented as a set of several printed circuit boards.

Let us take the macroprocessor architecture, considered in Chapter 3, as a basis for identification of requirements of architecture, design and technology of the modular-scalable MPCSs. In this case, the basic module structure may be presented as it is shown in Fig. 4.5.

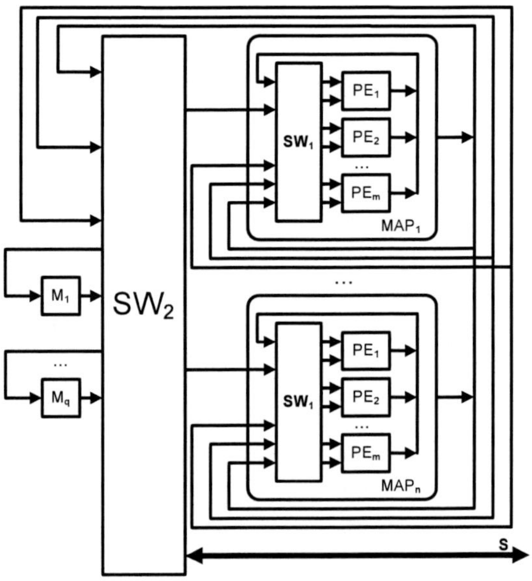

Figure 4.5. The outline structure of the basic module.

The basic module (BM) consists of several macroprocessors (MAP) and memory units (M), interconnected by a spatial communication system SW_2. Each macroprocessor contains m processing elements (PE), interconnected by a switch SW_1. The elements of the basic module, shown in Fig. 4.5, are interconnected by a two-level communication system. The switch of the first level interconnects the processing elements within the macroprocessor. The switch of the second level interconnects the macroprocessors within the basic module. The bus S interconnects the basic modules.

There are several approaches of performance augmentation. The first approach consists in connection of several basic modules, and creation of structures with higher performance by adding new levels of the communication system that interconnect larger sets of the basic modules (see Fig. 4.6).

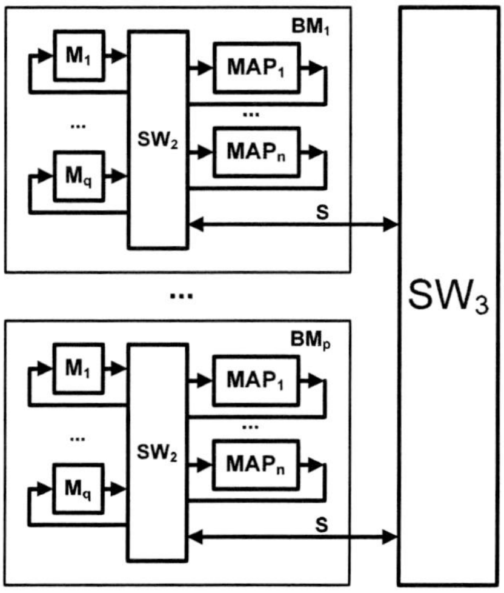

Figure 4.6. The generalized structure of a modular-scalable multiprocessor system with a hierarchical communication system.

The disadvantage of such approach is requirement of additional primitives for implementation of the highest levels of the communication system.

Another approach of performance augmentation consists in combination of several basic modules by the in-module switch SW_2 (see Fig. 4.7). In this case, it is possible to use the basic modules of only one type. However, the remote basic modules will be connected by relatively long nets, and informational channels that will get more and more narrow. Various combinations of the considered approaches are acceptable.

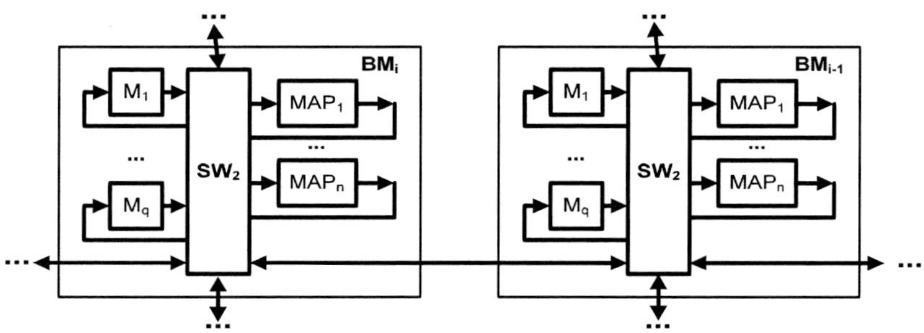

Figure 4.7. The generalized structure of a modular-scalable multiprocessor system on the base of an intermodule switch.

Figure 4.8 shows the standard structure of the basic module. The basic module consists of the macroprocessors MAP_1–MAP_n. Each macroprocessor contains the processing elements

PE_1–PE_m connected by the switches SW_1, and the distributed memory units, managed by a correspondent distributed memory controller (DMC).

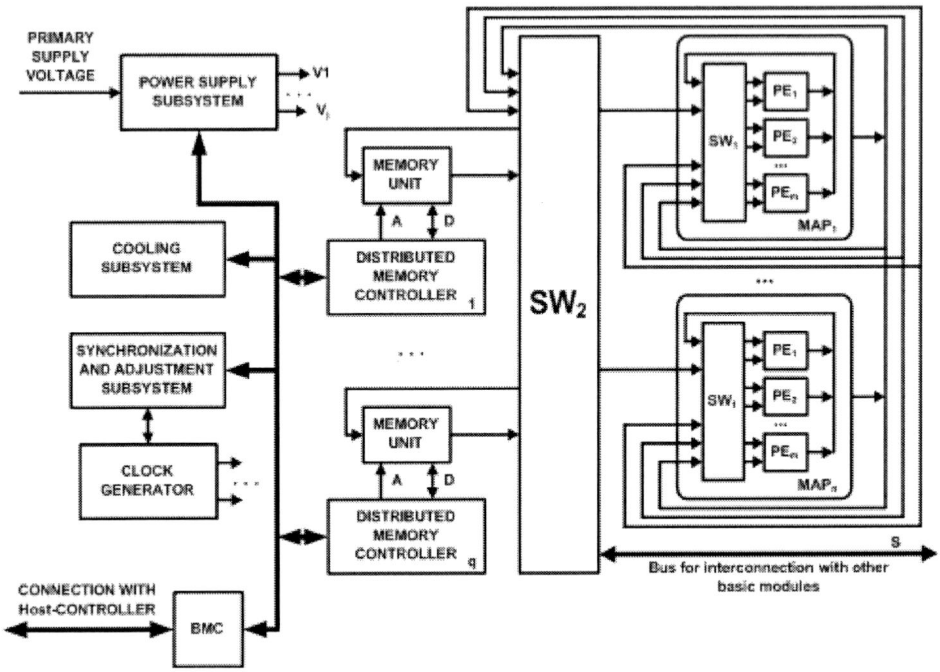

Figure 4.8. The standard structure of the basic module.

All components are connected to the spatial switch SW_2, which also provides connections with the external informational channels of the basic module. The macroprocessors, the distributed memory units, and the switch SW_2 are used for organization and execution of computations both in a single basic module and in several basic modules, combined for a single task solving.

All systems of the basic module are managed and controlled by a basic module controller (BMC). The basic module controller loads input data, unloads results from the distributed memory units, and loads an executable code into the distributed memory controllers. Besides, the basic module contains auxiliary subsystems of synchronization, power supply and cooling.

The distributed memory controllers play a leading role in organization of computations, and execute the fragments of the parallel program loaded into the memory units. Executing the parallel program, the distributed memory controllers adjust the processing elements for execution of all necessary operations and form the communication channels, adjusting the switches SW_1 and SW_2. The communication channels connect the processing elements and implement the multipipeline computational structure, similar to the basic subgraph of the task within the single basic module.

While executing the fragments of the parallel program, one part of the distributed memory controllers creates and synchronizes the data streams, that pass through the computational structures. Another part of the controllers receive the results of computations. Figure 4.9 shows the basic module with the implemented subgraph.

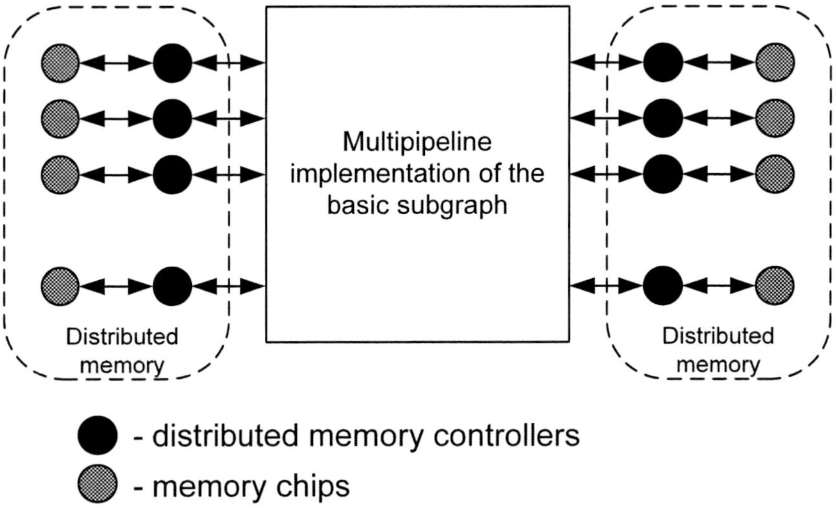

Figure 4.9. Data streams pass through the basic subgraph.

The distributed memory controllers of all the basic modules synchronously readjust the components of the system according to the structure of a new executable cadr. Synchronous reconfiguration is necessary for correct execution of the parallel program in all basic modules. Computations of every new cadr will not be started, while the current cadr is still being executed in any basic module.

The instruction set of the distributed memory controller effectively supports structural-procedural computations and provide flexible implementation of various program structures such as nested loops, recalculation of the loop parameters (an initial memory address, a memory address increment and a number of iterations), program calls, replacement of loop parameters with values from the memory.

The instruction set of the macroprocessor provides execution of a great variety of arithmetic and logical operations, signum functions and predicates, and a number of unary operations of format conversion. The communication system implements parallel access to the distributed memory channels of the basic module and data exchange between the components of the basic modules.

It is possible to cluster the basic modules by means of switches and programmable links. Figures 4.10 and 4.11 show structures of various combinations of the basic modules: blocks of the basic modules, and blocks of blocks (blocks of the second level).

According to the structure, shown in Fig. 4.10, the connections between the blocks are implemented by the communication system that connects the basic modules with similar numbers in different blocks according to full-accessible connection. Besides, the same connection is used for the basic modules within the computational blocks (an orthogonal communication system). According to the structure, shown in Fig. 4.11, the blocks are connected by a hierarchical communication system. The basic modules within each block are connected according to a full-accessible communication system.

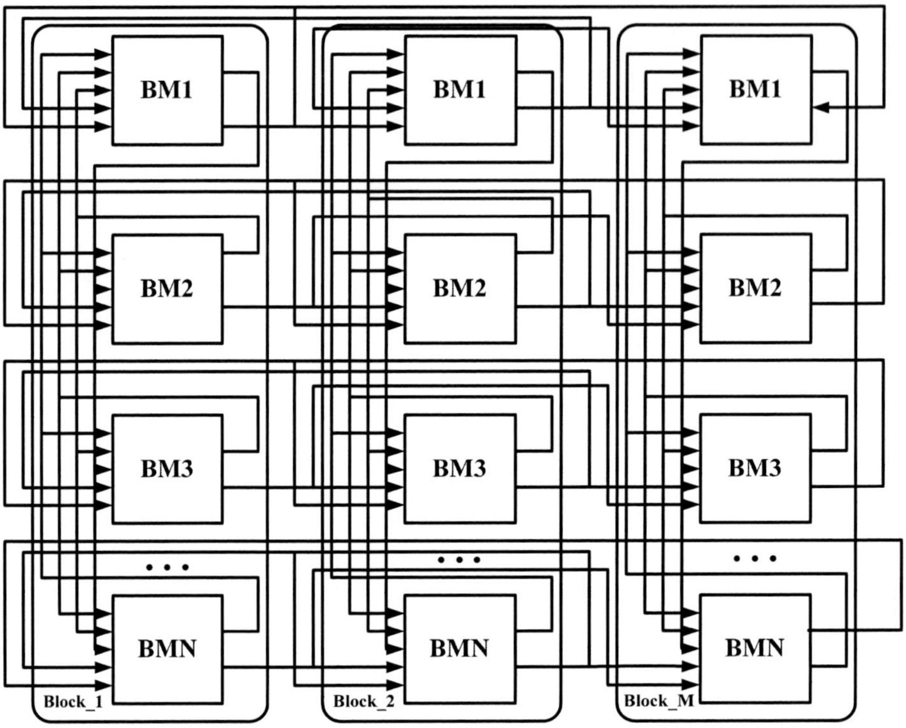

Figure 4.10. The blocks of the basic modules combined into the block of the second level by an orthogonal communication system.

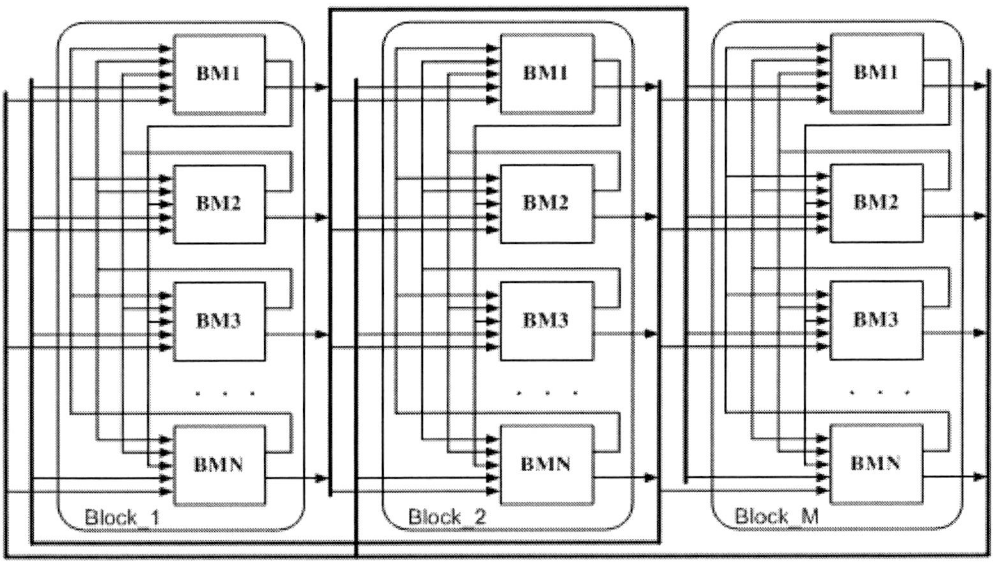

Figure 4.11. The blocks of the basic modules combined into the block of the second level by a hierarchical communication system.

Figure 4.12 shows the example of a modular MPCS with an orthogonal communication system [17] based on register links (RL).

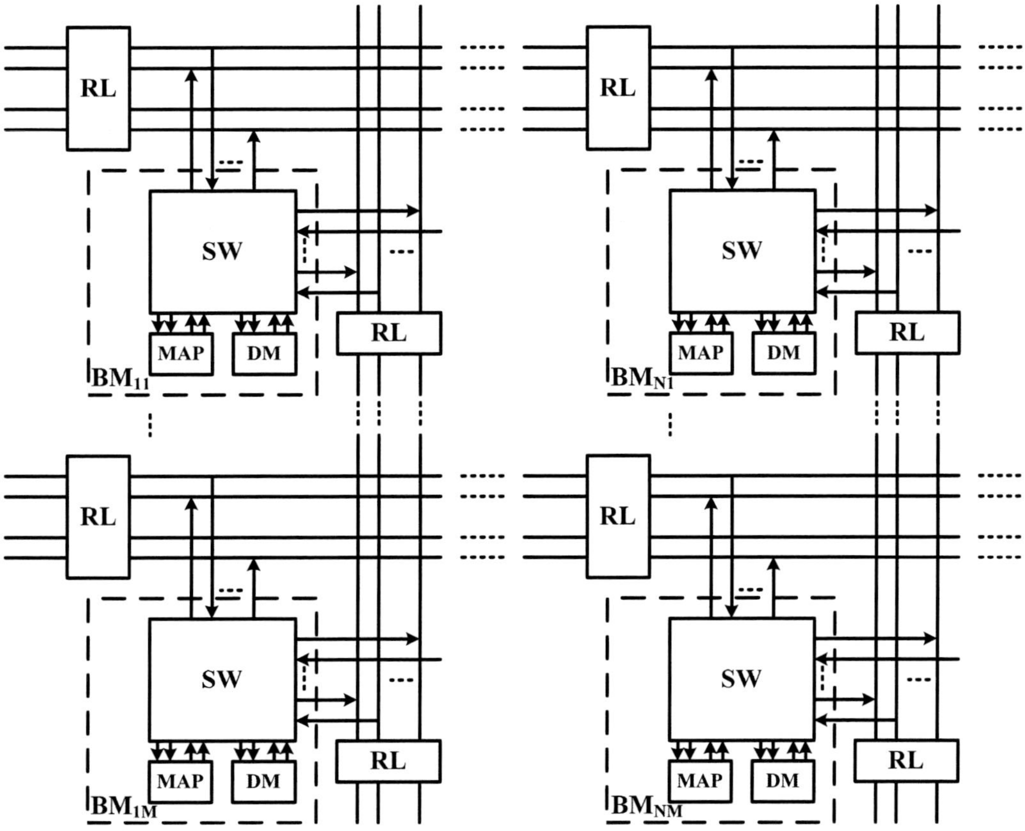

Figure 4.12. The structure of a modular MPCS built of the basic modules with an orthogonal communication system.

The basic modules (BM) are connected by *N* vertical and *M* horizontal *q*-channel data buses. In each basic module these buses are connected with an internal switch (*SW*) that connects the channels of the horizontal and vertical buses to the macroprocessors (*MAP*) and to the channels of the distributed memory (*DM*) of the basic module. Besides, the switch provides connection between the MAPs and the distributed memory channels within the basic module. In general case, *q* is defined according to the number of the macroprocessors, the distributed memory units, and the number of the distributed inputs and outputs of the basic module [22].

A switch that directly connects the outputs of the adjacent internal switches is another type of the orthogonal switch (see Fig. 4.13).

Such type of switch is more regular, and it is possible to increase clock rates of intermodule data exchange.

Owing to the principle of modular scalability, the performance of the MPCS may be increased when the number of its basic modules grows. As it will be shown in Chapter 6, resource independent and fault-tolerant programming will be also provided, and will depend on the number of the basic modules, allocated for task implementation, which is considered a variable parameter. The task can be solved on any configuration of the computational structure. Failure of one or several basic modules slightly decelerates the computation process [11] but does not abort it. Besides, in this case, there is no need of task recompilation.

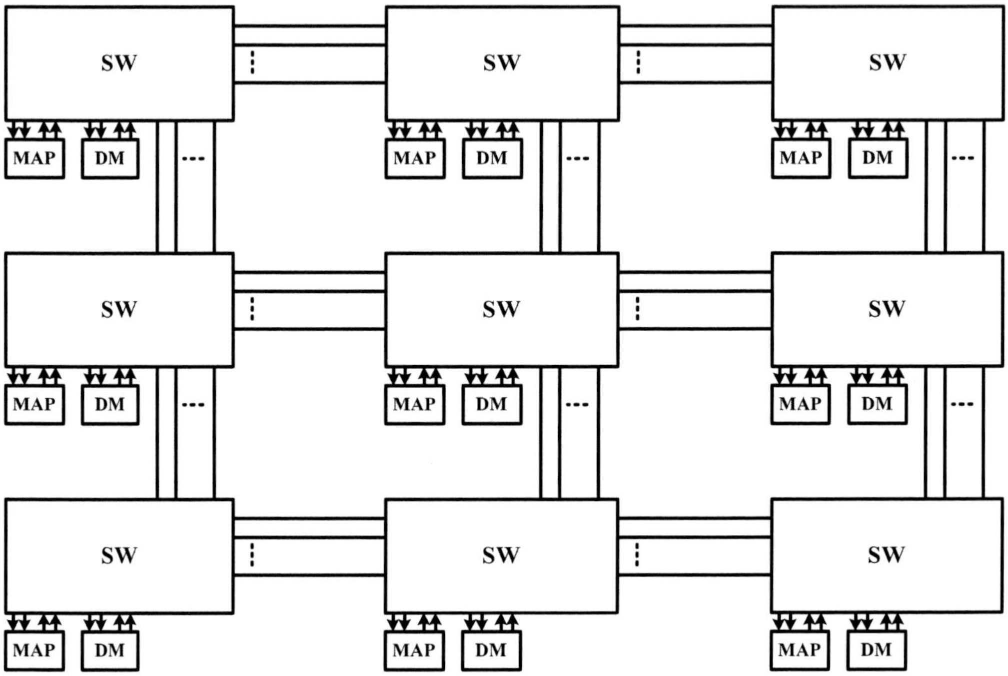

Figure 4.13. Direct connections between the basic modules.

4.3. Design Principles of Basic Modules of FGPA-Based Reconfigurable Multipipeline Computational Structures

Owing to FPGAs, used as principal elements of MPCS basic modules, reconfigurable multipipeline computing structures have specific features that are necessary for including of a set of functional units (see Fig. 4.4) into an FPGA field. Figure 4.14 shows the standard structure of the FPGA-based basic module.

The FPGA-based basic module possesses all features of a self-contained reconfigurable system [17]. At the same time, it can be combined with the similar basic module for performance increasing.

The main computing capabilities of the basic module are concentrated in the computation field that contains a number of FPGAs of large-scale integration. Figure 4.4 shows computational and communication structures, formed within the computation field: the macroprocessors MAP_1–MAP_n that contain several processing elements PE_1–PE_m each; the switches SW_1 that connect the processing elements; the intermodule switch SW_2; the distributed memory controllers for management of the distributed memory units. The distributed memory units are implemented on standard chips SRAM or SDRAM with the required size and performance.

Since the functional units can be implemented by means of several or even all chips of the computation field, it is necessary to reduce boundary effects between the FPGAs as much as possible. There are two approaches to this problem. The first approach consists in use of large-scale integration FPGAs for placement of large fragments of the functional units. The second approach consists in creation of regular and homogeneous connections between the FPGAs. The chips are placed in points of a 2D-lattice and direct links connect the adjacent chips according to an orthogonal communication system. Such communication system substantially simplifies the structure of the printed circuit board and improves its data transfer rate, because the length of connections between the adjacent chips does not exceed several centimeters. Transit channels are used for data transfer between the chips with no direct connections. The transit channels pass through the interjacent chips using the orthogonal communication system.

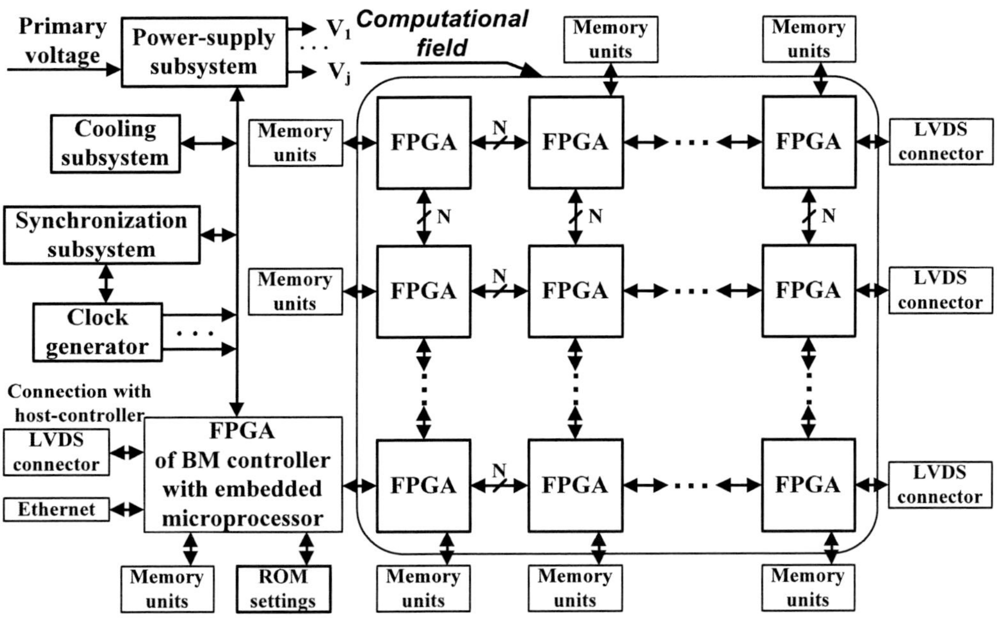

Figure 4.14. The structure of the FPGA-based basic module.

It is reasonable to use LVDS (Low Voltage Differential Signal) channels for connection of the FPGAs of the basic module computation field. Advantages of the LVDS are low power consumption of output drivers, low electromagnetic interference, inherent noise immunity, common-mode noise rejection, and availability of hardware resources for LVDS high data transfer rate in up-to-date FPGAs. Physically, each communication lane is two adjacent traces (a differential pair), connected to the special pins of the chips. The data transfer rate through each VLDS lane depends on implementation, and can amount to approximately several gigabits per second.

Implementation of the basic module is the next design level that requires specialized arrangements for reduction of boundary effects.

As far as computational resources of one basic module may become insufficient for implementation of multipipeline computing structures of large basic subgraphs, it is necessary to use two or more basic modules. Meanwhile, it is important to keep the high data transfer rate between FPGAs of different basic modules. Use of LVDS channels will keep the high data transfer rate, and will provide transfer of huge data streams between different basic modules, that are used for task implementation. In fact, the LVDS channels connect the in-module switches and provide various kinds of intermodule communications.

A standard interface for connection of the FPGAs, placed on one basic module or on several different basic modules, provides similar values of the data exchange speed, and reduces boundary effects within the whole circuit resources of multipipeline computing structures.

The distributed memory units are used for placement of the information vertexes of the basic subgraph, for reading of input data and writing of computational results. It is necessary to define an optimal proportion of the number of the FPGAs, distributed memory units and their size, for design of effective computing structures within the basic module. The proportion is different for various problems. At the same time, intermediate data storage is not necessary in multipipeline computing structures, because the data is transferred into the next pipeline stage for further processing. As a result, requirements to the total size of the basic module memory are reduced. The number of the distributed memory units is more crucial, than the size of each unit or the total memory size. Besides the distributed memory, implemented on the base of standard RAM chips, the FPGA on-chip memory is widely used in design of various computing structures.

Another feature of the FPGA-based basic modules is implementation of the distributed memory controllers (DMC) not as separate devices (see Fig. 4.9), but as functional units within the computation field (see Fig. 4.15).

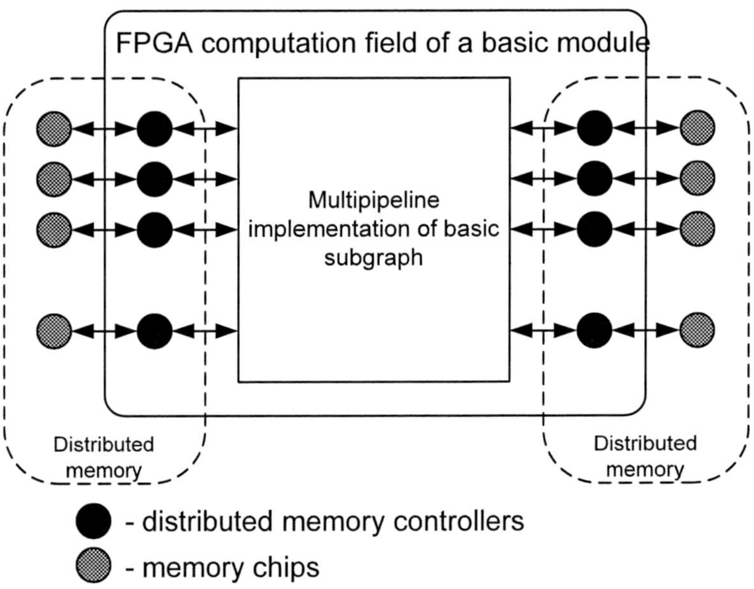

Figure 4.15. Implementation of the distributed memory in the FPGA-based basic module.

A large number of the FPGAs of various integration that are contained within the basic module, require a specialized engineering solution to the problem of acceleration of configuration bitstream loading. The FPGA-technology supports several modes of configuration bitstream loading: from a configuration ROM, from an external source, serial (bit after bit) or parallel (as a sequence of bytes). Various combinations of loading may be used.

As far as the FPGA computation field is configured directly by the basic module controller, so the controller itself is to be programmed from the configuration ROM. Pre-programming of the basic module controller starts immediately when power is on. As a result, all basic modules of the MPCS will simultaneously be ready for programming of the FPGA computation field.

The FPGAs of the computation field are to be programmed in a parallel mode, because such approach will accelerate programming in many times in comparison with serial loading of the configuration bitstream. Use of several parallel configuration channels will provide further acceleration of programming.

To fix the idea, let us assume that the computation field of the basic module contains $n \times m$ FPGAs, and the time of byte parallel programming of the configuration bitstream into one chip is t. If the configuration bitstream is simultaneously loaded into all chips in the parallel mode, then the loading time is t. If the configuration bitstream is loaded in each chip in turn through a single channel, then the loading time of all chips is $n \times m \times t$. If the set of the chips is divided, for example, into n groups containing m chips each, and the configuration bitstream for each group is loaded through n parallel channels, but through a single channel within each group, then the loading time is $m \times t$. The choice of the configuration bitstream loading method depends on many characteristics of the basic module, and on requirements of the readiness time and the MPCS hardware and software configuration.

4.4. Performance Increasing of Multipipeline Computing Structures

Nevertheless, onrush of the FPGA technology does not allow the circuit engineer to design a computing structure with the performance, high enough for solving large tasks during acceptable time, using only a single basic module. First of all, this is the result of design constraints of the basic module: the number of placed FPGAs is limited by the dimensions of the printed circuit board, the number of external links is limited by connectors, power consumption is limited by capabilities of the cooling system and by electrical constraints of conductors of the printed circuit board, etc. So, to increase the MPCS performance it is necessary to combine several basic modules into blocks.

Further increasing of the MPCS performance requires co-operation of several blocks combined into a rack (see Fig. 4.16). The racks may be combined into a large MPCS with sufficiently high performance, equal to tens or hundreds of teraflops. Taking into account relatively high real performance of MPCSs, which is usually not less than 50% from the peak one, the MPCS with the performance of approximately 80 Tflops may be included into the TOP-500 list (September 2011).

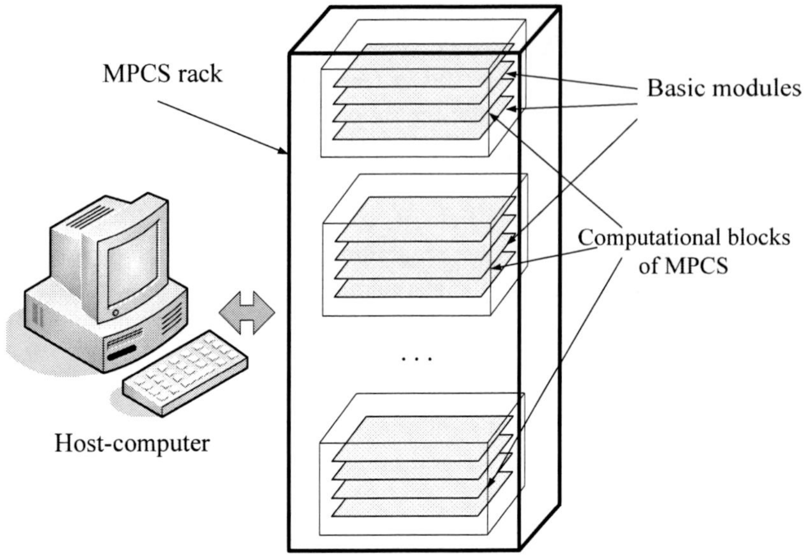

Figure 4.16. Design solutions of high performance MPCSs.

The FPGA-based basic modules possess all features of a self-contained MPCS. Owing to this, it is possible to create high performance systems, that will consist of tens and hundreds of basic modules, to provide software compatibility, and to keep unity of the architectural and design principles.

4.5. MPCS Basic Modules

It is evident, that design features of basic modules significantly influence MPCS characteristics, and, therefore, it is necessary to devote much attention to architectural solutions, that are used as a base for MPCS design. Let us analyze evolution of the engineering solutions that were used for design of FPGA-based basic modules at various periods of time.

4.5.1. The Structure of the First Generation Basic Modules

Let us consider the structure of the basic module of the first generation with a hierarchical communication system. A macroprocessor section (MAP) contains 8 FPGAs, connected by an internal switch. All processing elements of the MAP are connected by a full-accessible communication system [11]. A distributed memory unit is connected to the macroprocessor by switches SW_1 and SW_2. The inputs and outputs of the macroprocessor and the distributed memory unit are also connected by the full-accessible communication system. Selection of the output channels of the distributed memory unit and the macroprocessor for the third level communication is provided by a switch SW_3. The third level of communication is a level of intermodule communication, and it is implemented with the help of a switch SW_4

and connectors C_1, C_2 and C_3. Different basic modules may be connected according to the structure chart, given in Fig. 4.8.

The basic module was designed in SRI MCS (Taganrog, Russia) and was implemented in 1996 on the base of uncommitted logic arrays, and in 1998 – on the base of FPGAs [14]. Figure 4.17 shows the structure chart of the basic module of the first generation, designed on the base of the FPGA-technology.

To minimize hardware, that was necessary for the communication system, circuit engineers used MSB-first digit serial representation [71,101]. Owing to sequential code, used for data processing and transfer, it became possible to reduce hardware, that was required for organization of the communication system, in ten times. Nevertheless, implementation of the communication system requires approximately 60% of hardware resources.

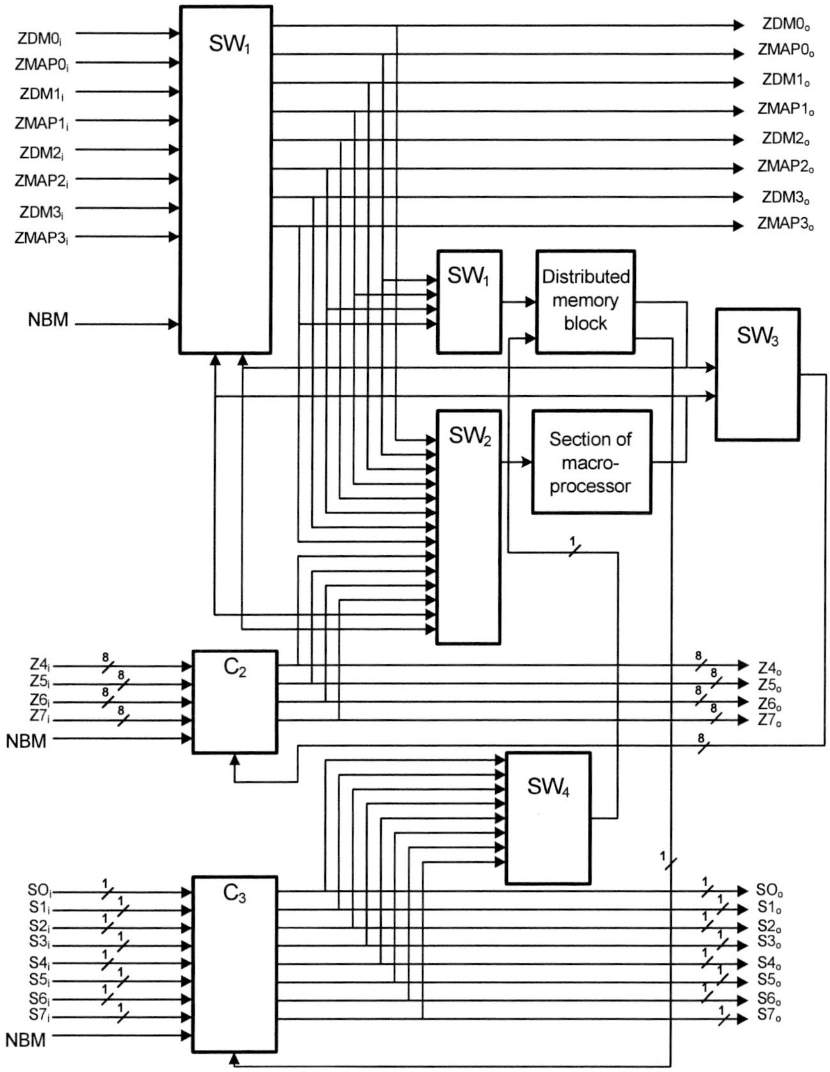

Figure 4.17. The structure chart of the basic module.

Figure 4.18 shows a photography of the basic module of the first generation.

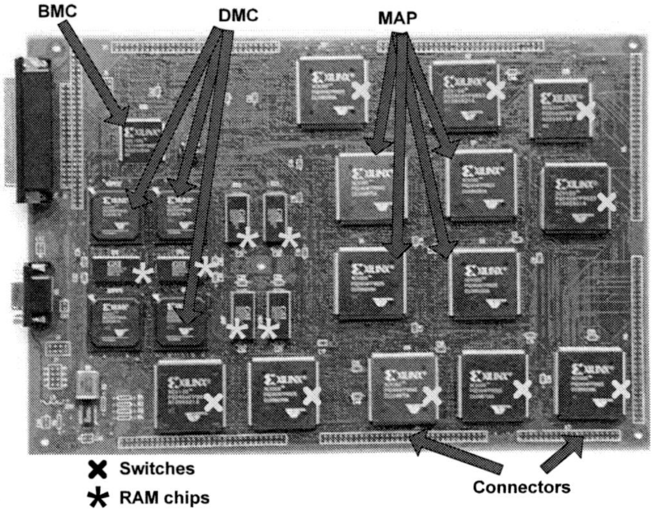

Figure 4.18. The FPGA-based basic module of the first generation.

It should be noted, that the result of the conflict between the sequential principle of processing and the FPGA architecture, oriented on parallel data processing, is performance degradation of the MPCSs, designed on the base of such basic modules.

Processing of data in digit serial representation requires much bigger amount of computational resources, than processing of the same data in parallel representation. Though by means of the FPGA-technology it was possible to create a computer on basis of up-to-date circuit components, its clock rate increased in 4 times, and its performance increased only in 2 times in comparison with the prototype, designed in 1996. As a result, for further designs it was planned to use traditional methods of data processing based on parallel codes.

4.5.2. The Structure of the Second Generation Basic Modules

Parallel data processing unavoidably leads to increasing of hardware resources that are needed for organization of a communication system. Usually hardware resources are proportional to a number of digits of connected channels. Therefore, the hardware resources, used for the communication system, had to be significantly reduced in the basic module, designed for parallel data processing. In this connection, the orthogonal communication system was selected. On the one hand, the orthogonal communication system provides full-accessible connection of the system elements, and on the other hand, it requires significantly smaller hardware in comparison with the hierarchical communication system, that was used in the basic modules of the first generation.

Predeterminacy of connections is another factor that was taken into account, while choosing the type of the communication system, according to which connections between vertices of basic subgraphs must be formed. Such predeterminacy is connected with determinacy of structural-procedural computational process, and is viewed in details in

Chapter 3. Owing to determinacy of structural-procedural computations, all connections that are necessary for basic modules and the whole MPCS, are defined during translation of the algorithm of some solving problem. As a result, it is possible to beforehand estimate correspondence between hardware communication capabilities and requirements of the task, and to use this information for partition of the information graph.

The orthogonal communication system, used for connection of the FPGAs within the basic module, is much simpler than the hierarchical one. It provides connections not only between the adjacent FPGAs, but also between any FPGAs by means of transit data transfer.

Such capability is provided by an internal switch, designed in each FPGA. The internal switch connects the internal elements of the FPGA to the data transfer channels, and connects all external informational channels of the FPGA, including the channels for transit data transfer.

The orthogonal communication system of the basic module has another additional advantage: physical connections between the FPGAs on the printed circuit board are rather short, they have almost equal length and electric parameters. Such features significantly simplify design of printed circuit boards, improve frequency characteristics of their channels and increase their transfer rate.

Figure 4.19 shows the standard structure chart of the basic module of the second generation, based on the orthogonal communication system. The basic module consists of:

- 16 macroprocessors (MAP); each MAP contains 4 processing elements and a switch that provides full-accessible connections between the processing elements and the external communication channels;
- the distributed memory, which contains 8 distributed memory controllers (DMC) with internal switches that provide full-accessible connections between the distributed memory channels and the external communication channels;
- static RAM chips;
- the basic module controller (BMC).

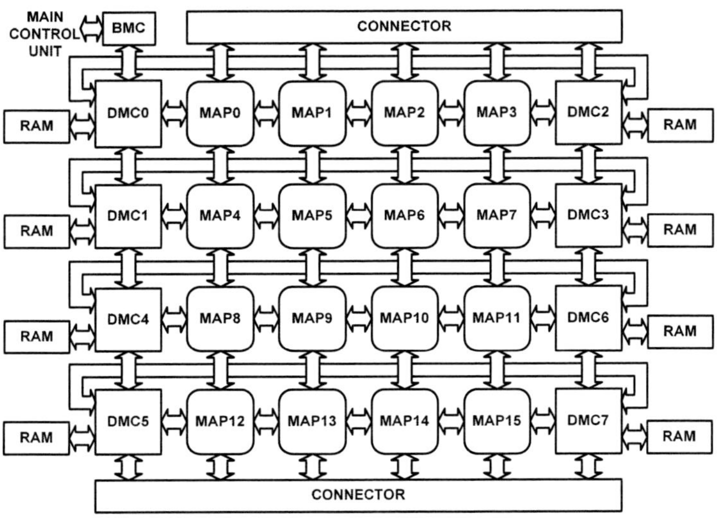

Figure 4.19. The structure chart of the basic module with the orthogonal communication system.

optimally similar to the structure of the solving task. The detailed description of the macroobject architecture is given in Chapter 5.

Besides architectural distinctions, the basic modules of the third generation have some design features:

- FPGAs of over 4 million equivalent gates and with the clock rate over 200 MHz;
- FPGAs with built-in PPC405 microprocessors;
- a large size dynamic memory DDR and DDR2 SDRAM for distributed memory organization;
- high speed LVDS channels for connection with other basic modules;
- a small-size secondary power supply on the basic module.

Several basic modules for various MPCSs were designed and produced in SRI MCS.

4.5.3.1. The third generation basic module 16V4-50

Figure 4.22 shows the basic module of the third generation that was designed in 2005 by scientists of SRI MCS.

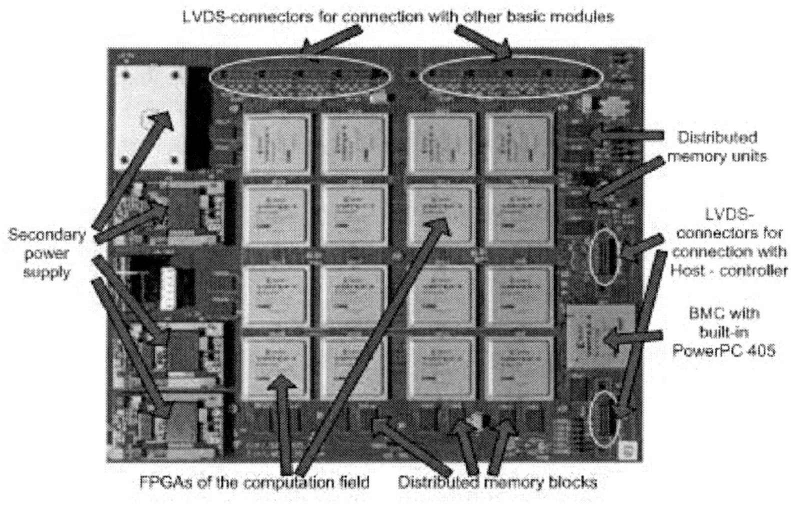

Figure 4.22. The basic module of the third generation designed on the base of Virtex-4 FPGAs.

The specification of the basic module:

Virtex-4 FPGA	17
Memory size, GB	1
Performance, Gflops	50
Basic module clock rate, MHz	200
LVDS	112
Total capacity of LVDS, Mb/sec	22400
Power consumption, KWatt	0.16
Dimensions, mm	233.35x280

The basic module 16V4-50 is based on the Virtex-4 FPGAs and is intended for solving problems of digital signal processing, linear algebra and mathematical physics. Each FPGA contains 4 million gates. The basic module controller contains a built-in processor PowerPC405. The basic module is implemented according to Eurocard 6U and has 112 data LVDS channels for data exchange with external devices. The total speed of data exchange is over 22 Gb/sec. The majority of the FPGAs are connected with the external distributed memory SDRAM.

The basic module 16V4-50 is a principal element of a modular-scalable reconfigurable multipipeline computational block (see Fig. 4.23).

Figure 4.23. A block of a modular-scalable MPCS.

The block is a self-contained MPCS; it may be used both as a standalone system, and as a part of complex systems with higher performance. The specification of the modular-scalable reconfigurable multipipeline computational block:

Basic module	4
Memory size, GB	4
Peak performance, Gflops	200
Power consumption, KWatt	0.9
Volume, l	39.9

To increase the performance, it is possible to combine several blocks into a single system. As a result, the total peak performance is increasing in proportion to the number of the basic modules of the system. Figure 4.24 shows a general view of the system, that is implemented in a 19" Schroff rack.

4.5.3.2. The Basic Module for Solving Tasks of Digital Signal Processing

The basic module is a standard node of a reconfigurable computer. It is a high-end computational resource with the performance up to 40 Gflops, oriented to structural-procedural computations. The basic module may be used as a standalone computer, as a component of general-purpose devices, and as a component of high performance reconfigurable multiprocessor computer systems.

There are two versions of the basic module: a mezzanine (the reference designation 8V2P-64) and a two-board. Figure 4.25 shows the structure of the mezzanine version. The performance of the mezzanine version of the basic module is 64 Gflops. The special connectors, placed on the printed circuit board of the basic module, are used for connection with the mezzanine 8V2P-64.

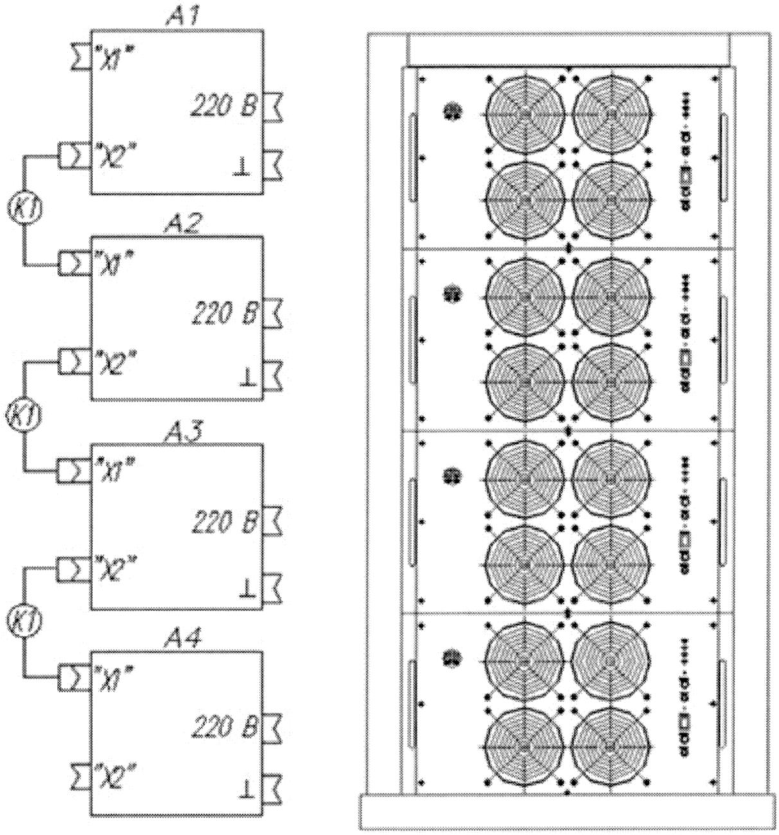

Figure 4.24. A general view of the MPCS implemented in a 19" Schroff rack.

Figure 4.26 shows the structure and connection of two basic modules of the two-board version. Here the reference designation of the basic module is 4V2P-32, and its performance is up to 32 Gflops.

Each basic module consists of 8 (or 4) XC2VP100-6FF1704C Virtex-II Pro Xilinx FPGAs. Besides 10 million logic gates, each FPGA contains two PPC405 RISC-cores. All Virtex-II Pro FPGAs of the basic module are connected by the orthogonal switch. The speed of data exchange between the FPGAs amounts to 3.2 GB/sec. Each XC2VP100-6FF1704C chip is connected with an SDRAM with capacity of 8M of 64-digit words. The basic module also contains a VME64X bus controller.

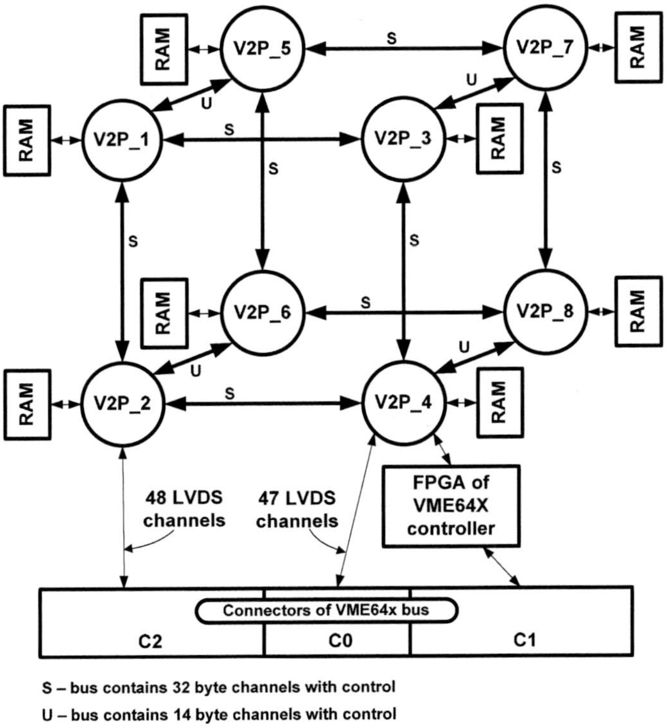

Figure 4.25. The structure of the mezzanine version of the basic module (8V2P-64).

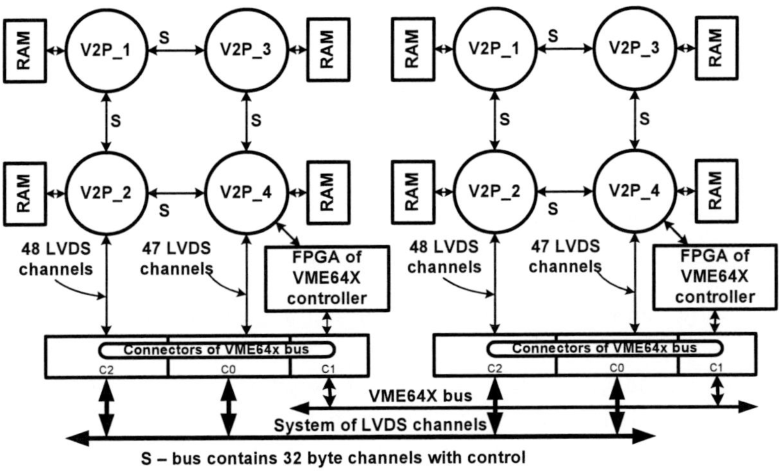

Figure 4.26. The structure of the two-board version of the basic module.

On the base of the general-purpose architecture of a single basic module, the application programmer can design various specialized and special-purpose virtual computer systems with the structure, similar to the structure of the solving tasks. The virtual special-purpose system is synthesized by configuring and adjustment of its components to large structural mathematical operations, to procedures of macroaccess to the distributed memory channels.

The communication structures, that connect the functional units and the distributed memory channels, are adjusted to the information graph of the algorithm. As a result, the special-purpose virtual architecture becomes similar to the structure of the solving task.

The basic module is designed according to Eurocard 6U and is intended for use in systems of digital communication monitoring. The basic module is embedded into the block with the VME64X control bus. Connection with other system components is provided by 95 LVDS channels.

The basic module contains the following subsystems:

- an informational subsystem of data transfer and distribution;
- a subsystem of computational process synchronization;
- a subsystem of adjustment to an executable cadr;
- a subsystem of FPGA configuring;
- a subsystem of FPGA configuring by JTAG;
- subsystems of power supply, consisting of two built-in secondary power supplies +1.5 V, 80 A and +2.5 V, 20 A, that get the primary voltage +48 V;
- a subsystem of data exchange with other MPCS modules through LVDS (95 channels in total).

To provide power supply for the secondary voltage supplies, the basic module is connected to the primary voltage supply SP-150-48 with the output voltage +48 V, 3.2 A.

The specification of the basic module 4V2P-32:

XC2VP100 FPGA	4
RAM size, MB	256
Real performance, Gflops	32
FPGA clock rate, MHz	200
Power consumption, Watt	85
LVDS	95
Speed of data exchange, Gb/sec	20
Maximum FFT size	2^{17}
Dimensions, mm	233.35×160

Figure 4.27 shows an exterior view of the basic module 4V2P-32 with heatsinks on the XC2VP100 Virtex-II Pro FPGAs.

Table 4.1 shows product profiles of modules of digital signal processing, manufactured by foreign enterprises and the product profile of the basic module 4V2P-32.

As follows from the table, the majority of parameters of the basic module 4V2P-32, running at the clock rate of 200 MHz, is inferior to none of foreign products, implemented on the base of such microprocessors as TigerSHARC ADSP-TS201S PowerPC MPC7455 running at the frequency of 500÷1,000 MHz. Such results were achieved owing to multipipeline structures that were implemented within the basic module 4V2P-32 and were similar to the solving tasks.

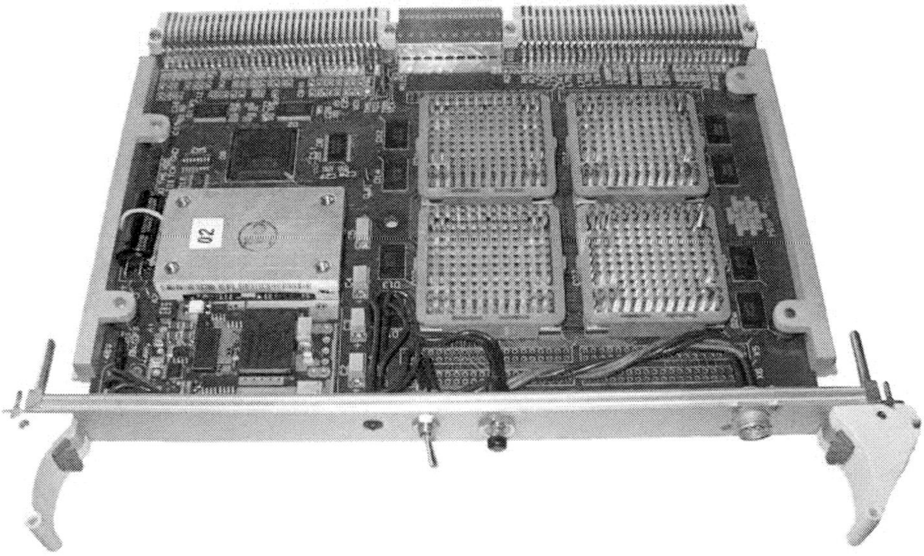

Figure 4.27. The basic module 4V2P-32 with heatsinks on the XC2VP100 Virtex-II Pro FPGAs.

Table 4.1. Product profiles of modules of digital signal processing

Parameter	TigerSHARC ADSP-TS101S	TigerSHARC ADSP-TS201S	PowerPC MPC7410	PowerPC MPC7455	4V2P-32
Clock rate (MHz)	250	500	500	1,000	200
Processors per module	8	8	4	4	4
Peak performance of the module, Gflops	12	24	16	32	32
Memory bus, bit/MHz	64/83.3	64/100	64/100	64/133	64/160
Input/output channels	2PMC+16Links	2PMC+16Links	2PMC	2PMC	95 LVDS
Peak speed of I/O channels, MB/sec	5,056	9,056	1,056	1,056	20,000
Capacity of I/O channels, Bytes/flops	0.42	0.38	0.07	0.07	0.64
FFT1024 per second (for one chip)	25,641	51,282	26,053	64,941	90,900
FFT1024 per second (for one basic module)	180,000	360,000	24,000	48,000	357,000

4.5.3.3. The Basic Module 4V4-25 of a Desk-Size Personal Computer Accelerator

A desk-size personal computer accelerator is intended for solving complicated fragments of tasks from various problem domains. The base of the accelerator is a module 4V4-25 implemented on an 18-layer 150×190 mm printed circuit board with double-sided placement of FPGAs. The computation field of the basic module contains 4 XC4VLX80-10FF1148 FPGAs. Each FPGA contains $8 \cdot 10^6$ equivalent gates. The basic module also contains an XC4VFX60-10FF1152 FPGA of a controller with a PPC405 microprocessor, FLASH-memory for storage of controller configuration data, 9 SDRAM chips of dynamic memory, a programmable clock generator with drivers, an LVDS-connector and an Ethernet-connector for connection with a PC, and other elements of surface mounting.

Figure 4.28 shows an external view of the basic module 4V4-25.

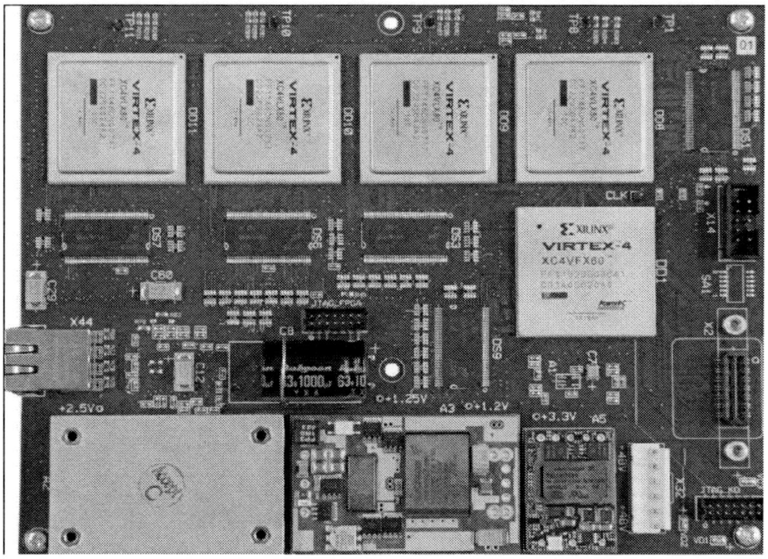

Figure 4.28. An external view of the basic module 4V4-25.

The specification of the basic module:

FPGAs of the computation field	4
Equivalent gates, million	32
Size of dynamic memory, MB	576
Performance, Gflops	25
Basic module clock rate, MHz	160
Speed of data exchange through Ethernet, Gb/sec	1
Power consumption, Watt	145
Dimensions of the basic module, mm	150×190

Figure 4.29 shows the general structure chart of the basic module 4V4-25.

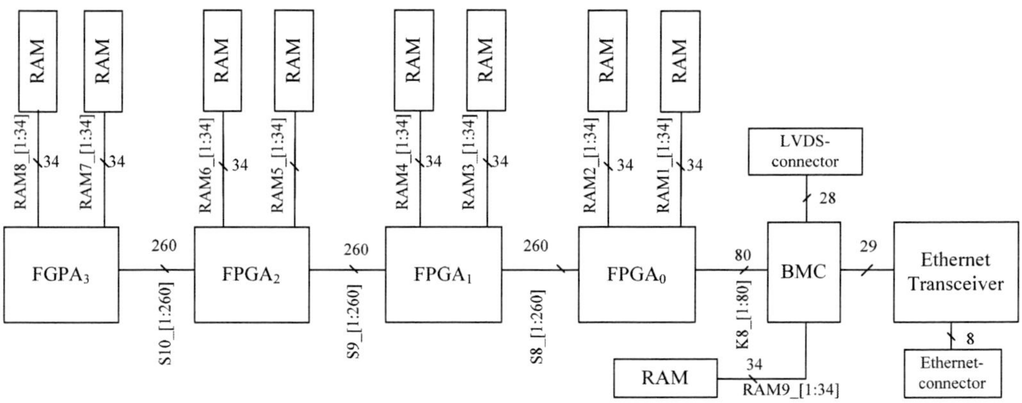

Figure 4.29. The general structure chart of the basic module 4V4-25.

The structure chart contains the following designations: BMC – a basic module controller; Ethernet-Transceiver – an Ethernet transceiver Marvell 8E1111.

The basic module controller is used for data exchange with the personal computer. The controller connects the personal computer and the chips of the computation field of the basic module by the Ethernet or LVDS channels, and implements an interface of FPGA configuration bitstream loading.

It should be noted, that the LVDS channel is an auxiliary element, and is used only for debug purposes and equipment checkout.

Figure 4.30 shows an external view of the desk-size personal computer accelerator in an open case. Heatsinks of different height are placed on the FPGAs of the computation field and on the BMC for heat removal. The dimensions of the accelerator case are 300×263×55 mm.

Figure 4.30. The desk-size personal computer accelerator.

4.6. Basic Modules of a Family of Reconfigurable Computer Systems

In 2009 scientists of SRI MCS SFU (Taganrog, Russia) designed and produced a family of reconfigurable computer systems (RCS) on the base of the MPCS theory, considered in Chapters 1 and 3.

The RCS family is oriented to solving complicated scientific and technical problems in such domains as computer network security, development of fundamentally new drugs and materials of new generation, real-time control of nuclear power engineering objects, aircrafts and spacecrafts, protein sequencing, optimization of oil field maintenance, tomographic research of near-surface layers of Earth using sonic and electromagnetic waves, neuromathematics.

The RCS family consists of:

- RCS-5 – a high-performance reconfigurable system with the performance over 6 Tflops;
- RCS-1R – a reconfigurable system with the performance over 1 Tflops;

- RCS-0.2-WS – a workstation with the performance over 300 Gflops;
- RAPC-50 and RAPC-25 – personal computer accelerators with the performance of 50 and 25 Gflops, respectively.

The RCS family is based on three types of basic modules: 16V5-75, 16V5-50 and 16S3-25. The basic module 16V5-75 provides the highest performance and is used in the high-end systems of the family – RCS-5, RCS-1R and RCS-0.2-WS. The basic modules 16V5-50 and 16S3-25 are used in the personal computer accelerators RAPC-50 and RAPC-25.

4.6.1. The Basic Module 16V5-75

The general principles of basic module design are similar, therefore, we will make a detailed analysis of the basic module 16V5-75, which has the following specification:

Performance (64 digits), Gflops	75
Performance (32 digits), Gflops	140
Power consumption, Watt	200
Size of distributed RAM, MB	1.25
XC5VLX110 FPGAs	16
Equivalent gates in FPGA	$8 \cdot 10^6$
Basic module clock rate, MHz	250
LVDS	224
Speed of data exchange between modules, Gb/sec	22

The computation field of the basic module 16V5-75 consists of 16 XC5VLX110-2FF1153 Virtex-5 Xilinx FPGAs. Each FPGA contains about $8 \cdot 10^6$ equivalent gates, placed in points of a 4×4 lattice and connected by the orthogonal communication system. Owing to such communication system, it is possible to significantly simplify the printed circuit board. Rather close connections (less than 4 cm) between the adjacent chips improve frequency characteristics of the board. Data is transferred between non-adjacent chips through the transit channels and pass through interjacent chips, using capabilities of the orthogonal communication system.

Connections between the FPGAs of the basic module computation field are based on the LVDS standard. Advantages of the LVDS are low power consumption of output drivers, low level of electromagnetic interference, inherent noise immunity, common-mode noise rejection, and availability of hardware resources for LVDS high data transfer rate in the FPGAs of the Virtex-5 family. Physically, each communication lane is two adjacent traces (a differential pair) connected to special pins of the chips. Special resources of the Virtex-5 FPGAs that maintain the algorithm of optimal bit and cadr synchronization, and that comply with all nuances of the communication line, provide robust data transfer through LVDS channels in the basic module 16V5-75.

The basic module contains the informational subsystem of data exchange with other MPCS modules through the high speed LVDS channels, that is intended for transfer of intermediate computational results from the FPGAs of the given basic module directly into the FPGAs of other basic modules. The data is transferred through special cables, connected to QTE-056 connectors (see Fig. 4.31).

A unified interface that is used for data transfer between the FPGAs of a single basic module, and between the FPGAs of different basic modules, provides equal exchange rates and reduces boundary effects within the united circuit resources of the RCS.

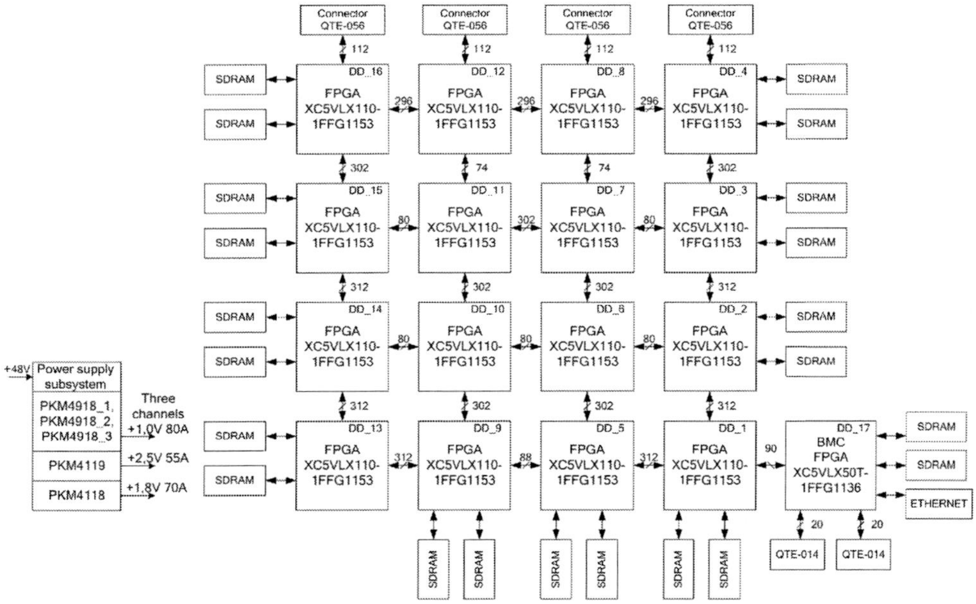

Figure 4.31. The structure of the basic module 16V5-75.

The peripheral FPGAs of the computation field, that are placed on the edges of the 4×4 lattice, are connected with DDR2 (Double Data Rate Two Synchronous Dynamic Random Access Memory) SDRAM chips of dynamic memory of the basic module 16V5-75. The distributed memory consists of 20 MT47H32M16BN chips of Micron Technology Inc. Each chip contains 2^{25} 16-digit words. The total size of the distributed memory is 1.25 GB. Besides, the basic module contains additional 128 MB in a memory chip, connected to the basic module controller.

The distributed memory is used for storage of input data, intermediate computational results, final results of data processing, and fragments of parallel programs.

The main functions of the BMC are management of the basic module subsystems and data transfer between the basic module and the host-controller (an IBM personal computer). The BMC also loads configuration bitstreams of all FPGAs of the computation field, implements some diagnostic functions of the basic module devices, loads input data into the distributed memory, unloads computational results, sets the synchronization subsystem, collects and transfers information about the current status of the basic module into the host-controller. The BMC is implemented in a XC5VLX50T-1FFG1136 Virtex-5 Xilinx FPGA with a built-in PPC405 microprocessor. The BMC is to be programmed from the configuration ROM that contains the configuration bitstream for the BMC. Pre-programming of the basic module controller starts immediately when power is on. The BMC is connected to the host-controller by LVDS channels, two QTE-014 connectors and an Ethernet-channel.

The BMC controls the temperature of the FPGAs of the computation field by polling System Monitor hardware units that are built in the FPGAs and measure their temperature. The subsystem of secondary power supply is placed directly on the basic module and contains portable voltage transducers Ericsson PKM with the input voltage +48V, received from the primary power supply. The complex cooling system removes heat and keeps the required temperature conditions for all FPGAs of the basic module. It contains heatsinks placed on the FPGAs of the computation field, and heat-sink fans.

So, the basic module 16V5-75 is a high-performance computational node with the performance over 75 (140) Gflops. It may be used as a component of computational blocks, containing from 1 to 8 basic modules with the performance from 75 to 600 Gflops. At the same time, the basic module is a stand-alone device which can be connected to the IBM PC as an accelerator and can be used for solving various tasks.

Figure 4.32 shows an external view of the basic module 16V5-75 printed circuit board with the placed electronic elements. Figure 4.33 shows the assembled basic module 16V5-75 with the cooling subsystem, its fixing part and the LVDS cables.

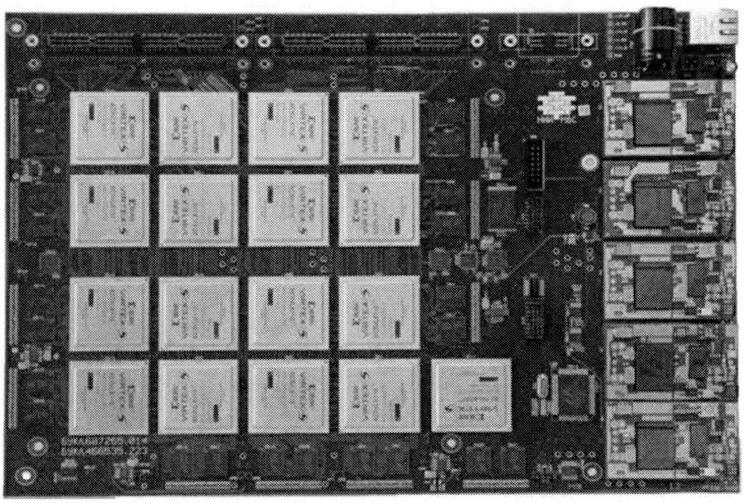

Figure 4.32. The printed circuit board of 16V5-75.

Figure 4.33. The assembled basic module 16V5-75.

4.6.2. The Basic Modules 16V5-50 and 16S3-25 of the Personal Computer Reconfigurable Accelerators

The basic module 16V5-50 is intended for kitting-up the personal computer accelerator RAPC-50.

The computation field of the basic module 16V5-50 consists of 16 XC5VLX110-1FFG1153 Virtex-5 Xilinx FPGAs. The basic module controller is implemented in a XC5VLX50T-1FFG1136 FPGA with a built-in PPC405 microprocessor. As far as the RAPC-50 contains the only basic module 16V5-50, it has no LVDS channels for connection with other basic modules in contrast to the basic module 16V5-75. Instead of LVDS-connectors for intermodule channels, the module 16V5-50 has auxiliary memory chips. Thereby, the total size of its distributed memory is extended to 1.5 GB. The performance of the basic module 16V5-50 is 50 Gflops.

Another distinction of the basic module 16V5-50 is its clock rate that was decreased from 320 to 250 MHz in comparison with the basic module 16V5-75. As a result of this, power consumption, and requirements for the primary power supply dimensions and the cooling system, which are crucial to the desk-size accelerator RAPC-50, were also reduced. All other parameters of the basic modules 16V5-50 and 16V5-75 are similar.

Figure 4.34 shows a general view of the 16V5-50 printed circuit board.

The basic module 16S3-25 is intended for kitting-up the personal computer accelerator RAPC-25. The computation field of the basic module 16S3-25 contains 16 XC3S4000-5FG1156 Spartan-3 Xilinx FPGAs with capacity about 4×10^6 equivalent gates. The basic module controller is implemented in a XC4VFX60-10FF1152C Virtex-4 FPGA with a built-in PPC405 microprocessor. The distributed memory consists of 24 DDR2 SDRAM chips with the total size of 1.5 GB. The performance of the basic module 16S3-25 is 25 Gflops.

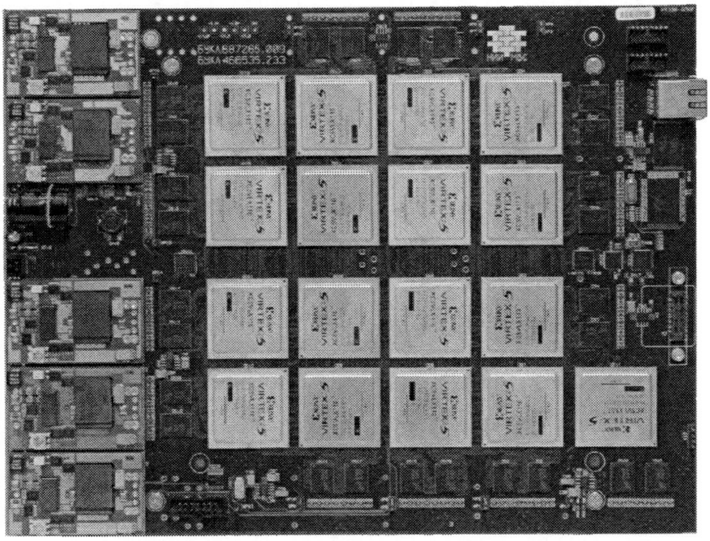

Figure 4.34. A general view of the 16V5-50 printed circuit board.

The clock rate of the FPGAs is 200 MHz, and therefore, adjacent FPGAs are connected by standard direct links, not by LVDS channels as in the basic modules 16V5-75 and 16V5-50. Figure 4.35 shows a general view of the 16S3-25 printed circuit board.

Figure 4.35. A general view of the 16S3-25 printed circuit board.

4.7. High-End Models of the RCS Family

High-end models of the RCS family are the workstation RCS-0.2-WS and the reconfigurable computer systems RCS-1R and RCS-5, designed on the base of the basic module 16V5-75. The computational block RCS-0.2-CB, whose architecture is sufficiently similar to the architecture of the workstation RCS-0.2-WS, may be used for kitting-up the systems RCS-1R and RCS-5.

In this connection, let us analyze the design principles of the workstation as a representative of the RCS family, and the design principles of the block RCS-0.2-CB as a standard architectural unit, that may be used as principal components for design of the systems RCS-1R and RCS-5, and computer systems with much higher performance.

4.7.1. The Workstation RCS-0.2-WS and the Computational Block RCS-0.2-CB

The basic module 16V5-75 is a principal component of the workstation RCS-0.2-WS and the computational block RCS-0.2-CB with the performance of 300 Gflops. These products are based on a multiprocessor computational block (MCB) that consists of 4 basic modules connected into a computational resource by high-speed LVDS and Ethernet channels.

The workstation RCS-0.2-WS is a representative of the RCS family. It is intended for solving problems of microelectronic design, real-time control of complex objects, modeling of complex technical and natural objects and processes, monitoring system design, remote sensing, tomography, etc.

The architecture of the block RCS-0.2-CB is practically similar to the architecture of the workstation RCS-0.2-WS, but the design of these two products is quite different. Design distinctions depend on the product functions: the workstation is a desktop version of a computer system intended for a stand-alone use; the computational block is an embedded element for kitting-up the racks ST-1R of the systems RCS-1R and RCS-5, and for creation of supercomputers of various configurations. The computational block RCS-0.2-CB has the same specification as the workstation RCS-0.2-WS, and is intended for solving the abovementioned tasks as a component of the systems RCS-1R and RCS-5.

The real performance of the workstation RCS-0.2-WS and the block RCS-0.2-CB at solving tasks of linear algebra, mathematical physics, digital signal processing is not less than 60% from the peak one.

Hardware and software tools of the systems RCS-0.2-WS and RCS-0.2-CB provide dynamical two-level reconfiguration of their architecture during task execution. The first level of architecture reconfiguration is so-called program level, i.e. the level of processing elements and distributed memory channels. It provides fast reconfiguration of the system for execution of the task from the given domain. The second level is a circuit level, i.e. the level of logic gates of FPGAs. It provides modification of the instruction set of the processing elements, and the high specific performance of the system, when it is used for execution of various tasks from different domains.

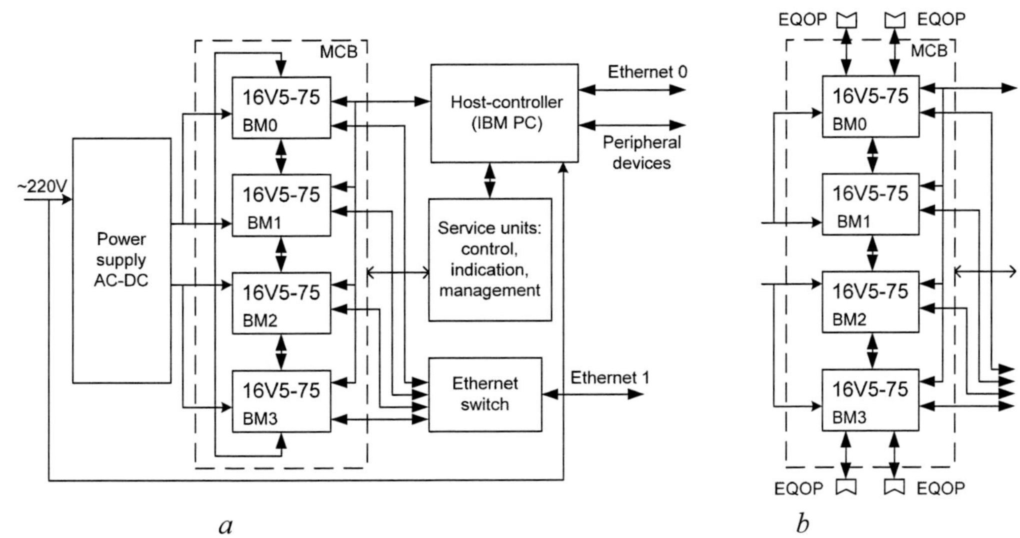

Figure 4.36. The structures of the workstation RCS-0.2-WS (a) and the computational block RCS-0.2-CB (b).

Architecture distinctions between the block RCS-0.2-CB and the workstation RCS-0.2-WS consist in special features of connection between the basic modules in the MCB. Figures 4.36a and 4.36b show the outline structures of the workstation RCS-0.2-WS and the block RCS-0.2-CB. The basic modules of the workstation are connected into a ring. In the computational block the basic modules BM0 and BM3 have external outputs for connection with other similar computational blocks, and for embedding into a computational structure within the rack ST-1R. Except for this, the architecture of the block RCS-0.2-CB is quite similar to the architecture of the workstation RCS-0.2-WS.

The host-controller, included in these both products, is a personal computer IBM PC embedded into the cases of the RCS-0.2-WS and RCS-0.2-CB and managed by Windows XP. The host-controller provides management of all components of the workstation and the computational block, data storage (initial data, intermediate and final results), translation of parallel programs from a high-level programming languages into an executable code, testing and diagnostics, connection with peripheral equipment, connection to local and global information networks by Ethernet.

The cooling systems of the workstation RCS-0.2-WS and the computational block RCS-0.2-CB have some distinctions concerning purpose and maintenance features. Since the workstation is intended for autonomous functioning, it is possible to use only copper heatsinks and extract ventilation heat-sink fans for cooling the FPGAs of its basic modules. This leads to minimization of the workstation dimensions in comparison with the block RCS-0.2-CB, and to decreasing of power consumption and noise level. In comparison with the workstation, operating conditions of the computational block RCS-0.2-CB are more severe, because of overheating of the neighboring blocks. In this case, the stream of cooling air is to pass not only through the case of the block, but through the whole rack, meeting additional resistance. Therefore, in the computational block RCS-0.2-CB it is necessary to use the cooling system with case extract ventilation and direct ventilation fans, placed on the copper heat-sinks of its FPGAs (see Fig. 4.29).

Figures 4.37 and 4.38 show the open cases of the workstation RCS-0.2-CB and the computational block RCS-0.2-CB.

Figure 4.37. The workstation RCS-0.2-WS.

Figure 4.38. The computational block RCS-0.2-CB.

4.7.2. The Reconfigurable Computer System RCS-1R

The reconfigurable computer system RCS-1R is intended for equipping scientific centres, dealing with research in the fields of physics, chemistry, biology, space, design of information control systems for potentially disastrous industry management, design of systems for solving problems from various domains such as aerospace, automobile and energetics. The peak performance of the system RCS-1R is 1,200 Gflops.

Figure 4.39 shows the structure chart of the system RCS-1R. The system RCS-1R is designed on the base of the computational rack SR-1R, which contains 4 computational blocks RCS-0.2-CB, a PC, an Ethernet switch, a power supply system and a cooling system.

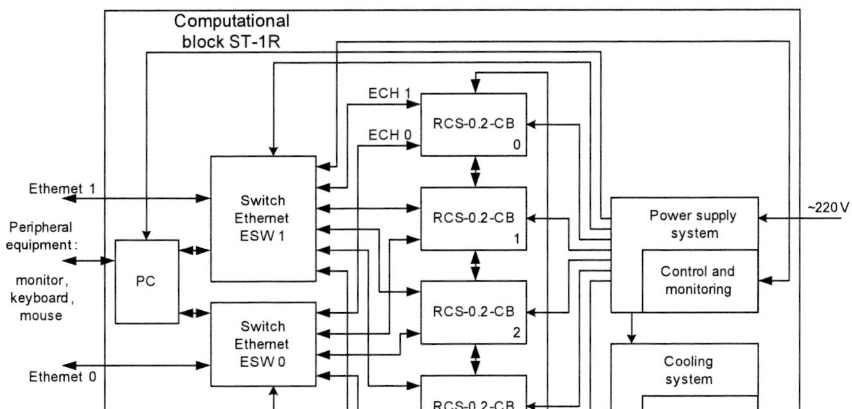

Figure 4.39. The structure chart of the system RCS-1R.

The four computational blocks RCS-0.2-CB are united by high-speed LVDS channels into a computational resource that contains up to 16 basic modules 16V5-75 and provides the peak performance of 1.2 Tflops.

The interblock connections are extensions of the intermodule connections and the connections between the FPGAs of the computation fields.

In total, such system of logic connections is a global LVDS for data transfer with a uniform speed within the rack ST-1R. Figure 4.40 shows an external view of the rack ST-1R of the computer system RCS-1R.

Figure 4.40. An external view of the rack ST-1R of the computer system RCS-1R.

Each computational block RCS-0.2-CB has 2 Ethernet channels ECH0 and ECH1 for connection with external networks. For management and monitoring, the host-controllers of the computational blocks are connected by a switch ESW1 through the outputs ECH1 under general control of the PC. The network outputs ECH0 of the block RCS-0.2-CB are united by a network switch ESW0 of the rack. The switch provides direct connections with any basic module included into the rack, omitting the host-controllers of the blocks.

Such connection of computational resources makes capabilities of reconfigurable components of the rack RCS-1R the most effective. On the one hand, the high-speed LVDS channels connect the computation fields of all 16 basic modules into a single computing environment, which allows the circuit engineer to create a multiprocessor computer system with structural-procedural organization of computations within 4 blocks RCS-0.2-CB, and to use all advantages of resource independent RCS software. On the other hand, owing to the communication system of the computational blocks RCS-0.2-CB, network technologies and the Ethernet switch, the computer system RCS-1R may possess features of a cluster supercomputer. Both the block RCS-0.2-CB and the module 16V5-75 can be used as clusters.

The embedded PC is an industrial computer like IBM PC, implemented in the special case for placing into the rack ST-1R. The PC is intended for management of 4 RCS-0.2-CB blocks, the Ethernet switches, the power supply and the cooling systems, and for connection with peripheral equipment (a monitor, a keyboard, a mouse manipulator, etc.), and also for connection of the reconfigurable computer system RCS-1R to local and global information networks.

4.7.3. The Reconfigurable Computer System RCS-5

The high-end model of the RCS family is the system RCS-5 with the peak performance of 6,000 Gflops. The system RCS-5 is intended for placing in scientific-research centres for solving such problems as development of fundamentally new drugs and materials of new generation, design of information modeling systems for long-term weather forecast, modeling of seismic activity, protein sequencing, tomographic research of near-surface layers of the Earth using sonic and electromagnetic waves, neuromathematics.

Figure 4.41 shows the simplified structure chart of the reconfigurable computer system RCS-5.

The system RCS-5 contains 5 ST-1R racks, a host-computer, a power supply system, a cooling system. A monitor, a keyboard, a mouse manipulator and other peripheral equipment are connected to the system RCS-5 for status control and interaction with the application programmer.

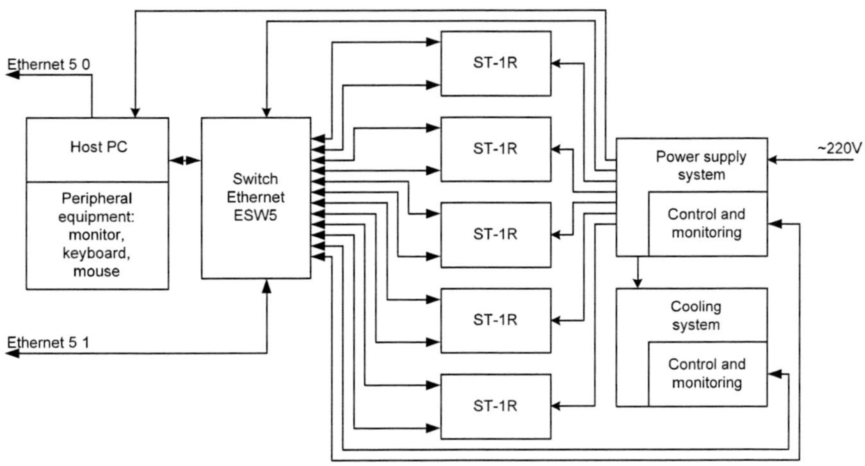

Figure 4.41. The simplified structure chart of the system RCS-5.

The main computational resources of the system RCS-5 are 5 ST-1R racks, connected by the Ethernet switches with the help of network technologies under general control of the host-computer. On the one hand, such hardware configuration makes possible to combine computational resources of the ST-1R racks. On the other hand, hardware may be as much as possible unified because the computational part of the system RCS-5 is assembled from the standard units, already used for creation of other representatives of the RCS family such as the 16V5-75 basic modules, the RCS-0.2-CB computational blocks, the ST-1R racks. Figure 4.42 shows the computer system RCS-5 that is placed in the computer room of the Scientific-Research Computing Centre at Lomonosov Moscow State University (SRCC MSU).

Figure 4.42. The computer system RCS-5 in the computer room of the SRCC MSU.

Figure 4.43 shows a possible structure of connection of the RCS-5 computational resources.

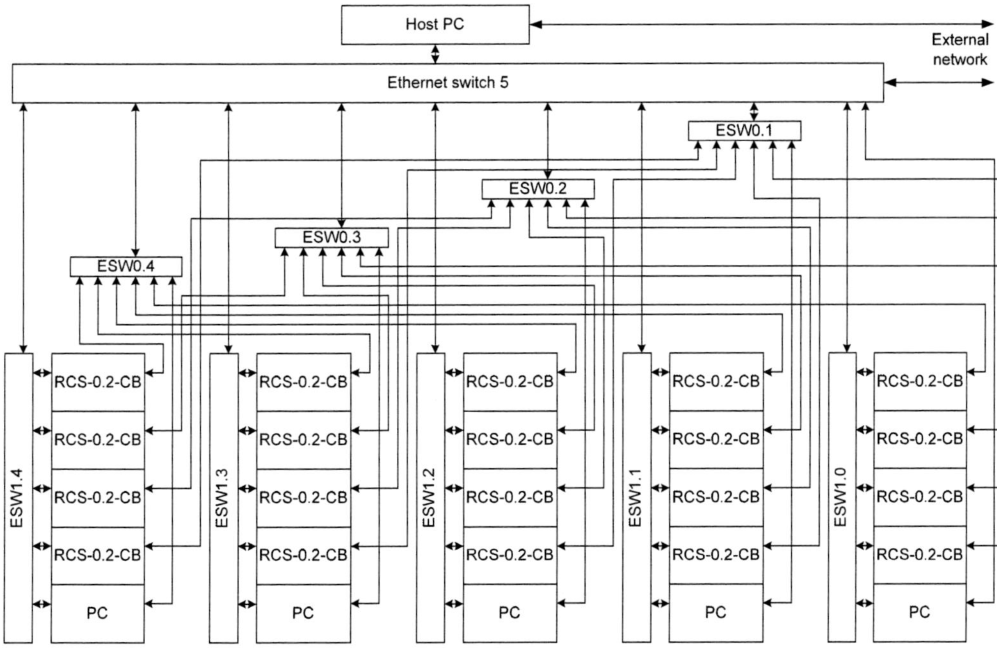

Figure 4.43. A network connection of the RCS-5 computational resources.

In this case, the Ethernet switch ESW0, contained in the rack ST-1R, is connected in order to create so-called "horizontal" connections between the RCS-0.2-CB blocks of the different racks of the system RCS-5.

The host-computer is a personal computer, like IBM PC, with a standard number of necessary devices and peripheral equipment. It is intended for management of computational resources of 5 ST-1R racks, the power supply and the cooling systems, and for connection with peripheral equipment (a monitor, a keyboard, a mouse manipulator, etc.), for connection between the reconfigurable computer system RCS-1R and local and global information networks.

4.8. The Reconfigurable Personal Computer Accelerators RAPC-50 and RAPC-25

The main purpose of the reconfigurable accelerators RAPC-50 and RAPC-25 is extension of capabilities of a personal computer like IBM PC, when it is necessary to solve computationally laborious and complicated problems such as problems of microelectronic design, mathematical modeling of complex technical and natural objects and processes, optimization of oil field maintenance, remote sensing, tomography, etc.

The accelerators RAPC-50 and RAPC-25 are additional computational resources with the peak performance over 50 and 25 Gflops, respectively.

The real performance of the accelerators RAPC-50 and RAPC-25 during execution of various tasks of linear algebra, mathematical physics, digital signal processing exceeds 60% from the peak one. Hardware and software tools provide dynamic reconfiguration of the architecture of the accelerators on program and circuit levels during execution of the tasks.

The accelerators RAPC-50 and RAPC-25 have similar structures, contain similar units and differ only by the types of the basic modules (see Fig. 4.44).

The accelerator RAPC-50 (RAPC-25) contains one 16V5-50 (16S3-25) basic module with an interface unit for connection with the PC, a voltage transducer AC-DC, a display panel and a cooling system.

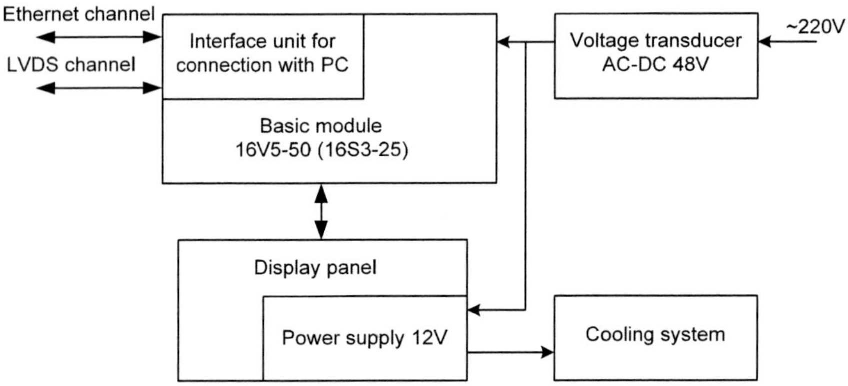

Figure 4.44. The simplified structure chart of the accelerator RAPC-50 (RAPC-25).

The specification and description of the basic modules 16V5-50 (16S3-25) are given in subsection 4.5.2. The main distinctions of the 16V5-50 (16S3-25) from the 16V5-75 are the following: the lower price and less power consumption with keeping the high performance.

Data and control information exchange between the accelerator RAPC-50 (RAPC-25) and the personal computer is implemented by the BMC through the LVDS or Ethernet channels.

The cooling system provides the required temperature of the components of the accelerator RAPC-50 (RAPC-25), and contains heatsinks and heatsink fans placed on the FPGAs of the basic module, and heatsink fans for forced ventilation of the whole case of the accelerator.

Figures 4.45 *a*, *b* show an external view of the accelerators RAPC-50 and RAPC-25. Figure 4.45 *c* shows the accelerator RAPC-25 (an open case).

Practical use of reconfigurable structures, designed according to the modular principle on the base of the FPGA-based modules, has shown that such structures provide the high real performance close to the peak one at solving various tasks from different problem domains, and have almost linear growth of the performance depending on growth of their number of basic modules. Successful development of computer-aided tools of design and programming of multipipeline computing structures will extend opportunities of design of new personal computer accelerators and supercomputers with performance characteristics that considerably exceed those of existing cluster supercomputers.

a

b

c

Figure 4.45. An external view of the accelerators RAPC-50 (a) and RAPC-25 (b), an external view of the open case of the accelerator RAPC-25 (c).

4.9. Basic Modules and Blocks with High-Density Assembly of FPGAs

At the moment of creation, the systems RCS-1R and RCS-5 had the highest performance among FPGA-based computer systems. The computation field of the ST-1R rack contains 256 XC5VLX110-2FF1153 Virtex-5 Xilinx FPGAs. The united computational resource of 5 computation fields of the system RCS-5 consists of 1,280 FPGAs. At the same time, it may be noted, that the considered systems are the experimental systems, and achievement of the maximum performance in a volume unity of a device was not the main purpose. In other words, the characteristic “compactness” is not critical for such systems. Here the characteristic “compactness” means a proportion of two important parameters of computer systems – the "real performance" and the "device volume". This characteristic is 3.5 Gflops/dm^3 for the RCS-0.2-CB block, and 0.9 Gflops/dm^3 – for the ST-1R rack (or for the systems RCS-1R and RCS-5).

To improve technical solutions and the characteristic "compactness", at the end of 2009 the team of specialists designed a basic module Orion (16V5-125OR), which contained a computation field fragment (16 XC5VLX110-2FF1153 Virtex-5 Xilinx FPGAs). The basic module has an advanced FPGA power supply system: each FPGA of the computation field has an individual secondary power supply that provides all necessary voltages. Figure 4.46 shows an external view of the Orion printed circuit board.

Figure 4.46. An external view of the Orion printed circuit board 16V5-125OR.

Owing to such assembly of the printed circuit board of the basic module, it was possible to create a computational block with the performance of 300 Gflops according to Eurocard 1U. Figure 4.47 shows an external view of the Orion computational block.

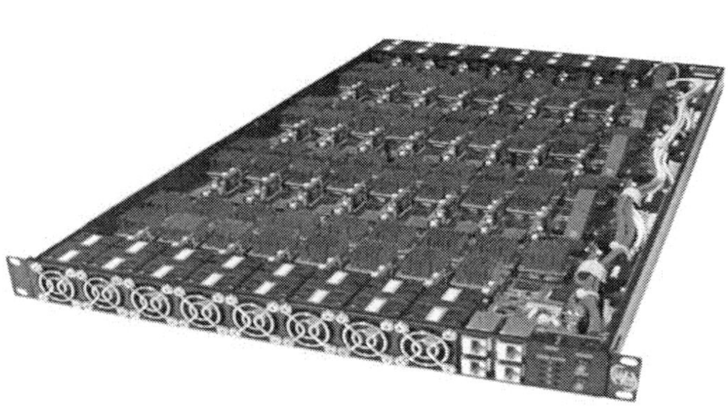

Figure 4.47. An external view of the Orion block (an open case) designed according to 19" Eurocard 1U.

Figure 4.48 shows the computation field of the Orion block. In contrast to the basic module 16V5-75, which contains only 4 FPGAs (out of 16) with channels for intermodule connections, all FPGAs of the basic module 16V5-125OR have LVDS channels for the computational field extension. It provides a higher speed of data exchange between the basic modules of the Orion block, and a higher performance in comparison with the computational block RCS-0.2-CB.

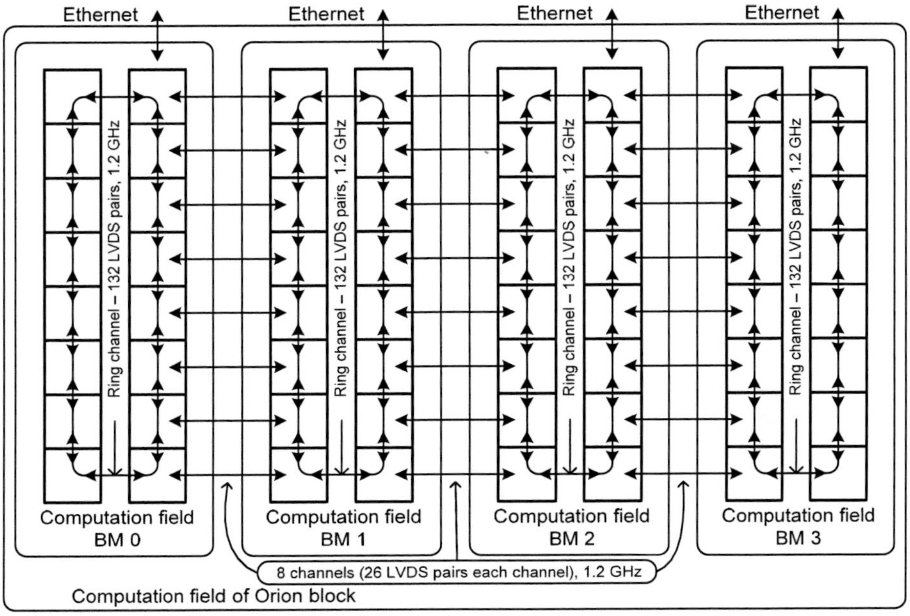

Figure 4.48. The united computation fields of the basic modules of the Orion block.

Table 4.1. Performance of basic modules, blocks and racks

Product	Basic module Pi_{32}/Pi_{64}*, Gflops	Block Pi_{32}/Pi_{64}, Gflops	Rack Pi_{32}/Pi_{64}, Gflops	Block, Gflops/dm^3	Rack, Gflops/dm^3
RCS-0.2-CB	16V5-75 200/75	RCS-0.2-CB 800/300	ST-1R 3,200/1,200	3.5	0.9
Orion	16V5-125OR 250/85	Block Orion 1,000/340	Rack Orion 24,000/8,160	22.0	6.1

*) Pi_{32} – peak performance (single precision);
Pi_{64} – peak performance (double precision).

Table 4.1 shows profiles of the basic modules, blocks and racks based on the Orion blocks and the computational blocks RCS-0.2-CB.

4.10. Basic Modules Based on Virtex-6 FPGAs

At present, the team of scientists of SRI MCS SFU produces RCSs of the new generation on the base of computational blocks, implemented using Virtex-6 FPGAs. The printed circuit boards of the new generations have an open scalable architecture [1], and are placed in the computational blocks of advanced embodiment, named Saiph and Rigel (the stars of Orion constellation). Figure 4.49*a,b* shows the printed circuit boards of the computational blocks, based on Virtex-6 FPGAs.

a)

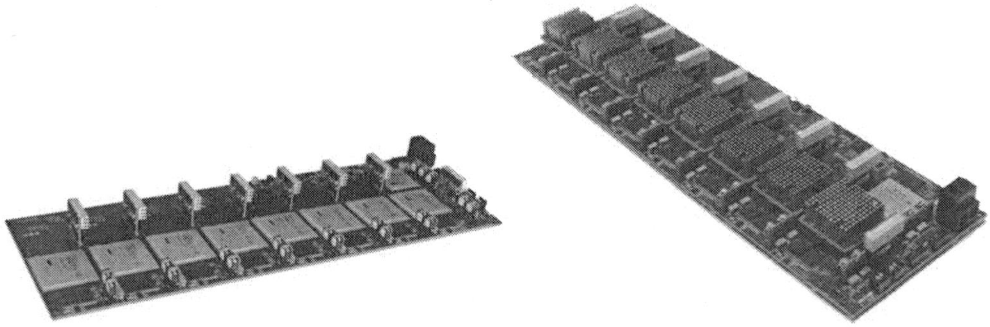

b)

Figure 4.49. The printed circuit boards of the computational blocks, based on Virtex-6 FPGAs; Saiph (a) and Rigel (b).

Table 4.2 shows specifications of the considered printed circuit boards of the computational blocks.

Table 4.2. The specifications of the computational block printed circuit boards

Printed circuit board of computational block	Number of Virtex-6 FPGAs	Number of equivalent gates in one FPGA, million	Interface and speed of data exchange between blocks, Gb/sec	Power consumption, Watt
Saiph, Rigel	8	24	Gigabit Ethernet, 1	300

The computational blocks, based on the Saiph and Rigel printed circuit boards, are designed according to Eurocard 6U and 1U, respectively, and are intended for placing into a standard 19" computational rack, that is a basic component of FPGA-based highly-performance complexes. Figure 4.50*a,b* shows the Saiph and Rigel computational blocks.

Use of Virtex-6 FPGAs as principal components for design of the Saiph and Rigel printed circuit boards makes possible to keep price of the computational block and to increase its performance in 1.5-2 times in comparison with the similar solution Orion, based on Virtex-5 FPGAs. So, the new generation computational modules can be used as principal components for design of RCSs of various architectures and configurations, and can provide considerable competitive advantages of the most part of performance characteristics of these systems, such as specific performance, power efficiency, etc.

Table 4.3 shows the values of peak performance and other technical parameters of the considered computational modules and racks. The performance values of the computational modules and 47U racks correspond to processing of data, represented in a single-precision (Pi_{32}) and a double-precision (Pi_{64}) format according to the IEEE-754 standard.

a b

Figure 4.50. The new generation computational blocks, based on Virtex-6 FPGAs.

Table 4.3. Performance of the RCS computational modules and racks on the base of Virtex-6 FPGAs

Computational module	Performance of computational module Pi_{32}/Pi_{64}, Gflops	Number of computational modules in a 19" rack 47U	Performance of a rack Pi_{32}/Pi_{64}, Tflops
Saiph	1,600/500	6	9/3
Rigel	1,600/500	24-36	34.5 – 51.8

Table 4.4 shows the values of total speed of data exchange between the FPGAs and the distributed memory units, and between the FPGAs within a single computational module, and between different computational modules.

Table 4.4. Speed of data exchange

Computational module	With distributed memory units, Gb/sec	Between FPGAs of computational field, Tb/sec	Between modules, Tb/sec
Saiph, Rigel	12.8	1.0	1.0

So, in comparison with the Orion-based RCSs, the new generation computational modules Saiph and Rigel, based on Virtex-6 FPGAs, provide perfect opportunity for design of higher performance computer systems with no increasing of their price. At the same time, the computational modules are sufficiently stand-alone, and can be easily combined with a personal computer (IBM PC) as accelerators for solving various problems.

The new generation computational modules Saiph and Rigel inherit the programming principles of the computational modules of the previous generations. The considered computational modules are programmed by means of an integrated software suit which supports structural-procedural methods of organization of computations.

Chapter 5

Reconfigurable Multipipeline Computing Structures with Macroobject Architecture

5.1. Design Principles of Reconfigurable Computing Structures on the Base of Macroobject Architecture

FPGA-based implementation of MPCS basic modules gave a new look and an original approach to the problem of rational use of computational resources of multipipeline computing structures.

As it was shown in the previous chapter, a basic module of a reconfigurable computing structure consists of a number of FPGAs, united into a computation field, i.e. the basic module may be considered as an analog of a superFPGA or an FPGA of the second level. For instance, the basic module, which consists of 20 FPGAs with integration about $(8–10)\cdot 10^6$ equivalent gates, can contain up to $1.6\cdot 10^6$ logic units, 6–8 thousand of memory units and about 1500 multipliers, etc., that may be used for design of various computing structures.

In highly-performance MPCSs that contain tens of basic modules, the number of logic units, potentially available to the circuit engineer, amounts to tens of millions. Owing to huge circuit resources, the circuit engineer can implement practically any computational nodes with any possible characteristics.

Within such huge hardware resources the application programmer can create various computing structures for execution of cadrs of solving problems, using FPGA elements of the lowest circuit level (such as logic units), memory units, multipliers, etc.) and FPGA CAD tools. It is clear, that such a solution is optimal according to use of the hardware resources and effectiveness of computational process, because a special-purpose computing structure will be designed within the MPCS computation field for every cadr of the solving problem. Therefore programming on the FPGA circuit level is such a winning approach.

However, several disadvantages restrict practical use of the approach. Firstly, programming of a large FPGA field is a very complicated and laborious process that requires high qualification of circuit engineers and special software tools. In addition, standard FPGA CAD tools allow circuit engineers to create only single-chip solutions, while implementation of parallel-pipeline computing structures may be considerably optimized during mapping on several FPGAs or on the whole computation field of the basic module. Certainly, it is possible to use library elements (IP-cores) of computational units for FPGA programming, but such approach will lead to degradation of computing structure effectiveness and only slightly will simplify the design process.

Secondly, according to structural-procedural organization of computations the problem is to be partitioned into a sequence of cadrs, and fast reconfiguration of the MPCS computation field is required for execution of each next cadr. Reconfiguration of even one FPGA takes rather long time which, in some cases, amounts to 10 seconds [11]. As a result, all advantages of high-performance multipipeline computing will be nullified by continuous cadr changing even in the case of pre-generated files of FPGA configuration. Therefore, reconfiguration of the computing structure, which requires reloading of the configuration files even for one FPGA of the computation field, is extremely unwanted when passing from one cadr to another. Loading of a new configuration is acceptable for single-cadr problems, i.e. for the problems whose information graphs can be entirely mapped on the hardware resources of the MPCS computation field. If the hardware resources of the computation field is insufficient for implementation of the entire information graph, such approach is inapplicable, and it is necessary to use the method of structural-procedural computations.

In this case, it is advisable to use a complex approach, based on two-level – circuit and system – programming of the architecture of the basic module. On the level of circuit programming the application programmer, using tools and capabilities of FPGA CAD software, can develop special-purpose functional nodes, oriented on the domain of the solving problem, and switching facilities that provide connection of the functional nodes into multipipeline structures [20]. In other words, on the level of circuit programming the circuit engineer can design special-purpose samples that may be used as components of future multipipeline structures, which will execute cadrs of applications of a certain problem domain. Then, during implementation of the problem from the given domain, the basic module will be efficiently programmed. The samples will be adjusted and connected by means of switching facilities into the multipipeline structures, which will perform the cadrs of the problem. Effectiveness of the multipipeline structures, synthesized according to such a method, is slightly lower than of those, designed according to the traditional circuit approach, but the cost and the time of design are essentially lower owing to pre-designed engineering solutions. The main thing is that the time of system reconfiguration for execution of every new cadr will only depend on the operating time of switching facilities.

Design experience of FPGA-based MPCSs proves, that for every class of problems it is possible to create a certain set of functional nodes, that can be used for future development of optimal multipipeline computing structures for the most effective solution of problems of the given class. The highest effectiveness of the structures is provided in the case, when all computing elements are placed in the field of logic units of all FPGAs of the basic module or in the MPCS computation field.

Let us define a notion of a macroobject for implementation of the suggested approach of reconfigurable multipipeline computing structure programming.

A macroobject is an architecturally indivisible set of functional nodes (objects), united by a spatial communication system and placed in one or several FPGAs. It is possible to vary the number of functional nodes of any type, parameters of the nodes (such as the size of operands, the number of data channels, the instruction set, etc.), but the functions of the nodes must remain unchangeable.

Each macroobject can contain various computational functional nodes, united by an internal spatial switch, which connects elements into a stand-alone device and provides correct interaction and data transfer between different macroobjects.

To use the macroobject in programming of various multipipeline structures within the MPCS computation field, the following conditions are to be fulfilled on the phase of the project definition:

1) a variety of functional nodes, that may be used for design of macroobjects of various problem domains, is to be defined;
2) a set of parameters of the functional nodes, which may be varied during design of the macroobject, are to be defined. The variable parameters of the functional nodes may be the following: the size of operands, the number of inputs/outputs, the types of external interfaces, the instruction set of the functional nodes, etc.;
3) a library of the functional nodes of the circuit level (IP-cores), designed as the FPGA logic circuits or as VHDL descriptions, is to be created;
4) a set of standard interfaces for interconnection of the functional nodes into multipipeline computing structures according to the information graph of the algorithm is to be created. The interfaces are to be designed as IP-cores;
5) structures of macroobjects for various problem domains is to be defined.

Hereinafter, the IP-cores are considered as pre-designed circuit solutions of the self-contained functional nodes, represented as VHDL program units and available for the application programmer for circuit programming of FPGAs.

So, the macroobject is a sample that may be redefined by the application programmer during development of some engineering solution. Then the sample may be replicated in FPGAs of the basic modules in some required quantity, and interconnected with the macroobjects of the same or various types into the multipipeline computing structures, similar to the structure of the solving problem.

From the application programmer's point of view, the macroobject is a virtual special-purpose device with programmable parameters, a sample which is used during MPCS architecture programming, and which may be implemented within the MPCS computation field without any modifications of its functions.

Examples of the macroobjects, designed for such problem domains as mathematical physics, digital signal processing, linear algebra, are described below. The examples give an overview of the MPCS macroobject architecture. In addition, subsection 5.7 gives a detailed description of so-called general-purpose macroobject, which may be used for solving problems from various problem domains. It is evident, that the general-purpose macroobject cannot provide the same effectiveness of special-purpose macroobjects. However, such macroobject has some advantages:

- the general-purpose macroobject is placed in one FPGA chip;
- the structure of the macroobject remains unchanged and does not depend on its placement in FPGAs of basic modules;
- there are formal methods of representation of the problem in the structural-procedural form for the general-purpose macroobject.

The listed features considerably simplify the programming process for those specialists, who are not capable enough in other, more effective methods of MPCS programming.

5.2. A GENERALIZED STRUCTURE OF A MACROOBJECT

Let us consider a generalized structure of a macroobject. Research shows that macroobjects, designed for various problem domains, generally contain the following objects: processing elements (PE), a distributed memory controller (DMC), a data converter (DC), a dynamic switch (DS), various memory units (M), a procedural ALU (PALU), an interface unit (IU) and a local control unit (LCU). Figure 5.1 shows the structure of the general-purpose macroobject.

The processing element performs processing of two operands according to an operation code, provides synchronization of operand flows and carries out operations with constant values.

The procedural ALU carries out calculations that cannot be effectively implemented in the structural form. The number of such operations is to be big enough to provide processing of data stream elements passing through the structural components of the macroobjects.

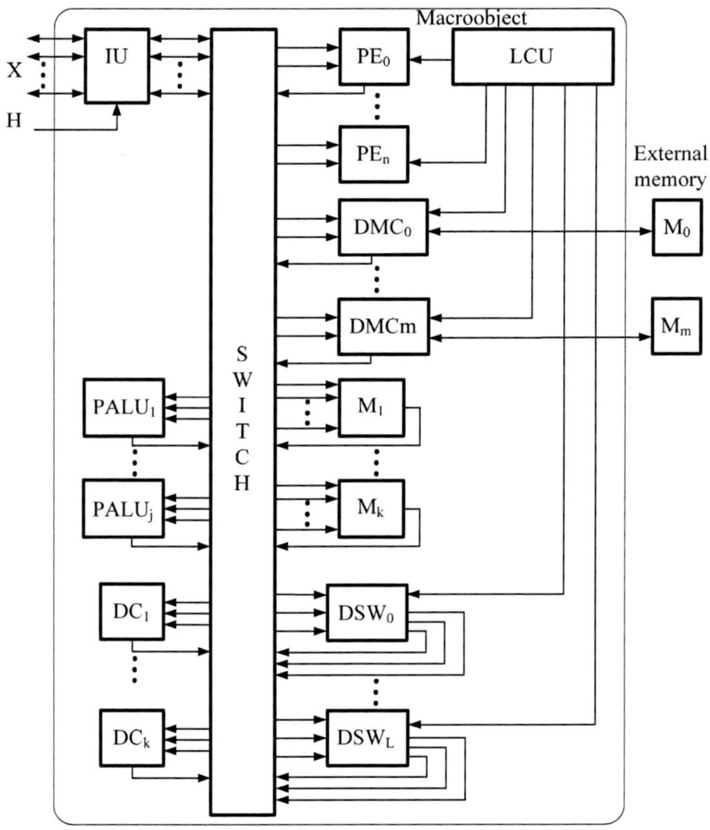

Figure 5.1. The structure of the general-purpose macroobject.

The number of the procedural ALUs may amount to several hundreds. Such number of the ALUs requires considerable reduction of the instruction set of such elements and hence, reduction of hardware resources needed for their implementation. The procedural ALUs are specially designed for each problem domain.

The data converter performs a fixed function. It should be noted, that functional transformation may be performed on a group of operands. Scalar product of vectors, numerical integration, matrix multiplication, solving a system of logic equations may be considered as examples. As a rule, synchronization of data streams is not provided in the data converter.

The set of the processing elements, adjusted to certain arithmetic-logical operations, and the set of the data converters, connected by the internal spatial switch according to the information graph of the solving problem, is the macroobject which processes data arrays according to the algorithm or the fragment of the problem. The macroobject may contain auxiliary nodes and units, for instance, data buffers for data stream synchronization or comparators-switches.

Data streams within the macroobject and the macroobject-based multipipeline computing structure are generated by the distributed memory controllers that organize high speed parallel access to the segmented internal and external memory of the macroobject. The internal memory of the memory controller is divided into a number of segments with certain functions. The operator segment stores a control program. The parameter segment contains parameters of operators of group access to the distributed memory units. The parameters are transferred and processed by address processors. The internal memory of the controller also contains the macroinstruction segment for computational unit adjustment. The macroinstruction segment contains instructions for the processing elements and functional nodes, and addresses of the static switches and operations of the dynamic switches. For execution of procedures, the internal memory of the controller contains the stack segment which stores call addresses of procedures and status words of the distributed memory controller.

The dynamic switches switch data streams between sources and targets. A source/target address may be arbitrarily modified for each data value. The set of the static and dynamic switches provides capability of creation of various computing structures for solving various problems from the problem domain that the macroobject was designed for.

Figure 5.2 shows a variant of the structure chart of the dynamic switch.

The dynamic switch contains multiplexers. Each multiplexer (MX) is controlled by a switching processor (SP) that operates according to an independent program. Functioning of the switching processors is synchronized by a synchronization control unit (SCU), which operates according to a single control program, stored in an instruction memory (IM).

The unit of the dynamic switch can control data streams in reserving modes. Analysis of data streams is carried out by a control unit (CU).

The macroobject contains the local control unit which controls the architecture of the reconfigurable parallel pipeline and provides universality of the system. Microprocessors Microblaze or PowerPC are suggested to be used as the control unit. These processors perform functions of a distributed operation system, load data into the macroobjects and eliminate the problem of "bottleneck" connection between a host-computer and the macroobjects.

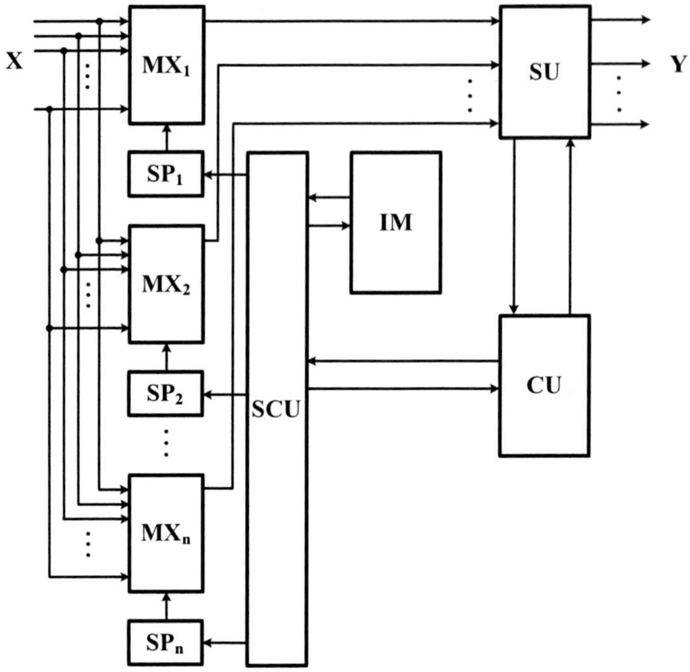

Figure 5.2. The structure chart of the dynamic switch.

5.3. COMPUTATIONAL UNITS OF MACROOBJECTS

5.3.1. Data Representation in the Macroprocessor

The functional nodes of the macroobjects may process data, represented in various formats. However, processing of floating-point data of single and double precision, represented according to the IEEE-754 standard, are of great interest to specialists dealing with effective hardware implementations of wide application. The IEEE-754 standard specifies the uniform data representation and processing methods for floating-point numbers in modern computers.

To fix the idea, let us analyze implementation of the functional nodes that carry out operations on floating-point data. According to the format of floating-point data representation a 32-bit number holds 1 bit of the significand sign, 8 bits of the exponent, 23 bits of the significand. The most significant bit of the significand is "implicit", hence, in fact, the significand holds 24 bits.

5.3.2. A Generalized Structure Chart of Floating-Point Functional Nodes

A generalized structure chart of floating-point functional nodes, that operate according to the IEEE-754 standard, is shown in Fig. 5.3 and consists of 3 units: a unit of input operand control, a unit of arithmetic operations and a unit of result.

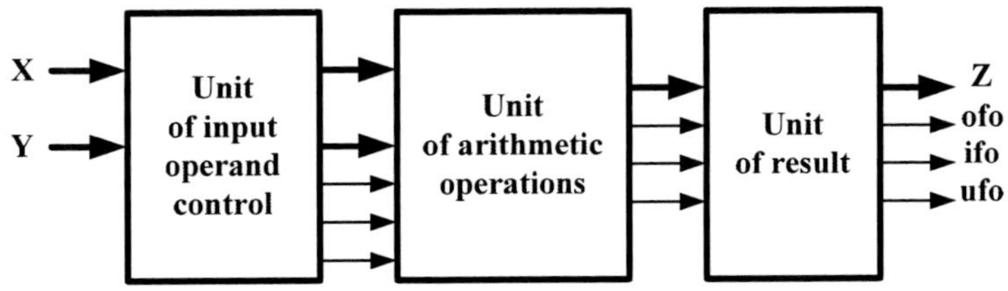

Figure 5.3. A generalized structure chart of floating-point functional nodes.

The unit of input operand control analyses and transforms input data according to the IEEE-754 standard.

As far as calculations are performed on 24-digit significands, and the significands of input data hold only 23 digits, then the most significant digit of the normalized significand is to be reset. The sense of resetting is the following. If the input number is normalized (the exponent is not equal to the minimum $(-127)_{10} = (00000000)_{\text{offset}}$), then the most significant digit of the significand is reset as unity. If the input number is denormalized and its exponent is equal to the minimum, then the most significant digit of the significand is reset as zero. Denormalized numbers may be results of such operations as addition, subtraction, multiplication. The unit of arithmetic operations carries out the floating-point operation and generates attributes of exponent overflow. The unit of result transforms the result number according to the IEEE-754 standard.

5.3.3. The Functional Node of Floating-Point Addition

Addition and subtraction of floating-point numbers is carried out by the functional node implemented as an IP-core. Figure 5.4 shows the simplified structure of the functional node of addition. Operands *A* and *B* are concurrently passing to the adder inputs, taking into account resetting of the "implicit" unity.

According to the algorithm of floating-point addition (subtraction), the unit of exponent matching calculates an exponent difference *DEN*. Taking into account the sign of the exponent difference and following the algorithm, the significand *MO* with greater exponent, and the significand *MD* with lesser exponent, and the exponent of the result *PZ* are selected.

The operation of exponent matching consists in denormalization of the significand *MD* with lesser exponent and is performed in the unit of exponent matching.

The significand of the result is obtained as follows. The significands *MO* and *MDEN* are in true representation. If their signs are equal, then addition of significands is carried out, and the sign of the result becomes equal to that of the input significands. If the signs of the significands are different, then the significand of the result value is obtained as the difference between the significands of the two input values. The significand with a greater absolute value becomes the minuend, and the one with a lesser absolute value becomes the deduction. The difference receives the sign of the significand with a greater absolute value. Then the result significand is normalized. The result value is rounded to the nearest integer value.

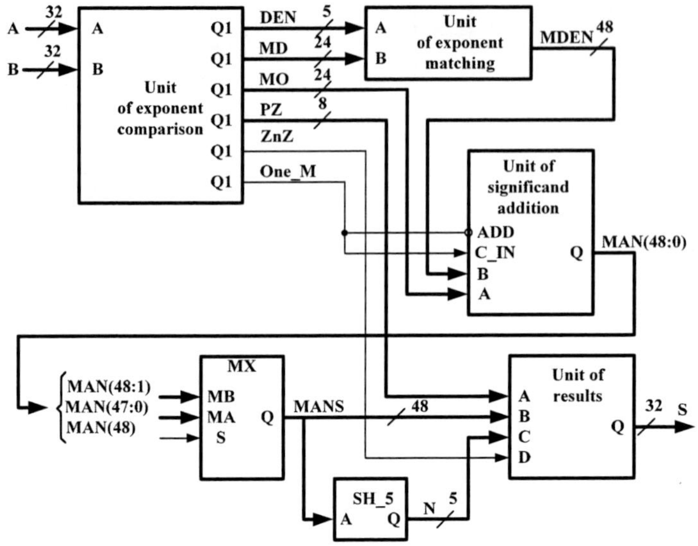

Figure 5.4. The structure chart of adder SMFP 32.

5.3.4. The Functional Node of Floating-Point Multiplication

Multiplication of floating-point numbers is carried out by a functional node implemented as IP-core. The operation of multiplication is performed solely on normalized numbers. Figure 5.5 shows the simplified structure chart of the functional node of multiplication.

The exponent of the result value is calculated as an algebraic sum of the exponents of the multiplicands with the help of the unit of exponent addition. Multiplication of two 24-digit significands is carried out by the unit of significand multiplication. The significand of the product holds 48 digits. Then normalization of the product significand and correction of the exponent are carried out. The rounding bit of the product significand is obtained in the unit of rounding.

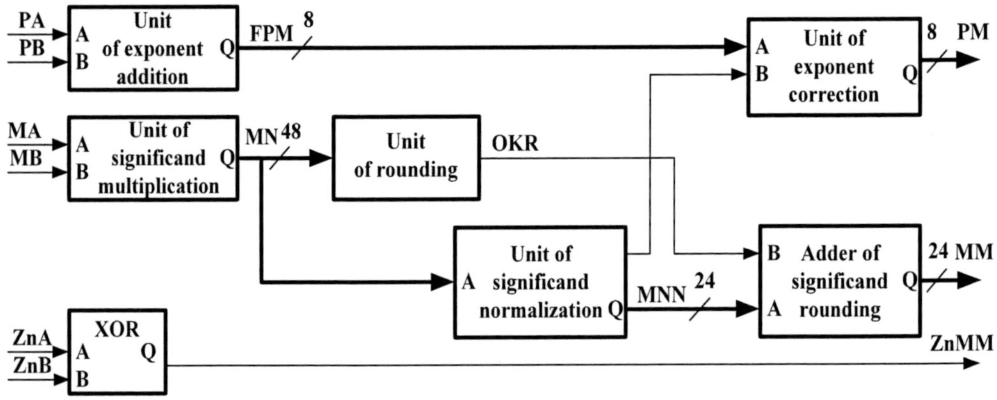

Figure 5.5. The structure chart of multiplier MULT 32.

The multiplier returns the sign of the significand, exponent and product significand.

The considered functional nodes of addition and multiplication, which process floating-point data of single precision, help to understand the principles of design of MPCS computational nodes. Therefore, consideration of functional nodes of division, square-rooting, etc., as well as analysis of functional nodes of double precision, based on the same design principles, are omitted.

5.4. The Macroobject for Solving Problems of Mathematical Physics by Means of Grid Methods

As an example let us consider a structure of a macroobject, oriented on solving problems of mathematical physics, and analyze use of the macroobject for creation of a problem-oriented structure within a basic module [121].

Any macroobjects, oriented on solving problems from one or another problem domain, are based on a fixed set of library functional nodes. The set of functional nodes is specified for each problem domain and reflects its distinctive features. The set contains functional nodes of data processing, units of internal RAM for intermediate data storage and time matching of data streams inside the macroobject, special functional nodes (for instance, a node of control of convergence of iterative processes), interface units for data exchange with other macroobjects, etc. In particular, the number and variety of the functional nodes of the macroobject, oriented on solving problems of mathematical physics, may be specified by calculations that are to be carried out in a node of a grid domain. Such macroobject contains functional nodes, a set of delay units, a unit of error estimation, a unit of boundary conditions, distributed memory controllers, and a communication structure for connection of the elements. All functional nodes are parameterized and their internal structure may be changed according to the solving problem. The number of the functional nodes and the communication structure are also variable and are specified by the parameter set. Figure 5.6 shows the structure of the macroobject for solving problems of mathematical physics.

The macroobject for solving problems of mathematical physics contains the following units:

- adders, multipliers, divisors, etc., $a_1,...,a_k$;
- delay units $DU_1...DU_{k1}$;
- the convergence unit CU;
- the unit of boundary conditions UBC;
- the dynamic data switches SW1, SW2;
- the distributed memory controllers DMC;
- the interface units of data receiving from other macroobjects $X_1...X_q$;
- the interface units of data transfer into other macroobjects $Y_1...Y_p$.

The computing structure for calculations in the grid node is created by means of the functional nodes $a_1,...,a_k$ and the switch.

The numbers of the functional nodes and connections between them are variable, and are specified by loadable parameters. The distributed memory controllers implement the structure

The equation (5.2) expresses Darcy's law, which describes process of fluid filtering in a porous medium. After substitution of (5.1) into (5.2), the following equation is obtained

$$\frac{\partial}{\partial t}\left(\phi(p)\frac{1}{B(p)}\right)+\nabla(K(x,y,z)\cdot\lambda(p)\cdot(\nabla p-\rho_{sc}g\nabla D))=q_{sc}, \quad (5.3)$$

where $\phi(p)$ is the porosity;

$B(p)$ is the coefficient of volume expansion of oil (is given as a table);

$\mu(p)$ is the oil viscosity (is given as a table), Pa·sec;

$K(x,y,z)$ is the tensor of absolute permeability, m^2,

$$K=\begin{pmatrix} k_x & 0 & 0 \\ 0 & k_y & 0 \\ 0 & 0 & k_z \end{pmatrix}, \quad (5.4)$$

$\lambda(p)$ is the oil mobility, 1/Pa·sec,

$$\lambda=\frac{1}{\mu(p)\cdot B(p)}, \quad (5.5)$$

ρ_{sc} is the oil density at standard conditions, kg/m^3;

g is the gravitation constant, m/sec^2;

$D(x,y,z)$ is the vector of the selected depth, m;

q_{sc} is the production of the oil well at standard conditions, m^3/day;

p is the pressure, Pa;

x,y,z are the coordinates of a three-dimensional space;

∇ is the Hamilton operator,

$$\nabla=\frac{\partial}{\partial x}i+\frac{\partial}{\partial y}j+\frac{\partial}{\partial z}k. \quad (5.6)$$

The oil porosity is obtained according to the equation

$$\phi(p)=\psi\phi_0\left(1+c_r\cdot p+\frac{c_r^2\cdot p^2}{2}\right), \quad (5.7)$$

where ψ is the grittiness coefficient (is considered to be a constant),

ϕ_0 is the porosity at a certain bottom-hole pressure p_{ref} (is considered to be a constant);

c_r is the rock compressibility at a certain bottom-hole pressure p_{ref} (is considered to be a constant), 1/Pa.

The boundary conditions in a domain Γ are defined according to the equation

$$p|_{\Gamma} = p_0, \tag{5.8}$$

and the initial conditions are defined according to the statistical equilibrium, which leads to equality of the oil velocity to zero. Taking into account the lack of impermeable girths ($k \neq 0$) with Dirichlet or Neumann boundary conditions, the initial distribution is determined according to the equation

$$\nabla(\lambda(p) \cdot (\nabla p - \rho_{sc} g \nabla D)) = 0. \tag{5.9}$$

The solution of the problem of modeling of fluid filtering in a porous medium is the solution of the equation (5.11). The equation may be solved by a finite difference method (or a grid method), which refers to methods of an approximate solution of differential equations. Universality and applicability to solving various equations of mathematical physics became the factors which determined the choice of the method. According to the finite difference method a domain of continuous variation of arguments is replaced by a finite (discrete) ensemble of points (nodes), called a grid. Functions of a discrete argument, which are defined in the grid nodes, and are called grid functions, are considered instead of functions of a continuous argument. Derivatives, which are contained in a differential equation, are replaced (or approximated) by difference quotients; the differential equation is replaced by a system of algebraic equations (a difference equation). The initial and boundary conditions are also replaced by difference initial and boundary conditions for the grid function. It is natural to require, that the obtained difference boundary problem is to be solvable and its solution, when the number of the grid nodes is increasing, is to approximate (to converge) to the solution of the initial problem for the differential equation. To solve the equation of mass balance, let us create an analytic grid $M+1 \times N+1 \times K+1$ in the given domain. Assume the grid steps in the line of coordinates x, y, z to be constant and equal to Δx, Δy, Δz, respectively. Let the values in the grid nodes be designated as $p_{i,j,k}$, where $i = 0,1,...,M$, $j = 0,1,...,N$, $k = 0,1,...,K$.

Let us rewrite the equation (5.1):

$$\nabla u = \frac{\partial u_x}{\partial x} + \frac{\partial u_y}{\partial y} + \frac{\partial u_z}{\partial z} = 0. \tag{5.10}$$

Then, using representation of partial derivatives in the finite-difference form, we represent the components of the equation (5.10) as follows

$$\frac{\partial u_x}{\partial x}=\frac{\lambda_{i+1,j,k}((p_{i+1,j,k}-p_{i,j,k})-\rho_{sc}g(D_{i+1,j,k}-D_{i,j,k}))}{2\Delta x^2}-$$
$$-\frac{\lambda_{i-1,j,k}((p_{i,j,k}-p_{i-1,j,k})-\rho_{sc}g(D_{i,j,k}-D_{i-1,j,k}))}{2\Delta x^2},$$
$$\frac{\partial u_y}{\partial y}=\frac{\lambda_{i,j+1,k}((p_{i,j+1,k}-p_{i,j,k})-\rho_{sc}g(D_{i,j+1,k}-D_{i,j,k}))}{2\Delta y^2}- \quad (5.11)$$
$$-\frac{\lambda_{i,j-1,k}((p_{i,j,k}-p_{i,j-1,k})-\rho_{sc}g(D_{i,j,k}-D_{i,j-1,k}))}{2\Delta y^2},$$
$$\frac{\partial u_z}{\partial z}=\frac{\lambda_{i,j,k+1}((p_{i,j,k+1}-p_{i,j,k})-\rho_{sc}g(D_{i,j,k+1}-D_{i,j,k}))}{2\Delta z^2}-$$
$$-\frac{\lambda_{i,j,k-1}((p_{i,j,k}-p_{i,j,k-1})-\rho_{sc}g(D_{i,j,k}-D_{i,j,k-1}))}{2\Delta z^2}.$$

After substitution of the obtained expressions of partial derivatives into the equation (5.10) and solving it for $p_{i,j,k}$, we obtain

$$p_{i,j,k}=\frac{1}{\dfrac{\lambda_{i+1,j,k}+\lambda_{i-1,j,k}}{2\Delta x^2}+\dfrac{\lambda_{i,j+1,k}+\lambda_{i,j-1,k}}{2\Delta y^2}+\dfrac{\lambda_{i,j,k+1}+\lambda_{i,j,k-1}}{2\Delta z^2}}\times$$
$$\times\left(\frac{\lambda_{i+1,j,k}(p_{i+1,j,k}-\rho_{sc}g(D_{i+1,j,k}-D_{i,j,k}))+\lambda_{i-1,j,k}(p_{i-1,j,k}+\rho_{sc}g(D_{i,j,k}-D_{i-1,j,k}))}{2\Delta x^2}+\right. \quad (5.12)$$
$$+\frac{\lambda_{i,j+1,k}(p_{i,j+1,k}-\rho_{sc}g(D_{i,j+1,k}-D_{i,j,k}))+\lambda_{i,j-1,k}(p_{i,j-1,k}+\rho_{sc}g(D_{i,j,k}-D_{i,j-1,k}))}{2\Delta y^2}+$$
$$\left.+\frac{\lambda_{i,j,k+1}(p_{i,j,k+1}-\rho_{sc}g(D_{i,j,k+1}-D_{i,j,k}))+\lambda_{i,j,k-1}(p_{i,j,k-1}+\rho_{sc}g(D_{i,j,k}-D_{i,j,k-1}))}{2\Delta z^2}\right).$$

To solve the system of linear algebraic equations (5.12) we use the method of simultaneous over-relaxation with the following iterative formula:

$$p_{i,j,k}^{(n)}=\omega F(p_{i+1,j,k}^{(n-1)},p_{i-1,j,k}^{(n-1)},p_{i,j+1,k}^{(n-1)},p_{i,j-1,k}^{(n-1)},p_{i,j,k+1}^{(n-1)},p_{i,j,k-1}^{(n-1)})+(1-\omega)p_{i,j,k}^{(n-1)}, \quad (5.13)$$

where F is the function, described by (5.12);

ω is an iteration parameter, $0<\omega<2$;

n is an iteration index.

The term "simultaneous" means that the obtained new values of the pressure $p_{i,j,k}$ will be used only in the next iteration. This leads to increasing of the number of iterations, but allows parallelization of computational process.

The inequality

$$\max_{i,j,k}\left|p_{i,j,k}^{(n)}-p_{i,j,k}^{(n-1)}\right|<\varepsilon, \tag{5.14}$$

is the condition of termination of calculations; here ε is the prescribed accuracy.

To implement the problem of fluid filtering in a porous medium on the MPCS, it is to be transformed in the structural-procedural form. Therefore, the equation (5.13) is to be represented as follows:

$$\begin{aligned}
p_{i,j,k} &= \frac{1}{C_1(\lambda_{i+1,j,k}+\lambda_{i-1,j,k})+C_2(\lambda_{i,j+1,k}+\lambda_{i,j-1,k})+C_3(\lambda_{i,j,k+1}+\lambda_{i,j,k-1})}\times \\
&\times(C_1(\lambda_{i+1,j,k}(p'_{i+1,j,k}-(D_{i+1,j,k}-D_{i,j,k}))+\lambda_{i-1,j,k}(p'_{i-1,j,k}+(D_{i,j,k}-D_{i-1,j,k})))+ \\
&+C_2(\lambda_{i,j+1,k}(p'_{i,j+1,k}-(D_{i,j+1,k}-D_{i,j,k}))+\lambda_{i,j-1,k}(p'_{i,j-1,k}+(D_{i,j,k}-D_{i,j-1,k})))+ \\
&+C_3(\lambda_{i,j,k+1}(p'_{i,j,k+1}-(D_{i,j,k+1}-D_{i,j,k}))+\lambda_{i,j,k-1}(p'_{i,j,k-1}+(D_{i,j,k}-D_{i,j,k-1})))),
\end{aligned} \tag{5.15}$$

where

$$C_1=\frac{1}{2\Delta x^2},$$

$$C_2=\frac{1}{2\Delta y^2},$$

$$C_3=\frac{1}{2\Delta z^2},$$

$$p'_{i,j,k}=\frac{p_{i,j,k}}{\rho_{sc}g}.$$

Transition from $p_{i,j,k}$ to $p'_{i,j,k}$ is called a procedure of normalization, when the coefficients are removed to minimize the number of arithmetic operations. In our case, if both parts of (5.9) are divided by the constant coefficient $\rho_{sc}g$, and $p_{i,j,k}$ is replaced by $p'_{i,j,k}$ and the obtained equation is solved for $p'_{i,j,k}$, then the equation (5.14) is obtained. After solving (5.14), it is sufficient to multiply each $p'_{i,j,k}$ by $\rho_{sc}g$ in each grid point. As a result, we get the final solution of the problem of calculation of initial distribution of pressure. Hereinafter, we will designate the normalized variable as $p_{i,j,k}$, assuming that $p_{i,j,k}$ is to be multiplied by $\rho_{sc}g$.

As it was mentioned above, for solving the system of equations (5.16) we will use the method of simultaneous over-relaxation, according to which the n-th approximation of the grid function $p_{i,j,k}^{(n)}$ in the grid points (x_i, y_j, z_k) is calculated according to the recurrent formula

$$p_{i,j,k}^{(n)} = \frac{\omega}{C_1(\lambda_{i+1,j,k} + \lambda_{i-1,j,k}) + C_2(\lambda_{i,j+1,k} + \lambda_{i,j-1,k}) + C_3(\lambda_{i,j,k+1} + \lambda_{i,j,k-1})} \times$$
$$\times (C_1(\lambda_{i+1,j,k}(p'_{i+1,j,k} - (D_{i+1,j,k} - D_{i,j,k})) + \lambda_{i-1,j,k}(p'_{i-1,j,k} + (D_{i,j,k} - D_{i-1,j,k}))) +$$
$$+ C_2(\lambda_{i,j+1,k}(p'_{i,j+1,k} - (D_{i,j+1,k} - D_{i,j,k})) + \lambda_{i,j-1,k}(p'_{i,j-1,k} + (D_{i,j,k} - D_{i,j-1,k}))) + \quad (5.16)$$
$$+ C_3(\lambda_{i,j,k+1}(p'_{i,j,k+1} - (D_{i,j,k+1} - D_{i,j,k})) + \lambda_{i,j,k-1}(p'_{i,j,k-1} + (D_{i,j,k} - D_{i,j,k-1})))) +$$
$$(1-\omega) p_{i,j,k}^{(n-1)}.$$

5.4.2. An Example of Solving a Problem of Mathematical Physics by Means of a Macroobject

Let us consider the solution of the problem of fluid filtering on the 3rd generation basic module 16V4-50 which contains 16 FGPA chips with integration of $4 \cdot 10^6$ equivalent gates. The macroobject, which was described earlier, may be implemented in 4 FPGA chips of the basic module.

There are two alternative implementations of the given computational structure on the basic module 16V4-50.

According to the first type of implementation 4 independent macroobjects are placed on the basic module. Each macroobject is placed into a separate quadrant. In this case the macroobjects operate independently and process, in general case, arbitrary nodes of the grid domain with no data exchange.

The second type of implementation provides shared use of the macroobjects, placed in 4 quadrants of the basic module. In this case 4 quadrants of the basic module simultaneously process 4 adjacent points of the grid domain.

Let us make a detailed consideration of the first type of calculations, according to which the problem is implemented in 4 FPGAs interconnected into the quadrant within the basic module. Let us also make a detailed analysis of the structure of each FPGA of the macroobject. Let the equation (5.16) be transformed into

$$p_{i,j,k}^{(n)} = c1 + c2 + c3 + c4 + (1-\omega) p_{i,j,k}^{(n-1)}, \quad (5.17)$$

where

$$c1 = \frac{\omega}{C_1(\lambda_{i+1,j,k} + \lambda_{i-1,j,k}) + C_2(\lambda_{i,j+1,k} + \lambda_{i,j-1,k}) + C_3(\lambda_{i,j,k+1} + \lambda_{i,j,k-1})},$$
$$c2 = C_1(\lambda_{i+1,j,k}(p'_{i+1,j,k} - (D_{i+1,j,k} - D_{i,j,k})) + \lambda_{i-1,j,k}(p'_{i-1,j,k} + (D_{i,j,k} - D_{i-1,j,k}))),$$
$$c3 = C_2(\lambda_{i,j+1,k}(p'_{i,j+1,k} - (D_{i,j+1,k} - D_{i,j,k})) + \lambda_{i,j-1,k}(p'_{i,j-1,k} + (D_{i,j,k} - D_{i,j-1,k}))),$$
$$c4 = C_3(\lambda_{i,j,k+1}(p'_{i,j,k+1} - (D_{i,j,k+1} - D_{i,j,k})) + \lambda_{i,j,k-1}(p'_{i,j,k-1} + (D_{i,j,k} - D_{i,j,k-1}))).$$

Figures 5.8 *a, b, c, d* show the generalized structures, programmed in 4 FPGAs to obtain the values $c1$, $c2$, $c3$, $c4$ according to the equation (5.17).

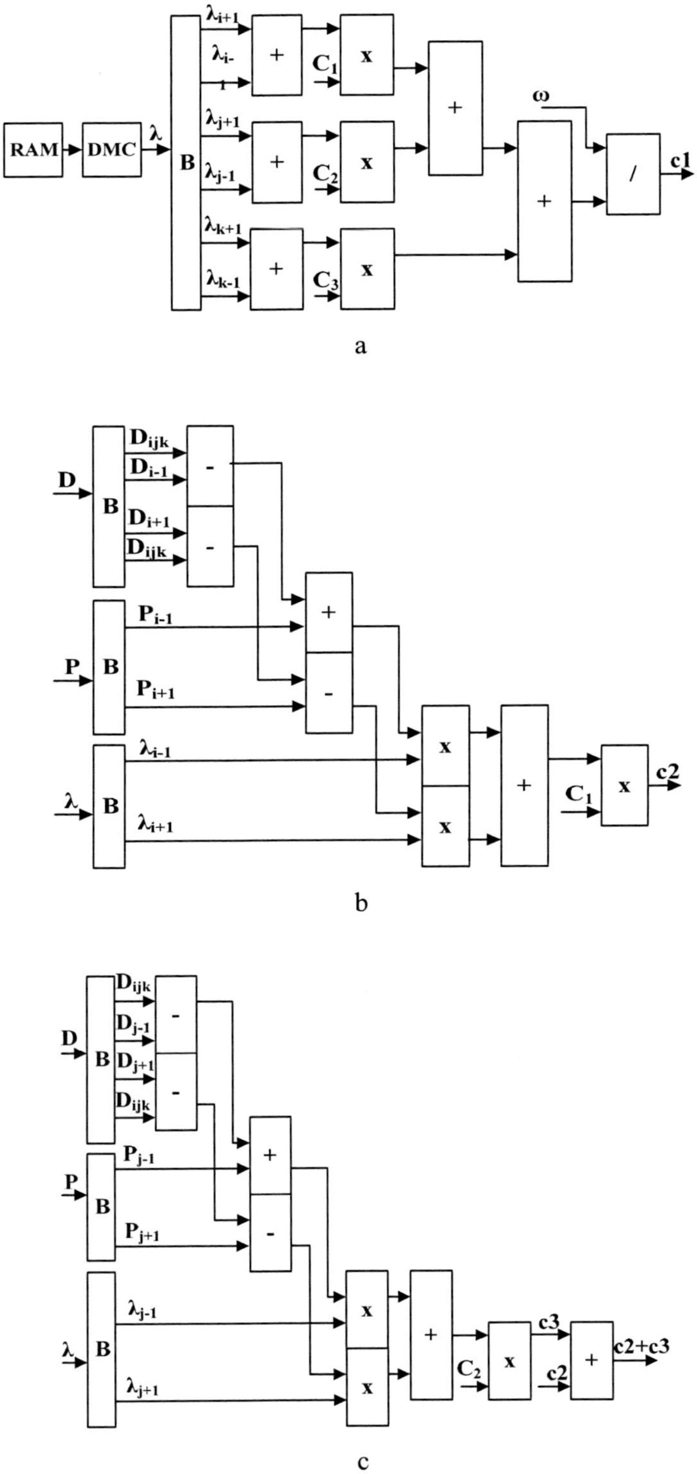

Figure 5.8. (Continued).

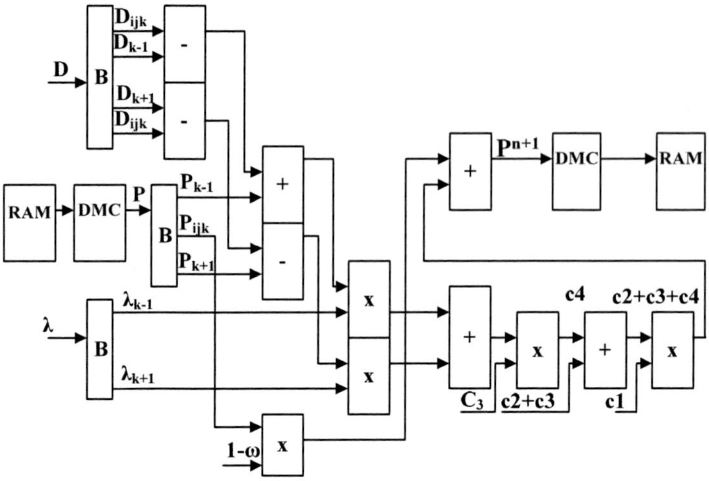

d

Figure 5.8. The structures for calculation of the values A, B, C, E according to the equation (5.5).

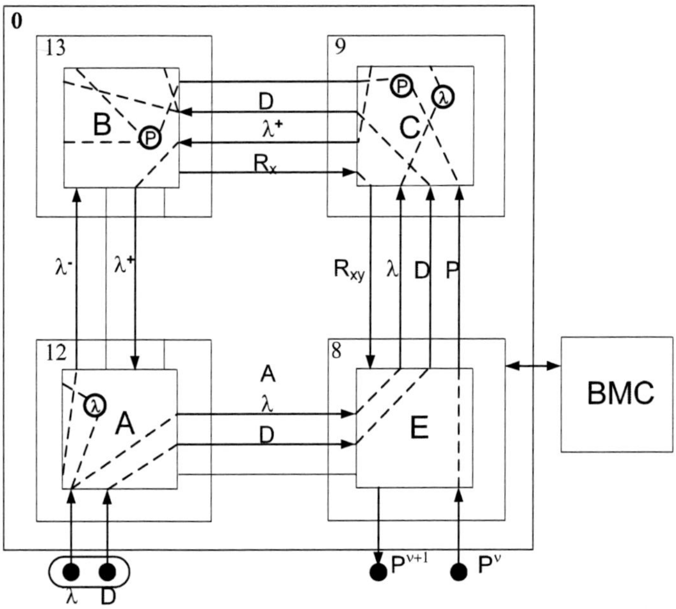

Figure 5.9. The generalized structure of the macroobject.

Figure 5.9 shows the structure of the macroobject that is created from FPGAs, used for the problem implementation.

Let us consider the second type of implementation, according to which 4 quadrants of the basic module simultaneously process 4 adjacent points of the grid domain.

Figure 5.10 shows the computational structure of the problem of interacting calculations, when the single problem is assumed to be solved by 4 interconnected quadrants.

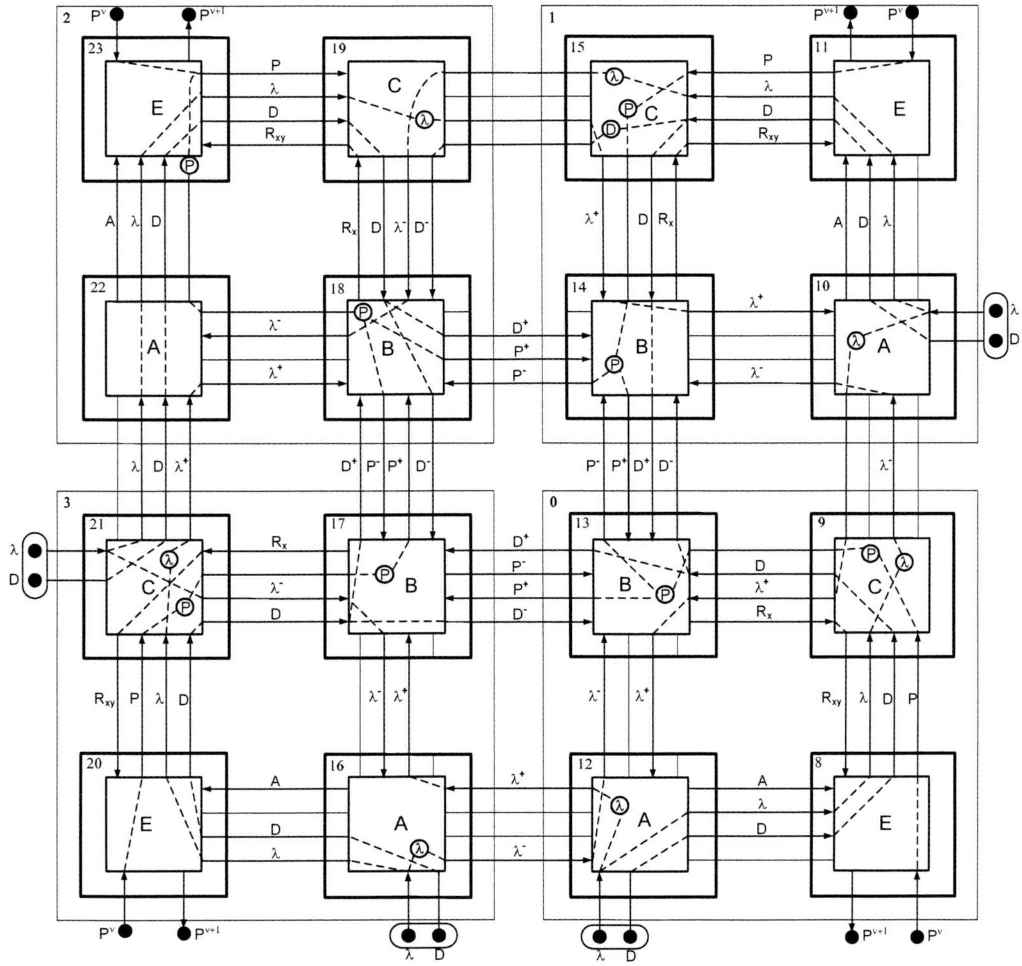

Figure 5.10. Implementation of interacting calculations in the basic module.

A great number of computing structures for solving various problems of mathematical physics may be created on the base of the macroobject. The application programmer may define a component set of the macroobject as well as modify its structure according to various computational algorithms and change properties of previously designed components, redefining library elements.

Design of hardware implementation of the computing structure, shown in Fig. 5.8, requires participation of highly-qualified circuit engineers, who implement and place it in the correspondent FPGAs of the basic module. However, other solutions, which need no circuit engineer for generation of FPGA configuration files, are also possible.

Chapter 6 contains the description of the solution of the problem by an application program written on a high-level language of description of structural-procedural solutions COLAMO. Owing to such approach, process of generation of configuration files can be automated.

5.5. The Macroobject for Solving Problems of Digital Signal Processing

The main algorithms of digital signal processing are an algorithm of digital convolution and an algorithm of discrete Fourier transform, because they are the base for various application algorithms, oriented on implementation of majority of special cases [70, 122, 111, 123].

A digital convolution with finite-impulse response is defined by the equation [111]:

$$y(n)=\sum_{k=0}^{M-1}h(k)x(n-k)\,,\ n=0,1,...,N-1. \tag{5.18}$$

Let us consider a pipeline implementation of convolution. Figure 5.11 shows the structure of the pipeline implementation of the algorithm of convolution (5.6).

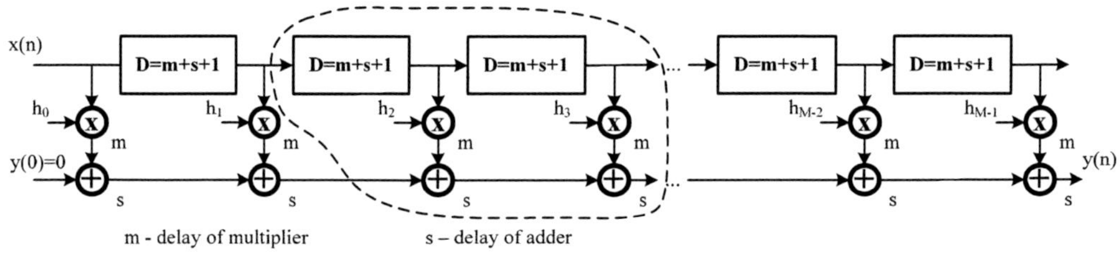

Figure 5.11. The structural implementation of the algorithm of convolution.

The algorithms of the discrete direct and inverse Fourier transform [120]

$$S(k)=\sum_{n=0}^{N-1}s(n)W^{n\cdot k}\,,\ W=e^{-j\frac{2\pi}{N}}\,,\ k=0,1,...,N-1\,, \tag{5.19}$$

$$s(n)=\frac{1}{N}\sum_{k=0}^{N-1}S(k)W^{-n\cdot k}\,,\ W=e^{-j\frac{2\pi}{N}}\,,\ n=0,1,...,N-1 \tag{5.20}$$

may be applied independently for calculation of a signal spectrum, or may be used in algorithms of fast convolutions, in algorithms of calculation of correlation functions, etc. The spectrum may be obtained by means of the discrete direct Fourier transform (DFT) according to (5.19).

The direct and inverse DFT may be calculated by means of effective algorithms, so-called fast algorithms of Fourier transform: the FFT for calculation of the direct DFT and the IFFT for calculation of the inverse DFT. Traditionally, the FFT algorithms are performed on a vector of input data. The length of the input data is powers of two: $N=2^p$, $p=1,2,3,...$. Hereinafter, the abbreviation FFT will contain the number N, which defines the size of transformation of the vector with the length N. In this case, the FFT algorithm, performed on the array with the length $N=2^9=512$, will be marked as FFT512.

Obviously, a repetitive fragment of the algorithm of convolution is a dotted part of the pipeline chart.

A specific feature of DSP algorithms is equality of the number of input data and the number of processing results, which is correct both for algorithms of filtering and for algorithms of fast orthogonal transformations. The statement is true, when the set of filter coefficients $\{h\}$, $i \in 0,1,...,M-1$ and the set of the coefficients $\{W^{n \cdot k}\}$, $n,k \in 0,1,...,N-1$ are stored or generated in the computing device, that performs the algorithm, but do not pass from external sources. The statement is also true for any self-contained fragments of the algorithms, mentioned above. The self-contained fragments of the algorithm are its parts, superposition of which provides performing of the whole algorithm. For the FFT512 algorithm such fragments are, for instance, the FFT2 and FFT8 algorithms.

Let us consider implementation of a modulo-2 FFT algorithm. The modulo-2 FFT basic operation with decimation in frequency may be performed according to the following algorithm:

$$\begin{aligned} &\dot{A}_{t+1} = \dot{A}_t + \dot{B}_t, \\ &\dot{B}_{t+1} = (\dot{A}_t - \dot{B}_t)\dot{W}_t, \\ &\dot{W}_t = \dot{W}_b \times \dot{W}_c \end{aligned} \tag{5.21}$$

The following equations may be individually written for real and imaginary parts:

$$\begin{aligned} &\mathrm{Re}\left[\dot{A}_{t+1}\right] = \mathrm{Re}\left[\dot{A}_t\right] + \mathrm{Re}\left[\dot{B}_t\right],\ \mathrm{Im}\left[\dot{A}_{t+1}\right] = \mathrm{Im}\left[\dot{A}_t\right] + \mathrm{Im}\left[\dot{B}_t\right], \\ &\mathrm{Re}\left[\dot{B}_{t+1}\right] = \left(\mathrm{Re}\left[\dot{A}_t\right] - \mathrm{Re}\left[\dot{B}_t\right]\right) \cdot \mathrm{Re}\left[\dot{W}_t\right] - \left(\mathrm{Im}\left[\dot{A}_t\right] - \mathrm{Im}\left[\dot{B}_t\right]\right), \\ &\mathrm{Im}\left[\dot{B}_{t+1}\right] = \left(\mathrm{Re}\left[\dot{A}_t\right] - \mathrm{Re}\left[\dot{B}_t\right]\right) \cdot \mathrm{Im}\left[\dot{W}_t\right] + \left(\mathrm{Im}\left[\dot{A}_t\right] - \mathrm{Im}\left[\dot{B}_t\right]\right), \\ &\mathrm{Re}\left[\dot{W}_t\right] = \mathrm{Re}\left[\dot{W}_b\right] \cdot \mathrm{Re}\left[\dot{W}_c\right] - \mathrm{Im}\left[\dot{W}_b\right] \cdot \mathrm{Im}\left[\dot{W}_c\right], \\ &\mathrm{Im}\left[\dot{W}_t\right] = \mathrm{Re}\left[\dot{W}_b\right] \cdot \mathrm{Im}\left[\dot{W}_c\right] + \mathrm{Im}\left[\dot{W}_b\right] \cdot \mathrm{Re}\left[\dot{W}_c\right]. \end{aligned} \tag{5.22}$$

The same equations may be used for basic operation of the inverse FFT algorithm [100], which requires use of the complex-conjugate coefficients $\dot{W}_t$. Complex conjugation is performed by reversal of the sign of the imaginary part of $\dot{W}_t$.

Let us consider calculation of $\dot{W}_t$ in details. The coefficient $\dot{W}_t$ may enter each basic operation from an external ROM with the capacity of $N/2$ complex words. For structural organization of computations, when several FFT basic operations are performed simultaneously, the number of various $\dot{W}_t$ must be the same. Therefore, each basic computing structure needs a ROM unit. As a result, this will lead to high hardware costs.

Other methods are based on calculation of $\dot{W}_t$. One of the methods, given in [124], consists in recursive calculation of the coefficients proceeding from the coefficient $W = e^{-j\frac{2\pi}{N}}$

with the minimum angle. However, the disadvantage of the recursive method is a rapidly accumulating error, therefore the method may be applied only for small N.

For structural-procedural implementations of the FFT the most suitable is a method, which is considered in [125, 126].

The coefficient $\dot{W}_t$ is obtained using basic $\dot{W}_b$ and auxiliary $\dot{W}_c$ according to the equation $\dot{W}_t = \dot{W}_b \times \dot{W}_c$. The coefficients $\dot{W}_b$ and $\dot{W}_c$ are also read from a ROM unit, therefore two tables – the table of basic and the table of auxiliary coefficients – are required for storage of $\dot{W}_b$ and $\dot{W}_c$, which are used for calculation of $\dot{W}_t$ in the N-size FFT algorithm. The coefficients, which are placed into the basic and auxiliary tables, may be obtained according to the following equations:

$$\dot{W}_b = \exp(-j2\pi pN_2 / N_M),\ \dot{W}_c = \exp(-j2\pi q / N_M),$$
$$p = 0,1,...,N_1 - 1,\ q = 0,1,...,N_2 - 1,\ N_1 \times N_2 = N_M / 2,$$

where N_M is the maximum possible size of the FFT which may be implemented in the macroobject.

In comparison with a traditional approach, such method of coefficient generation reduces the ROM capacity W in $N_M / (2(N_1 + N_2))$ times. For $N_M = 2^{16}$, $N_1 = 2^7$ and $N_2 = 2^8$ the ROM capacity is reduced in 85 times [124,125].

Since the DSP macroobject is oriented on pipeline execution of the FFT algorithm, then the sequence of execution of the basic operations, and, hence, the order of $\dot{W}_b$ and $\dot{W}_c$ selection is defined by the pipeline algorithm, described in [111].

Figure 5.12 shows the computing structure that corresponds to a classical basic operation, where $\dot{W}_t$ is generated by an external source.

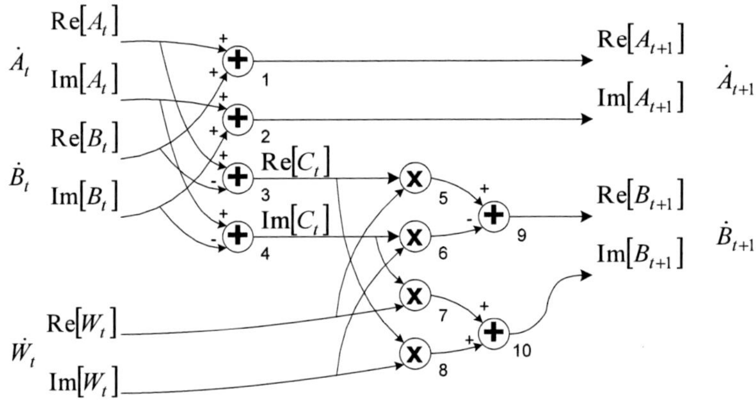

Figure 5.12. The computing structure of a classical basic operation.

Figure 5.13 shows the computing structure that corresponds to a modified basic operation, where $\dot{W}_t$ is calculated taking into account $\dot{W}_b$ and $\dot{W}_c$.

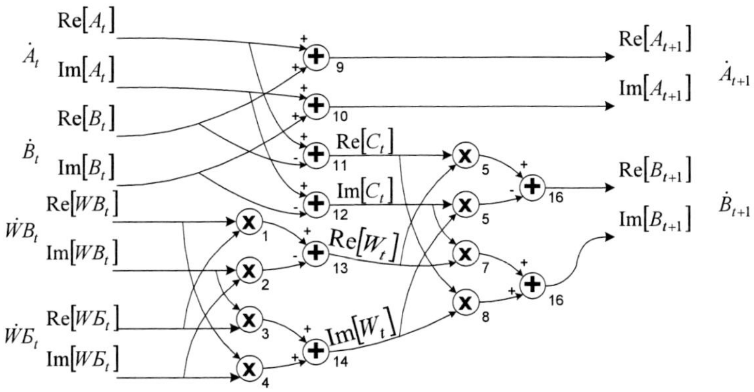

Figure 5.13. The computing structure of a modified basic operation.

Fast convolution, which uses the FFT and IFFT algorithms, assumes multiplication of the input signal spectrum by coefficients of convolution kernel spectrum. Since the FFT algorithm uses $\dot{W}_t = 1$ in the last iteration, then the basic operation of the last iteration of the FFT may be combined with multiplication by the coefficients of convolution kernel. The simplified FFT basic operation, combined with multiplication, is considered below.

The simplified basic operation with multiplication by the coefficients $\dot{H}_i$ and $\dot{H}_{i+1}$ of convolution kernel is obtained according to the equation

$$\begin{aligned} \dot{A}_{t+1} &= (\dot{A}_t + \dot{B}_t) \cdot \dot{H}_i, \\ \dot{B}_{t+1} &= (\dot{A}_t - \dot{B}_t) \cdot \dot{H}_{i+1}, \end{aligned} \tag{5.23}$$

or individually for the real and imaginary parts –

$$\begin{aligned} \mathrm{Re}\left[\dot{A}_{t+1}\right] &= \left(\mathrm{Re}\left[\dot{A}_t\right] + \mathrm{Re}\left[\dot{B}_t\right]\right) \cdot \mathrm{Re}\left[\dot{H}_t\right] - \left(\mathrm{Im}\left[\dot{A}_t\right] + \mathrm{Im}\left[\dot{B}_t\right]\right) \cdot \mathrm{Im}\left[\dot{H}_t\right], \\ \mathrm{Im}\left[\dot{A}_{t+1}\right] &= \left(\mathrm{Re}\left[\dot{A}_t\right] + \mathrm{Re}\left[\dot{B}_t\right]\right) \cdot \mathrm{Im}\left[\dot{H}_t\right] + \left(\mathrm{Im}\left[\dot{A}_t\right] + \mathrm{Im}\left[\dot{B}_t\right]\right) \cdot \mathrm{Re}\left[\dot{H}_t\right], \\ \mathrm{Re}\left[\dot{B}_{t+1}\right] &= \left(\mathrm{Re}\left[\dot{A}_t\right] - \mathrm{Re}\left[\dot{B}_t\right]\right) \cdot \mathrm{Re}\left[\dot{H}_{t+1}\right] - \left(\mathrm{Im}\left[\dot{A}_t\right] - \mathrm{Im}\left[\dot{B}_t\right]\right) \cdot \mathrm{Im}\left[\dot{H}_{t+1}\right], \\ \mathrm{Im}\left[\dot{B}_{t+1}\right] &= \left(\mathrm{Re}\left[\dot{A}_t\right] - \mathrm{Re}\left[\dot{B}_t\right]\right) \cdot \mathrm{Im}\left[\dot{H}_{t+1}\right] + \left(\mathrm{Im}\left[\dot{A}_t\right] - \mathrm{Im}\left[\dot{B}_t\right]\right) \cdot \mathrm{Re}\left[\dot{H}_{t+1}\right]. \end{aligned} \tag{5.24}$$

The simplified basic operation differs from the classical one by $\dot{W}_t$ which is equal to unity. Therefore, instead of carrying out $\dot{W}_t = \dot{W}_b \times \dot{W}_c$ and multiplication by $\dot{W}_t = 1$, multiplication by the coefficients $\dot{H}_i$ and $\dot{H}_{i+1}$ of convolution kernel may be carried out. The simplified basic operations with multiplication by $\dot{H}_i$ and $\dot{H}_{i+1}$ are used only in the algorithms of fast convolution, and unite execution of the basic operation on the last iteration of the FFT algorithm and multiplication by the coefficients of convolution kernel.

Figure 5.14 shows the computing structure, which corresponds to the simplified basic operation with multiplication by $\dot{H}_i$ and $\dot{H}_{i+1}$.

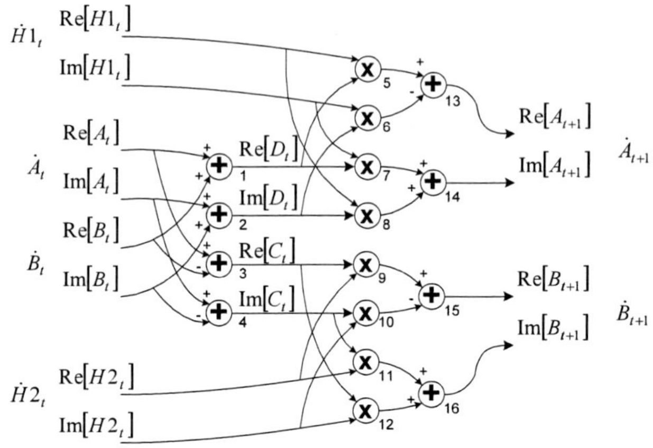

Figure 5.14. The computing structure of the simplified FFT basic operation.

Let us consider organization of computational process in the pipeline computational unit (PCU), which consists of 2 adders and 2 multipliers and carries out calculations according to (5.22) and (5.24). Assume that:

- the source of input data is a 32-bit RAM, in which real and imaginary parts of complex numbers are stored in a sequence of cells. The address of a complex number is the address A of its real part. In this case, the address of its imaginary part is $A+1$;
- the computing structure of the basic operation receives the input data $\mathrm{Re}\left[\dot{A}_t\right]$, $\mathrm{Im}\left[\dot{A}_t\right]$, $\mathrm{Re}\left[\dot{B}_t\right]$ and $\mathrm{Im}\left[\dot{B}_t\right]$ during 4 cycles;
- the computing structure of the basic operation returns the result data $\mathrm{Re}\left[\dot{A}_{t+1}\right]$, $\mathrm{Im}\left[\dot{A}_{t+1}\right]$, $\mathrm{Re}\left[\dot{B}_{t+1}\right]$ and $\mathrm{Im}\left[\dot{B}_{t+1}\right]$ during 4 cycles;
- $\dot{W}_t$ are calculated in the pipeline unit taking into account $\dot{W}_b$ and $\dot{W}_c$, and according to (5.22). The coefficients $\dot{W}_b$ and $\dot{W}_c$ are read from an internal ROM unit;
- $\dot{H}_i$ and $\dot{H}_{i+1}$ are read from an internal RAM.

In this case, during 4 cycles the pipeline computational unit receives the two complex numbers $\dot{A}_t$ and $\dot{B}_t$ (the initial data for the FFT basic operation), and returns the 32-digit floating-point results $\dot{A}_{t+1}$ and $\dot{B}_{t+1}$. By means of two channels it is possible concurrently to receive the four complex numbers $\dot{A}_t$ and $\dot{B}_t$, and to return the results of the simplified basic operation during 4 cycles. So, the data format and the order of data exchange through the external channels of the pipeline unit specify requirements to the structure of the computational part of the unit. The pipeline unit is to carry out calculations according to (5.22) and (5.24) during 4 cycles. This will match the time of pipeline calculations, and the time of input data transfer, and the time of result unloading through the external channels of the unit. Such computing structure is optimal, because data processing and data transfer through the external channels are equal and need no auxiliary hardware structures of matching and synchronization.

The computational part of the pipeline unit contains 2 adders and 2 multipliers, which carry out floating-point operations. Both these adders and multipliers are pipeline computing devices, that carry out addition and multiplication during 1 cycle in a dataflow mode. In addition to the adders and multipliers the unit contains registers, dual-port RAMs, multiplexers, an internal switch and ROMs. Figure 5.15 shows the structure chart of the pipeline computational unit.

Taking in account that each basic operation with calculation of $\dot{W}_t$ (5.22) and the simplified basic operation with multiplication by the coefficients $\dot{H}_i$ and $\dot{H}_{i+1}$ of convolution kernel (5.24) contains 8 operations of addition and 8 operations of multiplication, it is possible to claim that pipeline execution of each basic operation takes 4 cycles.

Creation of the pipeline FFT structures requires pipeline RAMs, placed between the units, that perform sequential operations of the FFT algorithm [111, 125]. These RAMs insert delays between the stages of the pipeline. The pipeline RAM may be implemented on the base of block memory of the Virtex family FPGAs as dual-port memory banks RAM AB. The complex results of calculations enter the RAM AB and are written into the addresses, generated by an addressing unit. After a certain number of cycles, which is defined by the pipeline FFT algorithm, the complex output data is read from the RAM AB according to the addresses, generated by the addressing unit of the RAM AB. The depth of the pipeline memory may be programmed according to the number of an FFT iteration.

Without loss of generality, assume that the depth of the pipeline adder is 4 cycles, and the depth of the pipeline multiplier is 3 cycles. Figure 5.16 shows the timing chart of calculations, that are carried out by the pipeline computational unit (see Fig. 5.15) according to (5.10).

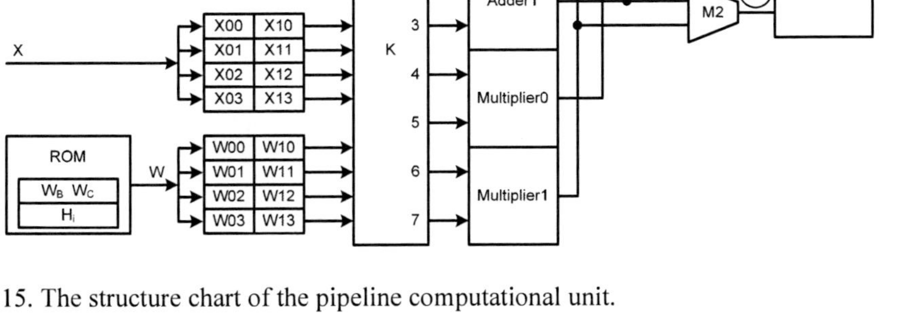

Figure 5.15. The structure chart of the pipeline computational unit.

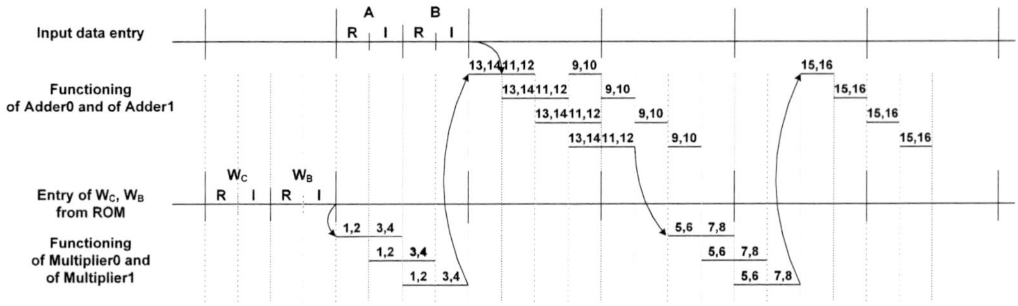

Figure 5.16. The timing chart of calculations.

In this Figure long scale marks denote 4-cycle intervals of transfer of the real and imaginary parts of $\dot{A}_t$, $\dot{B}_t$, $\dot{W}_b$ and $\dot{W}_c$ into the pipeline unit. Dotted scale marks denote cycles of functioning. Numerals denote the numbers of the arithmetic operations according to Fig. 5.13.

Control of functioning of the pipeline computational unit is provided by a microprogram control unit.

Microinstructions are cyclically read from the RAM of microprograms of the control unit. Read cycle, as it may be seen from Fig. 5.16, takes 22 cycles. The microprogram starts with writing of the real and imaginary parts of $\dot{W}_b$ and $\dot{W}_c$, and ends by returning results, obtained according to the informational graph (see Fig. 5.13), into the RAM AB.

Figure 5.17 shows the timing chart of the simplified basic operation with multiplication by the complex coefficients of convolution kernel, performed according to (5.24) and according to the informational graph (see Fig. 5.14).

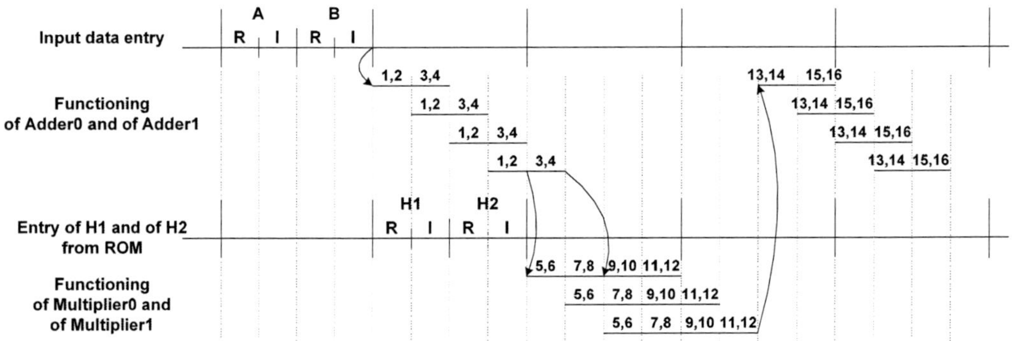

Figure 5.17. The timing chart of the simplified basic operation.

The timing charts (see Fig. 5.16 and 5.17) have a 4-cycle latency when a static pipeline with single-type operations is used. Besides, all stages of the pipeline are completely loaded and data conflicts are ruled out.

The microinstructions are cyclically read from the RAM of microprograms of the control unit. According to Fig. 5.17, the read cycle of the simplified basic operation takes 19 cycles. The microprogram starts with writing of the real and imaginary parts of $\dot{A}_t$ and $\dot{B}_t$, and ends by returning results, obtained according to the informational graph (see Fig. 5.14), into the RAM AB.

The pipeline computational unit, shown in Fig. 5.15, carries out the following standard operations of digital signal processing:

1) the FFT basic operation with $\dot{W}_t$ according to the informational graph, shown in Fig. 5.12;
2) the FFT basic operation with calculation of $\dot{W}_t$ on base of $\dot{W}_b$ and $\dot{W}_c$ and according to the informational graph, shown in Fig. 5.13;
3) the simplified FFT basic operation with multiplication by the coefficients $\dot{H}_i$ and $\dot{H}_{i+1}$ of convolution kernel according to (5.24);
4) the simplified FFT basic operation with calculation of square of the absolute value of the complex numbers $\dot{A}_{t+1}$ and $\dot{B}_{t+1}$;
5) the two-section fragment of the non-recursive filter according to Fig. 5.11;
6) a recursive section of a first-order filter;
7) a recursive section of a second-order filter. Its pipeline implementation may lead to instability of the filter and, as a result, special actions are required to keep stability of the filter [126, 127].

So, the pipeline computational unit (see Fig. 5.15) is a general-purpose computational node which performs a number of algorithms of digital signal processing. Therefore, with the help of this block, the structure of the macroobject of digital signal processing may be represented as it is shown in Fig. 5.19.

A brief description of the macroobject nodes is given below.

The controller of external connections is intended for creation of data exchange channels and transfer channels of control and adjusting information between the macroobjects of the basic module.

The unit of code transformation matches data representation forms used in various functional nodes.

As is known, modern analog-digital converters (ADC) do not generate data in the IEEE-754 floating-point format, but represent the data in offset binary codes or in the true form. The purpose of such data representation is to increase the ADC performance and to simplify the ADC structure. Some nodes of digital signal processing use the fixed-point radix complement form of different digit capacity for data representation.

The fixed-point format is applied in digital signal processing to increase the performance and to simplify the structure of functional nodes. Digital capacity of processing data may vary in certain limits to provide the required precision of calculations and the dynamic range of data representation on various stages of processing.

The unit of coefficient generation is intended for generation and storage of coefficients that are necessary for digital signal processing algorithms. The coefficients may be constant (FFT coefficients, coefficients that reflect features of hardware and data transfer channels, etc.) and non-constant (coefficients of adaptive filters, coefficients that depend on the state of the environment, etc.). As a rule, the constant coefficients are stored in ROM units. The variable coefficients change according to processing conditions, and either come from some external sources, or may be generated within the macroobject and stored in a RAM unit. Therefore, different macroobjects may contain different units of coefficient generation. In addition, the unit of coefficients contains hardware addressing facilities for data storage units.

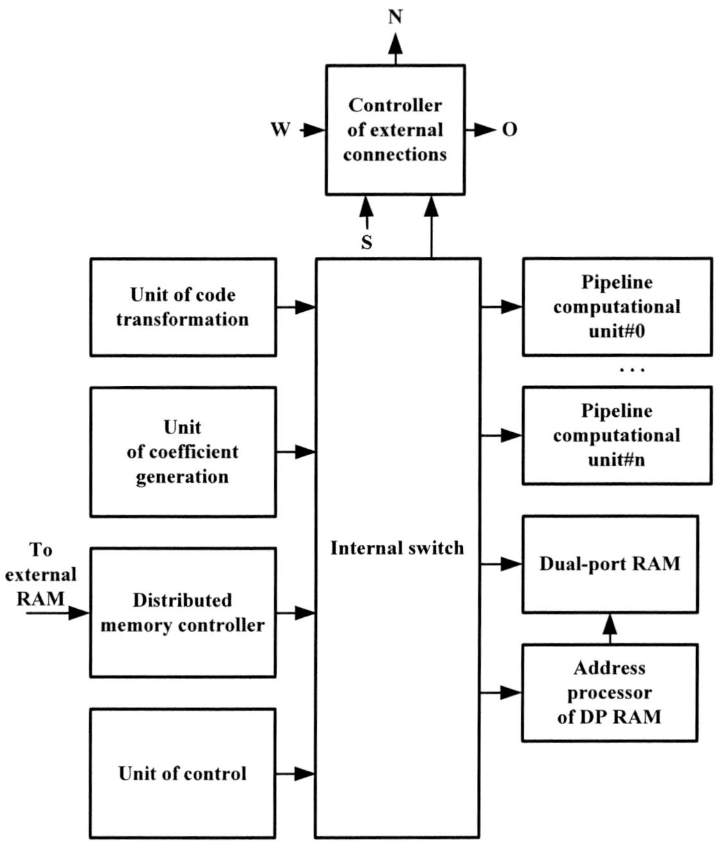

Figure 5.19. The structure of the macroobject of digital signal processing.

The distributed memory controller is the main active node which generates data streams in multipipeline computing structures. Use of the memory controller as a standard node for interaction with the memory keeps succession of software, when the described macroobject is used for solving problems of digital signal processing.

The pipeline computational unit is necessary for execution of basic computing operations of algorithms of digital signal processing. The principles of organization of the pipeline unit of the DSP macroobject are given above. In order to provide the required performance, the DSP macroobject contains several pipeline computational units.

The internal switch is intended for creation of data channels between all components of the macroobject.

The unit of control is used for transfer of control and adjusting information into all components of the macroobject, for synchronization of their functioning, and for organization of computational process.

As an example we consider the structure of the DSP-7.1 macroobject, which is implemented in the XC2VP100 Virtex II Pro FPGA and contains 5 pipeline units.

Characteristics of the basic modules 8V2P-64 and 4V2P-32, in which the DSP-7.1 macroobjects were implemented, are described in subsection 4.4.3.2.

Figure 5.20 shows the structure of the DSP-7.1 macroobject.

The dual-port RAM and the address processor of the dual-port RAM are intended for effective storage of intermediate results of calculations. Memory with a dynamic principle of data storage, such as SDRAM, DDR, DDR2, provides space-efficient storage of large amount of data, and is frequently used as an external RAM. Nevertheless, block access to data in dynamic RAM does not allow nonstandard types of data addressing that are used in algorithms of digital signal processing for permutation of elements of numerical sequences. For such purposes the macroobject contains a RAM with an arbitrary access, which, together with the address processor, can perform all required rotations of elements of numerical sequences [128, 129].

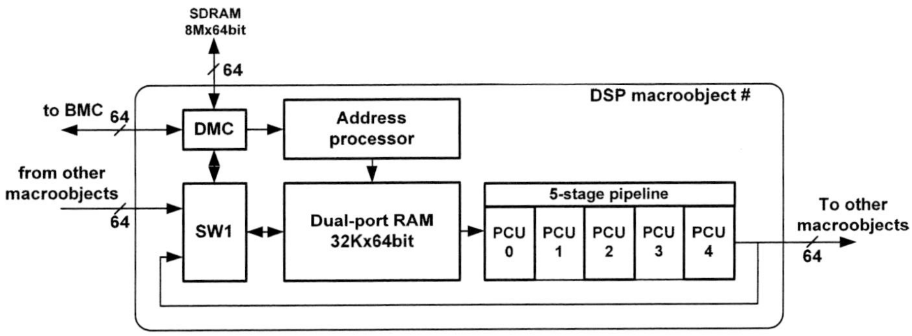

Figure 5.20. The structure of the DSP-7.1 macroobject.

The macroobject contains the following objects: the distributed memory controller - DMC; the data channel switch – SW1; the address processor; the dual-port RAM with the capacity of 32K of 64-bit words; the computational pipeline of 5 pipeline units – PCU0…PCU4.

The address processor is oriented on interaction with the dual-port RAM within the macroobject. The main functions of the address processor are:

- data exchange between SDRAM and the dual-port RAM under control of the distributed memory controller;
- various modes of data addressing, including modes, oriented on execution of digital signal processing algorithms.

Figure 5.21 shows the outline structure chart of the address processor.

The FIFO addressing unit is intended for generation of address streams for reading and writing data from/into the dual-port RAM during execution of the FFT, the IFFT, the fast convolution algorithm, the irredundant FFT algorithm. The algorithms of digital signal processing, implemented on pipeline computing structures, require rather complicated calculations for data addressing. Sometimes the procedural method of address generation takes tens of cycles for calculation of each address. Therefore, the addressing unit provides hardware generation of addresses, returning a new address during each cycle. Figure 5.22 shows the functional chart of the addressing unit, functioning in a FIFO mode. For saving space of the dual-port RAM, the results of data processing may be written to the same addresses, from which the data was previously read in the current cadr. Owing to the switch SW (see Fig. 5.22), the address processor can generate various types of addressing, that may be obtained by switching of address digits, and that provide execution of all fast DSP algorithms.

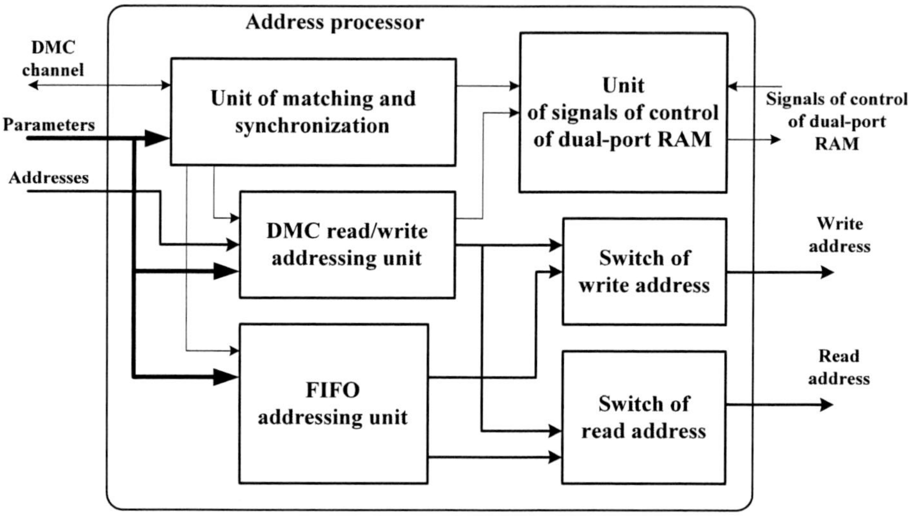

Figure 5.21. The outline structure chart of the address processor.

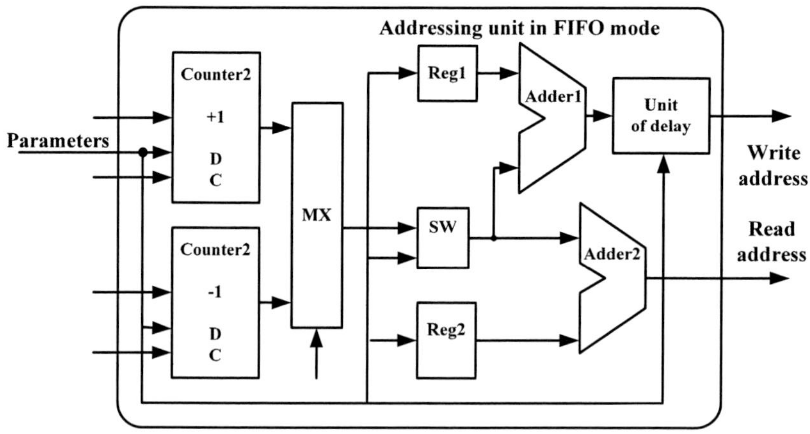

Figure 5.22. The functional chart of the addressing unit in a FIFO mode.

Theoretical justification of the method of address generation in the fast algorithms of digital signal processing is given in [129]. Since the suggested structure of the address processor allows writing of the processing results to the same addresses, from which initial values were read, then it allows double reduction of the capacity of the dual-port RAM of the macroobject. The performance of the designed DSP-7.1macroobject, running at the clock rate 200 MHz, is about 8 Gflops for the FFT1024 algorithm. Several DSP-7.1 macroobjects may be combined into a single computing structure with proportional increasing of its performance.

5.6. The Macroobject for Solving Problems of Linear Algebra

5.6.1. The Structure of the Macroobject for Solving Problems of Linear Algebra

The distinctive feature of problems of linear algebra is standard computing structures, such as a sum of pair products, and a sum of so-called pivot entries of matrices, that are needed in each iteration of an algorithm.

Figure 5.23 shows the structure of the macroobject for solving problems of linear algebra.

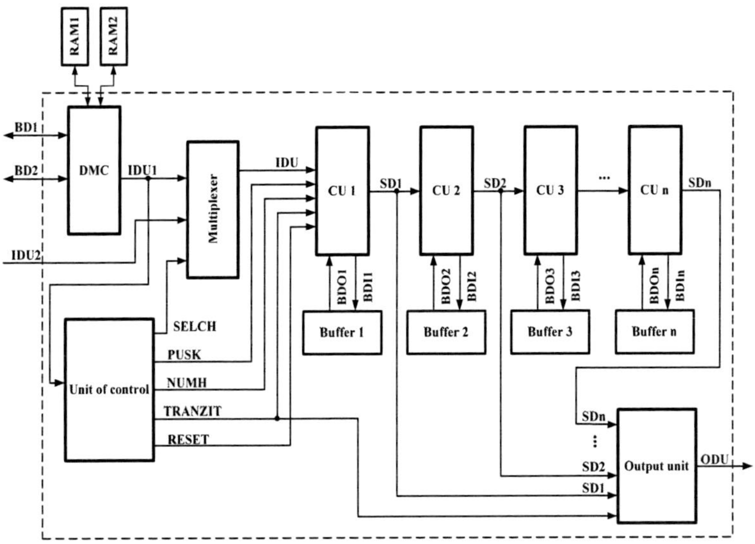

Figure 5.23. The structure of the macroobject for solving problems of linear algebra: CU1...CUn – computational units; DMC – a distributed memory controller.

The computational units perform basic functions of data processing on data, which passes from the RAM1 or RAM2. Figure 5.24 shows the structure of the computational units that are used in algorithms of linear algebra.

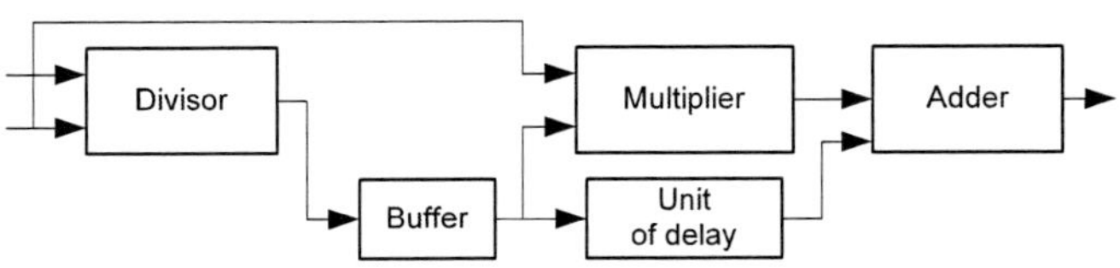

Figure 5.24. The computational unit for solving problems of linear algebra.

The main operation of problems of linear algebra is summation of two products [42], performed by the computational units. The number of the computational units, involved in execution of the algorithm, may vary. The properties of the computational units, that depend on program parameters, may also be changed by the program.

Data streams are generated by the distributed memory controller, which also provides interaction of the functional nodes between themselves and the memory. As a rule, the distributed memory controller is connected to 2 units of external RAM. One RAM unit stores the initial data, the other one stores the results of calculations. The data, that enters the inputs of the computational units, leaves the output of the multiplexer, controlled by the unit of control. The buffers are FIFO units that store pivot entries.

5.6.2. Implementation of the Algorithm for Solving a System of Linear Algebraic Equations by Means of Gaussian Elimination

In this subsection we analyze an example of a pipeline implementation of an algorithm for solving a system of linear equations by means of Gaussian elimination. To fix the idea, we use a basic module 16V4-50 of the third generation. Figure 5.25 shows the configuration of the basic module.

All fragments of the algorithm, such as forward Gaussian elimination, backward Gaussian elimination, normalization of a pivot row, processing of coefficients of succeeding rows, transit of pivot row coefficients within the computing structure, were implemented one by one in FPGAs of the computation field.

The macroobject operates in 3 modes.

Mode 1. Forward Gaussian elimination or transformation of an augmented matrix into an upper triangle form.

In the zero moment of time, an initial $N \times (N+1)$ matrix is located in the distributed memory. In each pair of memory units one unit contains the initial data, and is the source unit, and the other one is the target unit. The target unit stores the result data, obtained after execution of one iteration of the algorithm. After that, under control of the memory controller, the source unit and the target unit change places, and the next iteration of the algorithm begins. The number of iterations is defined by the number K of stages. Consequently, the number of iterations may be obtained as

$$I = \frac{N \times (N+1)}{K}. \quad (5.25)$$

During each iteration, K pivot rows and $N-K$ guided rows are written into the target memory unit. Figure 5.26 shows data location in the memory units before starting of the algorithm, during data processing, and after execution of 8 iterations. As it may be seen from Fig. 5.26a, during execution of the algorithm the data moves from the RAM1 into the RAM2, passing through the processing units. At first, the memory receives the pivot rows M1, and then – the guided L1. The obtained pivot rows are not used in further forward Gaussian elimination. Free space in the memory array is reserved for further use during execution of the next iterations of the algorithm.

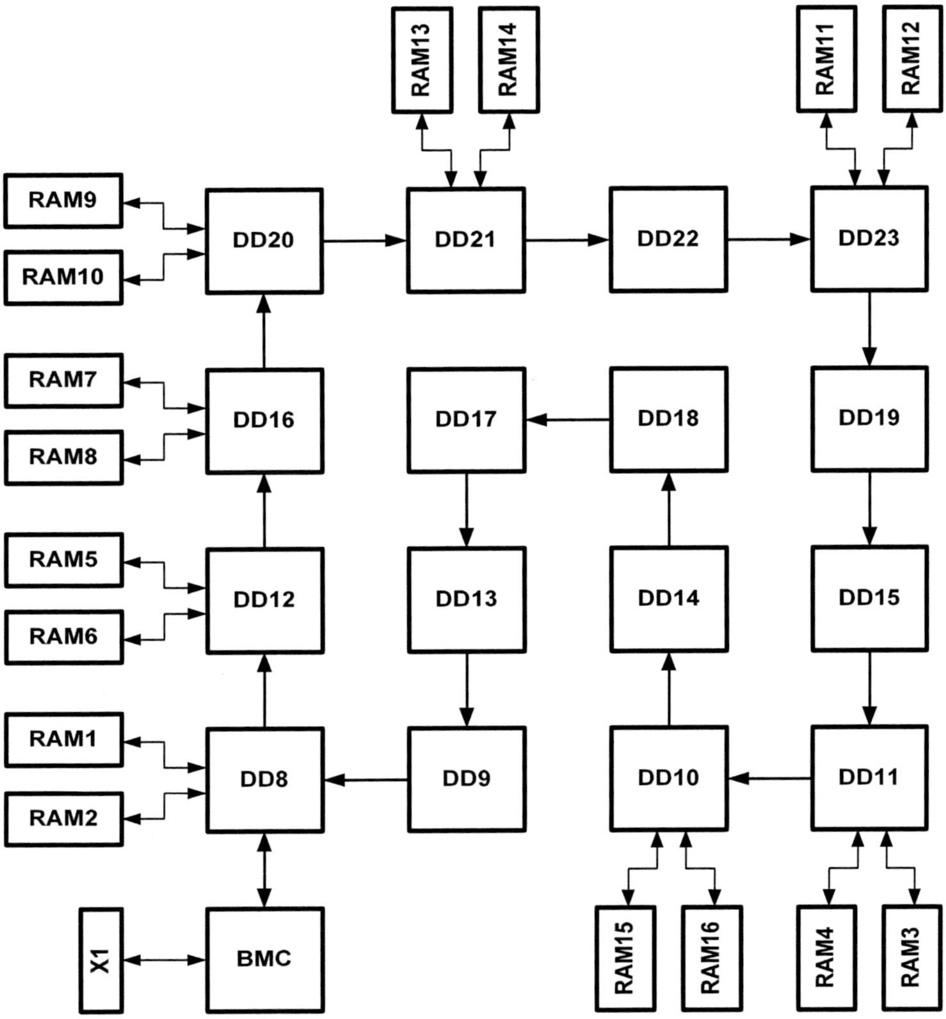

Figure 5.25. The hardware configuration of the basic module 16V4-50.

Mode 2. Backward Gaussian elimination or transformation of a triangle matrix into a diagonal form.

The flow chart of backward Gaussian elimination is similar to the forward algorithm except the only distinction: the memory units keep the desired coefficients, but not the data arrays (see Fig. 5.26b). During backward Gaussian elimination the obtained pivot rows, starting with the last one, pass through the processing units, and are transformed into the desired coefficients. The coefficients are the results of the algorithm. Such approach is applied for both the only pair and several pairs of memory units.

Mode 3. Transit of coefficients of pivot rows.

The mode is needed for processing of the matrix, whose rows are located in more than one pair of RAM. For instance, suppose the one half of the matrix is stored in the RAM1, and the second one – in the RAM5. Since the whole matrix, which is split and stored in two or more memory units, is to pass through the computing structure during one iteration, then it is necessary to provide a correspondence between the coefficients of the obtained pivot rows and the data, stored in the RAM5. It may be achieved by transit passing of the data in a circular buffer. Figure 5.27 shows the structure of the circular buffer.

The units BUF1-BUF4 are dual-port memory units. Data exchange is provided through the multiplexers MX, that connect transit buses. During forward or backward elimination the buffers write and store data, obtained in the processing units. In transit mode the multiplexer switches the buffers to the transit buses DTR and ODU, looping them into the circular buffer. So, the required quantity of data may be simultaneously transferred from all buffers of the DD1 into the DD2, etc. The obtained coefficients of the pivot rows are located in the required buffers, and control over all stages of the algorithm is transferred to the next unit, that continues processing of its own part of the data. The memory size, which is necessary for execution of the algorithm, is $2N^2$ 4-byte words. If $N = 10^4$, then the memory size is 800 Mbytes. If $N = 10^5$, then the memory size is 80 GBytes.

When $N = 10^4$ and the frequency is 160 MHz, the time of execution of the problem on the 16V4-50 basic module is 49 sec. The problem of SLAE with the dimension $N = 10^5$ takes 80 Gbytes of distributed memory. The time of execution of the SLAE problem with the dimension $N = 10^5$ on the single 16V4-50 basic module is 13.5 hours.

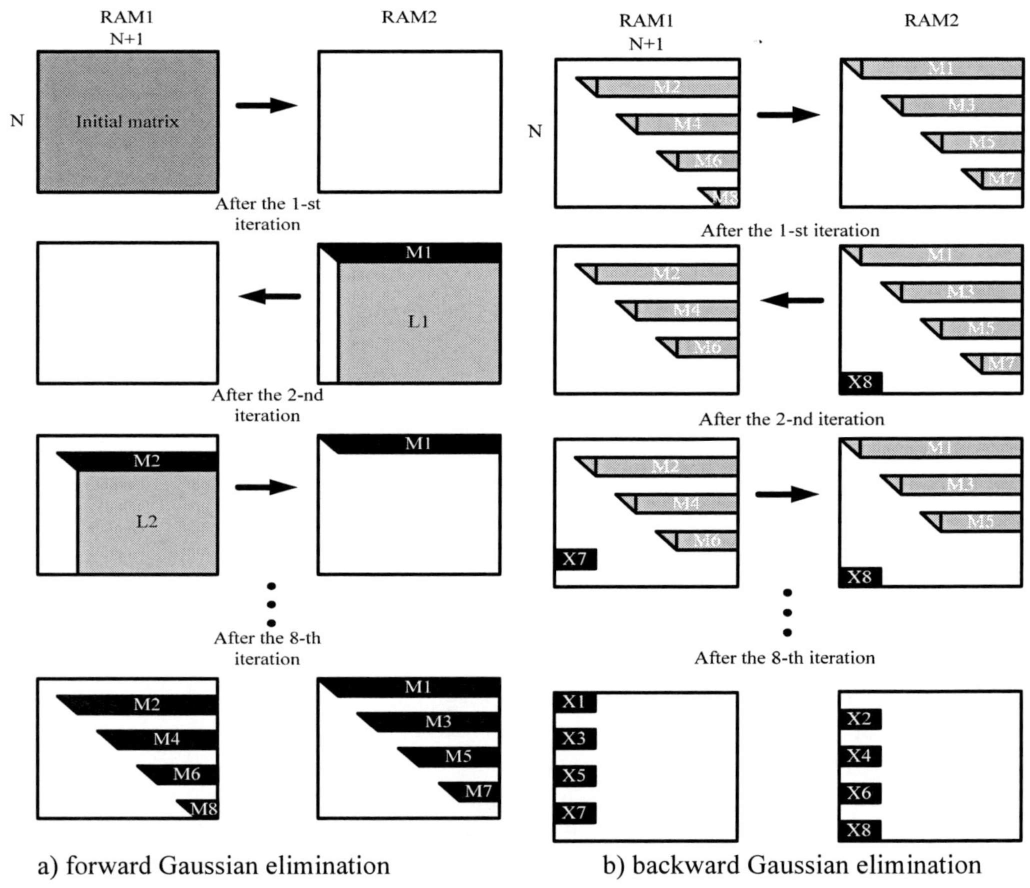

Figure 5.26. Data location in the memory units.

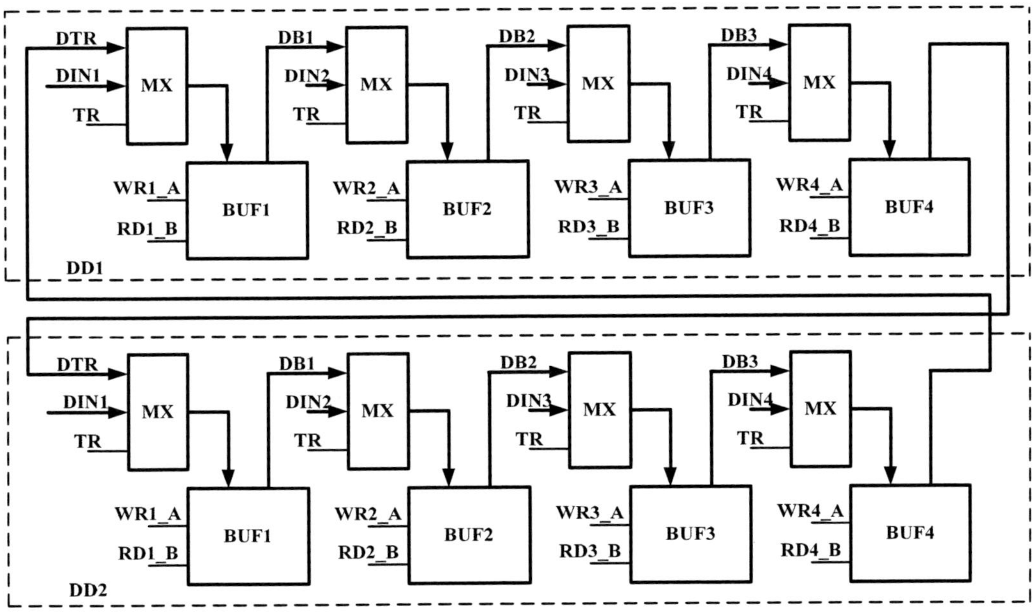

Figure 5.27. The structure of the circular buffer of the DD1 and DD2 FPGAs.

5.7. A General-Purpose Macroobject

A general-purpose macroobject or, in other words, a macroprocessor (MAP) takes a particular place among problem-oriented macroobjects.

For convenient statement of the material of the previous chapters, architecture of computation fields of MPCS was considered as the main computing element in terms of macroprocessor use. In turn, the macroprocessor is a set of processing elements combined by a switch into a single computational resource. It is assumed, that all processing elements are similar arithmetic-logical units, that carry out a set of elementary multidigit operations.

Obviously, usage of such general-purpose single-type processing elements (PE) leads to hardware redundancy of reconfigurable multipipeline computing structures. The processing element is to contain circuit resources, sufficient to perform the entire set of operations in spite of the operation that is being performed at the moment. As a result, such requirement leads to degradation of the specific performance of the whole MPCS.

The problem may partly be solved by using specialized processing elements. For instance, the macroprocessor may contain processing elements of two types: *additive* processing elements (APE) and *multiplicative* processing elements (MPE), shown in Fig. 5.28.

The additive processing element is intended for such operations as addition, subtraction, comparison. The multiplicative processing element may be used for multiplication, division, square-rooting, etc. In addition, each type of the processing elements is to perform elementary logical and stream operations.

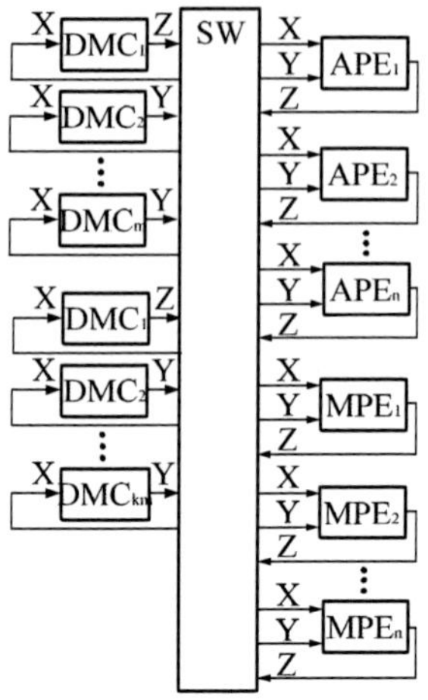

Figure 5.28. The macroprocessor with two types of processing elements.

Of course, it is difficult to imagine a computational problem, purely consisting of addition or multiplication. Therefore, the processing elements of both types are to be used evenly in the structure of the macroprocessor. Even if the computational problem purely consists of one-type operations, for example, addition of array elements, then the specific performance of the macroprocessor, that contains processing elements of various types, will remain practically equal to that of the macroprocessor, that contains the general-purpose processing elements. In the first case the multiplicative processing elements remain idle. In the second case the hardware multiplicative components of the general-purpose processing elements cannot be used. In the case of even use of addition and multiplication, the specific performance of the macroprocessor, that contains the processing elements of two types, practically increases in two times.

The specific performance of the macroprocessor and, consequently, of the entire RMCS may be increased even more owing to separation of functional units into a wider spectrum of objects. Nevertheless, there is no need of ultimate differentiation of the functions of the processing elements, because, in any case, some components of hardware resources will remain idle during execution of the most part of problems, or it will be necessary to reconfigure the structure of the processing elements.

The general-purpose macroobject has certain features and limitations, that distinguish it from a variety of problem-oriented and specialized macroobjects:

- the structure of the macroprocessor contains a fixed number of computational cores - processing elements (PE);
- parameters of the processing elements (a size of operands, an instruction set, a set of input and output buses) remain invariable in spite of the problem domain;
- parameters of the internal switch of the general-purpose macroprocessor remain invariable in spite of the problem domain;
- from the point of view of the application programmer, the macroprocessor is program-indivisible, i.e. when a macrooperation is changed, instructions are simultaneously changed for all processing elements and the switch.

The structure of the macroprocessor remains constant and does not depend on placement into one or another FPGA of the basic module. Invariability of the macroprocessor structure is compensated by the general-purpose instruction set of the processing elements and the programmable switch. As a result, effective parallel-pipeline computing structures may be created for various classes of problems. Of course, effectiveness of such structures is slightly less than of those that are created from macroobjects with no limitations listed above. However, lower effectiveness is compensated by rather a small number of configuration files required for FPGAs of the basic module and for the entire MPCS. As a result, price of design is reducing.

5.7.1. The Structure of the Macroprocessor

The general-purpose macroprocessor is intended for structural implementation of large self-contained fragments of problems. In the general case, the macroprocessor consists of internal processing elements and a programmable channel switch, which provides connection between the processing elements, and between the processing elements and the outputs of the macroprocessor. A detailed description of the processing element is given in subsection 5.7.4.

There are various alternative designs of the internal programmable switch of the macroprocessor. It is obvious, that crossbar communications can be implemented in the macroprocessor with a limited set of the processing elements.

Here we analyze the macroprocessor, in which the internal elements and the outputs are connected by a crossbar switch. Let the macroprocessor contain 4 processing elements and 1 input/output for each orthogonal direction. In this case the communication graph contains 16 vertices. Figure 5.29 shows the functional chart of the macroprocessor with crossbar communications.

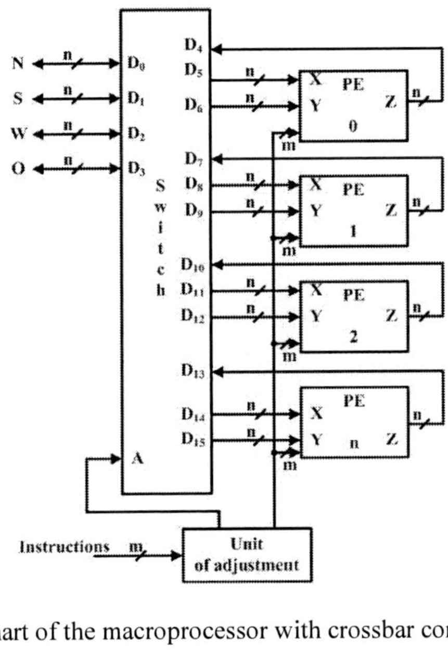

Figure 5.29. The functional chart of the macroprocessor with crossbar communications.

The switch provides connections between:

- the inputs and outputs of the processing elements;
- the macroprocessor outputs (in the general case each output is bidirectional);
- the macroprocessor outputs, that are programmed for data input, and inputs of the processing elements;
- the outputs of the processing elements and the macroprocessor outputs, that are programmed for data output.

The macroprocessors with crossbar communications were implemented in the basic modules of the second generation, because FPGAs that were used for their implementation, had limited circuit resources, insufficient for implementation of a macroprocessor with a large number of processing elements.

5.7.2. A Macroprocessor with a Cascade Communication Structure

Up-to-date FPGAs that contain several million of equivalent gates allow circuit engineers to design macroprocessors, that consist of 16 to 40 processing elements and have up to 8 32-bit channels for data transfer in each direction. In this case, the communication graph contains over 150 vertices. At present, creation of a crossbar switch that can connect more than 150 vertices by 32-bit channels is impossible because of the technological level of modern FPGAs. Therefore, the switch complexity is to be limited. A cascade switch, that provides internal connections between the elements of the macroobject according to the certain rules, may be used as an alternative to the crossbar switch. Assume that the macroprocessor contains 16 processing elements and 16 external data channels: 4 channels to each orthogonal direction.

It is suggested to group the processing elements of the macroobject into cascade lines L0-L3. Each line contains 4 processing elements: 2 additive and 2 multiplicative. The cascades of the macroprocessor are connected sequentially into a ring, therefore the outputs of the elements of the cascade line L3 are connected with the inputs of the elements of the line L0. The external outputs of the macroobject are assigned in fours to each direction N, S, W, O.

Since the communication structure of the macroobject is limited, and elements of disjoint cascades cannot be connected directly, then an internal element of the macroobject such as a transit bus is required. Each cascade line of the macroobject contains 4 transit buses T0-T3 that provide data transit between the cascades and the outputs of the macroobject. Depending on a performed macrooperation, each output of the macroobject may be programmed as both input and output. By means of the transit buses data may pass through the macroobject without any digital processing. The switch cascades SW0-SW3 provide connections between the processing elements, the transit buses and the outputs of the macroobject.

Figure 5.30 shows the functional chart of the macroprocessor with no distributed memory controller.

If an FPGA, that contains the macroprocessor, is connected with RAM microchips of distributed memory, then the RAM microchips are also connected with the external data channels of the corresponding direction. In this case, the macroprocessor contains a number of distributed memory controllers, which is equal to the number of RAM microchips. As a rule, for basic modules of the 2nd and the 3rd generations the number of RAM chips, connected with one FPGA, does not exceed 2. Figure 5.31 shows the functional chart of the macroprocessor with 2 distributed memory controllers.

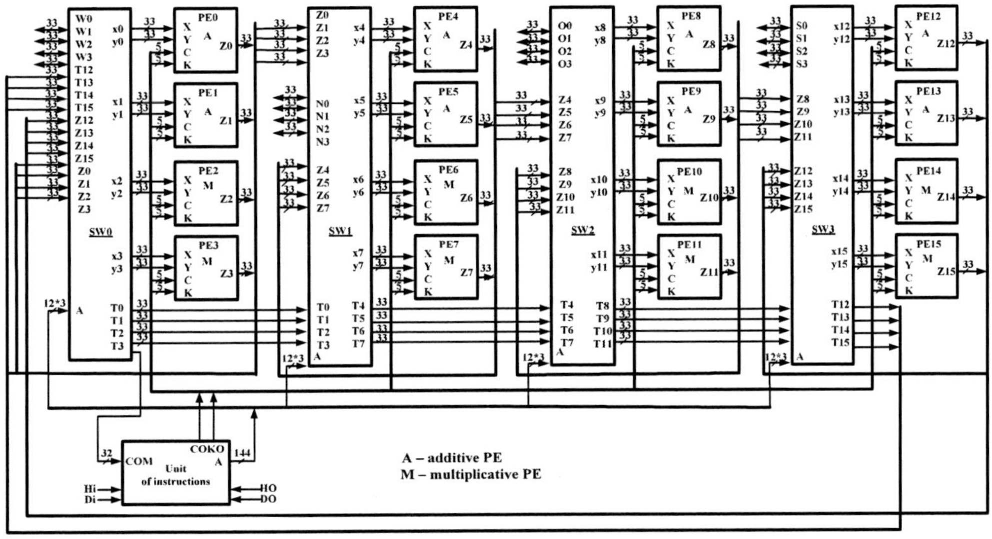

Figure 5.30. The functional chart of the macroprocessor with no distributed memory controller.

The macroprocessor consists of:

- 2 distributed memory controllers DMC0-DMC1;
- 12 processing elements PE0-PE11;
- 4 switch cascades SW0-SW3;
- a unit of instructions.

The RAM is connected to the macroprocessor structure by means of the distributed memory controller.

Owing to the switch cascades, data from the RAM may be transferred through the controller into all components of the macroprocessor and to the output buses as transit data.

5.7.3. The Cascade Switch of the Macroprocessor

Functionally the macroprocessor switch may be represented as a unit, which consists of 3 types of 33-bit multiplexers.

The first type multiplexers provide connections between the outputs of the processing elements and the outputs of the macroobject. The inputs of the multiplexers accept data from the outputs of the processing elements and transit buses.

The second type multiplexers provide connections between outputs of various elements of the macroprocessor and the inputs of the processing elements of the considered switch cascade of the macroobject. The inputs of the multiplexers accept data from the corresponding inputs of the macroobject, from the outputs of the processing elements of the previous cascade line, from the transit buses that belong to the previous cascade line, and from the output of the corresponding processing element which belongs to the considered cascade line.

The third type multiplexers provide connections between outputs of various elements of the macroprocessor and the transit buses of the considered switch cascade of the macroobject. The inputs of the multiplexers accept data from the corresponding inputs of the macroobject, from the outputs of the processing elements that belong to the previous cascade line, and also from the corresponding transit buses.

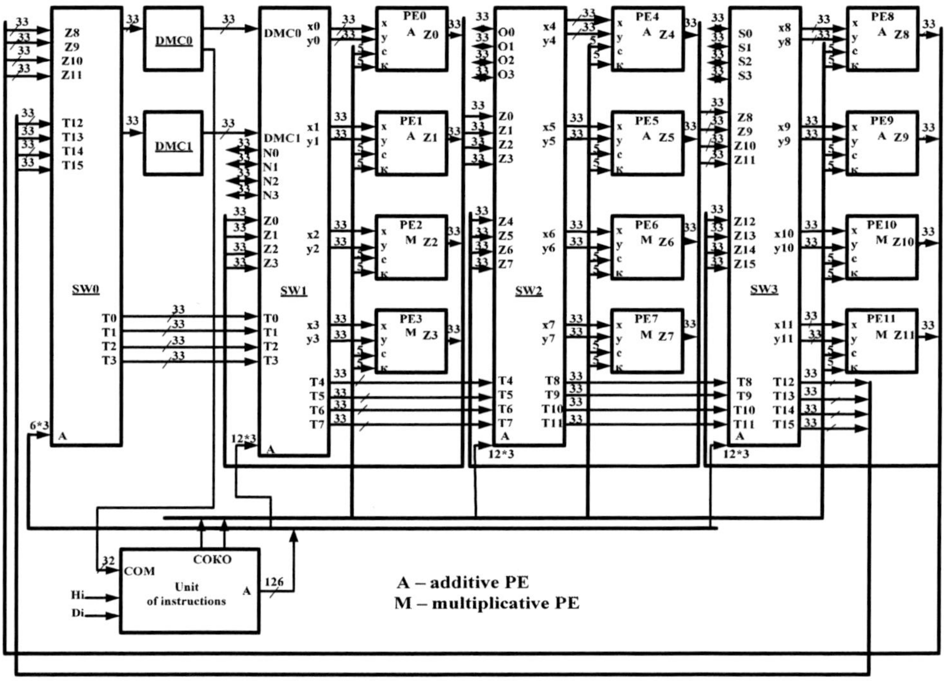

Figure 5.31. The functional chart of the macroprocessor with 2 distributed memory controllers.

In total, the switch of the macroprocessor contains 16 multiplexers of the first type, 32 multiplexers of the second type and 16 multiplexers of the third type. In comparison with the crossbar communication structure, the suggested cascade communication structure requires considerably less resources for the switches that connect the elements of the macroobject. As a result, larger resources may be used for calculations. Nevertheless, it should be noted that programming of the macroobjects with the cascade communication structures is rather complicated.

5.7.4. The Processing Element

The processing element is intended for carrying out arithmetic and logical operations on two operands that pass to its inputs in a 32-digit floating-point format. Each operand has a bit marker. The marker acknowledges transfer of a number through a data channel. In fact, the data is transferred in the data channel as 33-digit words. The processing element consists of an operand unit, an arithmetic-logical unit (ALU) and a converter. Figure 5.32 shows the structure chart of the processing element.

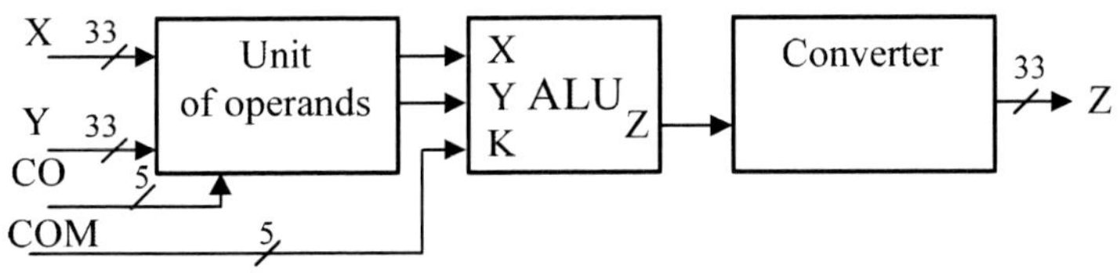

Figure 5.32. The structure chart of the processing element.

The ALU is the main computational node of the processing element. The ALU is intended for carrying out arithmetic and logical operations, and calculation of operands of comparison predicates, that pass to its inputs. The list of operations, that are carried out by the ALU, may be placed into 4 groups: arithmetic operations, logical operations, comparison operations and special operations.

The list of arithmetic operations contains: *addition, subtraction, multiplication, division, residue of division, count of units in a value.*

The list of logical operations contains: *bitwise conjunction, bitwise disjunction, modulo 2 addition, logical shift, cyclical shift.*

The list of operations of comparison: *greater than, not less than, equal to, not equal to.*

The list of special operations: *scaling, data synchronization, clearing of buffer memory.*

Using the given list of operations of the processing elements, the application programmer can develop macrooperations from various problem domains, modify and augment the list of operations that are carried out by the arithmetic-logical unit of the processing element.

The main purpose of the unit of operands is to match data streams on the input of the ALU. During calculations, operand streams, that pass to the inputs of the processing elements, may pass the routs with different operational delays. This leads to advance or delay of the one operand stream in comparison with the other. On the basis of analysis of the markers of the both operand streams, the operand unit delays the advance operand stream for a required number of cycles. As a result, the operand streams pass to the ALU inputs simultaneously. The unit of operands synchronizes the mismatched operand streams with the help of the buffer memory that accumulates advance operands. There are two independent subsystems of synchronization of calculations in the computation field of the MPCS: the system of cadr synchronization and the system of data stream synchronization. The system of cadr synchronization is implemented on a program level (see Chapter 6). The current chapter gives a detailed analysis of the synchronization of operand streams in multipipeline structures, based on the processing elements.

5.7.5. Synchronization of Calculations in MPCS

The system of data stream synchronization is implemented in the units of operands (see Fig. 5.32) that are contained in each processing element. Mismatch between data streams on the input of the processor occurs on account of different operational delays during calculation

computationally intensive fragments of tasks from various problem domains on an arbitrary number of FPGA-based basic modules. In addition, complexity of MPCS application development must be approximately similar to programming of cluster multiprocessor systems. So, it is necessary to create user-friendly tools for development and debugging of applications on programming languages of structural-procedural computations.

The system software that simplifies programming of complicated problems for reconfigurable multipipeline computing structures, may be placed into three groups according to its functionality:

- tools of application development;
- administrative tools of computational resources;
- utility programs and drivers.

Figure 6.1 shows the structure of the MPCS system software.

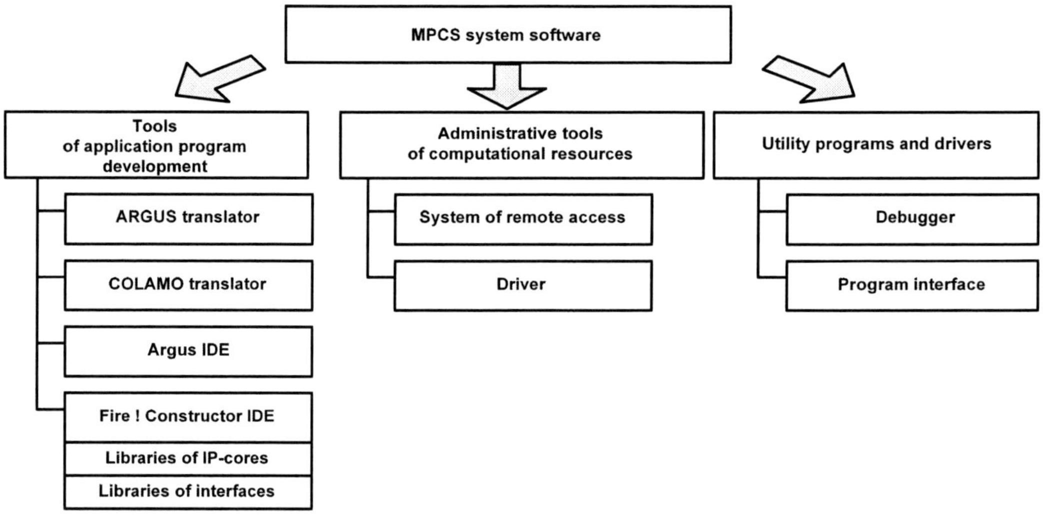

Figure 6.1. The structure of the MPCS system software.

A low-level language of structural-procedural programming Argus v.3.0 is an MPCS assembler which provides low-level access to basic module hardware. The language Argus v.3.0 is necessary for circuit implementation control and development of applications that use topological and architectural features of MPCS basic modules. At the same time, it should be noted that the application programmer, who uses the language Argus, has to precisely realize hardware capabilities of computing structures, just like any programmer who uses the assembler of a cluster MCS. Owing to use of a high-level programming language COLAMO, the application programmer has no need in detailed knowledge of MPCS hardware structure.

The high-level structural-procedural programming language COLAMO allows the application programmer, who has poor information about architectural features of the MPCS, to develop effective MPCS application programs. The language is based on an implicit description of parallelism. Algorithms are mapped on the MPCS architecture by a translator, which syntactically supports reconfiguration of the hardware platform, using elements of a library of scalable IP-cores.

An integrated development environment Argus IDE is intended for interactive development of parallel programs by means of the COLAMO and Argus. The Argus IDE provides effective development of scalable parallel programs for the MPCS and contains the COLAMO and Argus translators.

A synthesizer Fire!Constructor maps the information graph of a problem, generated by the COLAMO translator, on the specific MPCS hardware platform. Computing structures are designed on the base of a pre-developed library IP-cores of functional units and interfaces.

Circuit engineers create the IP-core libraries for various problem domains, and for various methods of data processing, for instance, fixed-point data, floating-point data, double precision floating-point data, etc. The Fire!Constructor uses the IP-core libraries for synthesis of computing structures. Besides, circuit engineers also create interface libraries that provide interaction of the various structural components, used in developed applications, and connections between the MPCS and peripheral devices.

By means of a debugger of parallel programs Argus Debugger, the user can interactively debug applications on various basic modules of the MPCS. The Debugger provides interactive access to each distributed memory controller of any MPCS basic module, including the registers of the controller and memory segments.

Owing to the unified program interface, which provides program access to the basic modules, the application programmer may use the functions of loading/unloading and the functions of computational process control, that are developed using traditional programming languages, such as C++, Object Pascal, Java, etc., and are available owing to the interface libraries that are plugged in an application project. The driver of the basic module provides program support of the hardware interface of access to the basic modules.

Remote access to the basic modules is implemented by a system of multiuser access, and application layer protocols.

On the level of direct access to computational resources, application processes load configuration and executable modules into distributed memory controllers, transfer instructions of computational process control, perform data input/output in/from distributed memory channels. The level of direct access provides a single application process with sole rights to interact with the components of the basic modules of the reconfigurable structure.

On the level of client requests the operating system schedules functions, and allocates the needed hardware resources according to the configuration and current occupancy of the basic modules by other problems. The level of client requests provides simultaneous interaction of the systems components with several application processes. To provide such interaction, a task scheduler orders processing of client requests, imparting functional properties of a multichannel queueing system to the operating system, which consider the basic module as a resource unit.

The application programmer may develop static, parameterized and scalable parallel programs for the MPCS.

The static parallel program is implemented on a fixed set of the basic modules, that remains unchanged during program execution. To execute each cadr, a fixed number of the basic modules is allocated, and their interaction is provided. During execution of the program, the procedures of access to distributed memory channels and to external communication channels are performed in each basic module, involved into computational process. Hence, the data stream passes through the computing structure which consists of several basic modules.

The parameterized parallel program differs from the static one in the of parameters that specify its execution and quantities of data streams. As a result, the parameters of execution of the problem may be changed by variation of the size of processed data.

The scalable parallel program also operates with data streams with a parameterized quantity specification. However, in contrast to the parameterized parallel program, the scalable program may be executed on any arbitrary number of the MPCS basic modules, that may dynamically vary according to releasing hardware resources.

Development of scalable structural-procedural programs consists in creation of structure components of parallel algorithms, procedures of access to distributed memory channels, rules of initial data generation in memory channels, functions of executable module generation according to allocated resources.

Development of parallel program components consists of several steps:

1) determination of data stream structures, the general parallel algorithm, the rules of data allocation in distributed memory channels;
2) partition of the algorithm into a set of components;
3) development and specification of the missing components, that are added to the database;
4) generation of the cadr form of the problem;
5) verification of the developed program. If execution of the program is impossible because of hardware restrictions, then the general algorithm is corrected and all previous steps are performed again;
6) debugging of the program.

The initial step of development is oriented on analysis of the algorithms of the solving problem, and on determination of the required data structures. In the next steps, the algorithm of the problem is created according to possible hardware platform restrictions: computational resources, memory resources, capacity of its data transfer channel, capabilities of its communication system. Methods of transformation of static solutions into the effective cadr form are considered in [66, 68, 72, 85]. According to the suggested methods, descriptions of parallel algorithms are generated. Then they are used as the base for execution of parallel programs of various configurations with multilevel organization of architecture programming.

The COLAMO-translator generates four basic components of the application for the reconfigurable structure:

- a structural component, represented as a virtual information graph or an ordered sequence of information subgraphs;
- a stream component, represented as a set of programs, written in the assembler Argus, for distributed memory controllers;
- a procedural component, which provides an ordered change of cadrs and structural components, if the parallel program is represented as a set of procedural or general-purpose macroobjects;
- a control component, developed for the host-computer. It performs the procedures of MPCS adjustment, start, stop and other control and service functions.

Let us make a detailed consideration of development of the structural component of the application, because it is the most difficult to MPCS users.

Figure 6.2 shows interaction of the system software components in the process of development of the structural component of some application for reconfigurable computer systems.

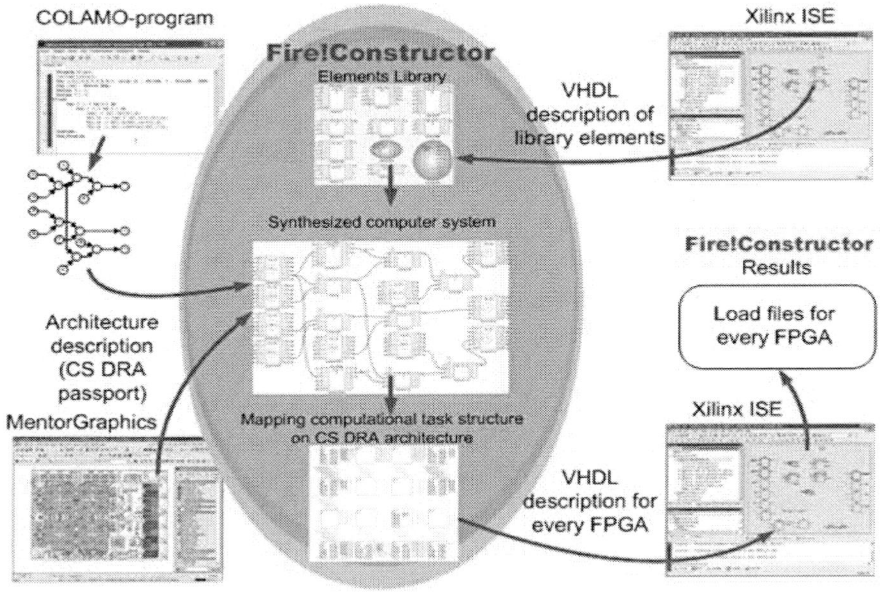

Figure 6.2. Interaction of the system software components of the reconfigurable computing structure.

Translating the parallel COLAMO-program, the translator generates the structural component as the virtual information graph of the problem. If the initial code of the parallel program contains errors, then the translator generates a list of errors and warnings, and returns it to the Argus IDE for further correction by the application programmer.

To implement the structural component, obtained after translation of the COLAMO-program, the Argus IDE calls another environment – the synthesizer Fire!Constructor, which is intended for creation of application computing structures on the base of the specified hardware platform.

As a result, the elements of the synthesized virtual information graph are replaced by the functional elements from the library of IP-cores (the library of elements). It should be noted, that computing structure creation may be based on several element libraries.

Using the information graph and the libraries of IP-cores and interfaces, the Fire!Constructor implements the structural component as a cadr computing structure. The computing structure is created from the information graph by replacement of its vertices with the library IP-cores of the computational nodes, and by replacement of its edges with the correspondent library interfaces. It is necessary to take into account the passport (description) of the hardware platform, for which the cadr computing structure is created. The passport contains information about the number and types of the FPGAs, about the configuration and the number of connections between the FPGAs and other components of the basic module, about the RAM chips and their types, about the number and types of external communication channels, etc.

The file of the MPCS architecture description is generated by a specialized tool – an architect, which converts the descriptions of the printed circuit boards of various basic modules, generated by Mentor Graphics, into the form, which may be used by the Fire!Constructor. One of the main functions of the Fire!Constructor is automatic distribution of a single computing structure among the FPGAs of the basic module. Besides, the Fire!Constructor translates the fragments of the computing structures, which were generated for separate FPGAs, into the language of hardware description VHDL. Then standard synthesizers of FPGA configuration, such as Xilinx ISE, generate configuration files for all the FPGAs of the basic module. A specialized loader tool loads the FPGA configuration.

Such two-level translation provides portability of the application, written in the high-level programming language COLAMO, to MPCSs of various architectures and configurations, designed on various basic modules. Owing to the complicated procedures, performed by the Fire!Constructor, the time of development and laboriousness of MPCS applications are considerably reduced. The list of these complicated procedures is the following:

- matching of inputs/outputs (ucf-files) of co-functioning FPGAs;
- automatic synchronization of data streams during placement of functional units in different FPGAs;
- automatic balanced placement of functional units into various FPGAs.

Theoretically, automatic synthesis of the computing structure and synthesis of multichip computing structures within the single basic module allows that highly qualified circuit engineers, who are usually involved in creation of circuit solutions for each FPGA of the basic module, may be excluded from development of the MPCS parallel program. The stream component of the structural-procedural program is generated as an executable module (out-file) during translation of the COLAMO-program. The executable module contains a sequence of statements for the MPCS distributed memory controllers, that control the RAM chips, connected to the controllers, and perform the procedures of reading of input data streams and writing of computation results, obtained during execution of the cadrs. The executable modules are loaded into the memory of the distributed memory controllers by the debugger or the specialized loader, and provide control of the memory controllers during functioning of the MPCS. Let us make a detailed consideration of the main components of the MPCS system software.

6.2. The Assembler

6.2.1. General Information

The programming language ARGUS v.3.0 is a low-level language, intended for programming of reconfigurable structures. The language of structural-procedural programming ARGUS v.3.0 supports the reconfigurable hardware platform of the computing structure owing to the specialized language structures and parameters of configuration, included during translation and generation of the executable code.

A program, written in the ARGUS v.3.0, is a text file, that contains language statements. The program may consist of several modules. In the simplest case, the program may solely contain the main file. In the general case, the program consists of the main program file and several units with description of structural components of the program. Owing to this, program development becomes more flexible and simple.

The main program file contains declarations of variables and the cadr sequence, which is determined by control operators. Each cadr consists of a number of computational instructions, that are carried out concurrently and asynchronously. The text of the main program, written in the ARGUS v.3.0, is bounded by the keywords PROGRAM and End_Prog.

The unit contains description of a cadr computing structure, descriptions of macroaccesses, special circuit solutions (configuration files) and connections between them. The macroaccesses and special circuit solutions may be written in the file in any possible sequence with no restrictions. The text of the unit is bounded by the keywords UNIT and End_Unit.

Owing to such "crop marks", the application programmer may structure the program, and place the library of macroaccesses and special circuit solutions into separate units, that may be plugged into the project, if it is necessary.

User's variables and statements of the ARGUS v.3.0 may be declared by pre-defined symbolic names or symbolic names, defined by the user. The pre-defined symbolic names are mnemonic designations of the operators and directives of the language. The user-defined symbolic names are the following:

- labels (local and global);
- names of macroaccesses;
- data identifiers;
- names of basic modules.

To define the symbolic names, the user may use alphanumeric symbols according to the following conditions:

- the first symbol cannot be a numeral;
- no separators are allowed.

A label is a symbolic name of a cell in the program memory of some distributed memory controller, which stores a machine equivalent of some operator of the memory controller, which follows the label. According to the format, the symbol ':' marks the end of the label name.

A sentence, written in the ARGUS v.3.0, may take several lines. The end of the sentence is marked by ';'. Usually, the sentence consists of three fields:

- a label field;
- an operation field;
- a field of operands.

There is also a field of comments, which may be placed in any part of the program with no restrictions.

The comment field and the label field are not obligatory.

The field of operands may be missing.

Any comment starts with '/*' and ends with '*/'. Nested comments are also available.

The ARGUS v.3.0 contains the following data types:

- *Number* is an integer;
- *Vector* is an array of *n* elements;
- *Stream* is an array of *n* elements;
- *Array* is a multidimensional array;
- *ParamAddress* is an address of a memory cell.

Arithmetic expressions in the ARGUS v.3.0 consist of user-defined symbolic names, numbers, operators and built-in functions.

6.2.2. The Elements of the Language

The format of the variable declaration:

Var name_of_variable : type_of_variable;

The format of the variable initialization:

Define name_of_variable = expression;

The elements of a vector may be defined according to two basic forms of an assignment statement.

The Form 1:
Define name_of_variable = [expression];

The Form 2:
Define name_of_variable = [expression, ..., expression];

If the number of the defined elements is less than the vector length, then:

- all missing elements will be replaced by computer zeros, if the elements of the vector are defined for the first time;
- all missing elements will remain unchanged, if the elements of the vector were defined before.

Peculiarities of the vector arithmetic consist in several rules. If one operand is a vector, and a second operand is a number, and a result is a vector, then the arithmetic operation is performed on each element of the vector-operand, and on the number-operand.

Describing Macroaccesses

Macroaccesses are used for adjustment of elementary connections between the channels of the distributed memory controller and other components of the reconfigurable structure. There are two types of macroaccesses in the ARGUS v.3.0: real and virtual. A real macroaccess adjusts two communication channels.

The name of a channel must start with the keyword CHAN. The number of the channel must follow next, and must belong to the range [0, 1].

There are six types of macroaccess instructions (see Table 6.1).

All instructions of one macroaccess must be of the same type. The macroaccess is to be concluded with the ENDMACROMEMORY statement, which indicates the end of the instruction sequence of the macroaccess. This statement is necessary because the instructions, that adjust the channels, may be written not in an ascending order of numbers of the distributed memory controller channels, but in an arbitrary one.

The following example shows a real macroaccess, which adjusts the distributed memory channels to a data read function:

```
RealMacromemory m00(out: alpha0,beta0);
alpha0<<chan0;
beta0<<chan1;
EndMacroMemory;
```

The macroaccess adjusts two channels to data read.

The following example shows the use of a virtual macroaccess:

```
VirtualMacroMemory m00(In : alpha0,beta0; out : alpha0,beta0);
EndMacroMemory;
```

alpha0, beta0, alpha0, beta0 are the logical names of the used distributed memory controllers.

A Data Read Instruction

According to this instruction, a data stream is read from the distributed memory according to the calculated addresses. The starting address is defined by the vector of starting addresses of the statement READ or READ_MOD. The statement is to refer to the same distributed memory page as the one in which the macromemory is loaded. The increment of the addresses is defined by the vector of address increments of the instruction. The read data follows to the output A of the macromemory channel.

A Data Write Instruction

According to this instruction, a data stream is written in the distributed memory according to the calculated addresses. The starting addresses are defined by the vector of starting addresses of the statement WRITE or WRITE_MOD. The statement is to refer to the same distributed memory page as the one in which the macromemory is loaded. The increment of the addresses is defined by the vector of address increments of the instruction. The data enters the input A of the macromemory channel.

Table 6.1. Types of macroaccess instructions

Name of instruction type	Instruction format	Graphic representation
Read of data	output_name<<channel_name;	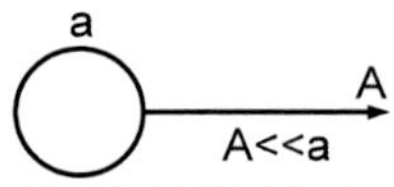
Write of data	channel_name<<input_name;	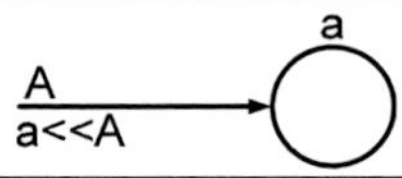
WHILE-read of data	output_name<<channel_name (input_name);	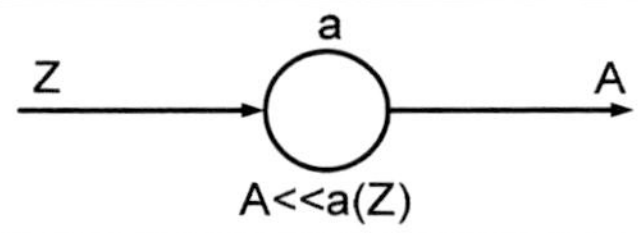
WHILE-write of data	channel_name<<input_name (input_name);	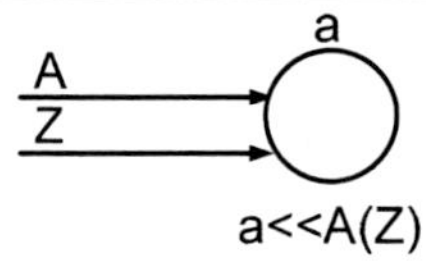
Read of data according to external address (external increment)	output_name<<channel_name [input_name; input_name];	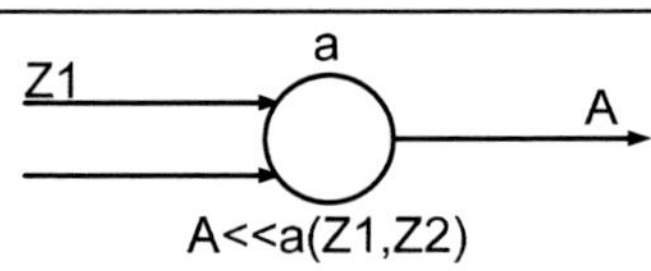
Write of data according to external address (external increment)	channel_name<<input_name [input_name; input_name];	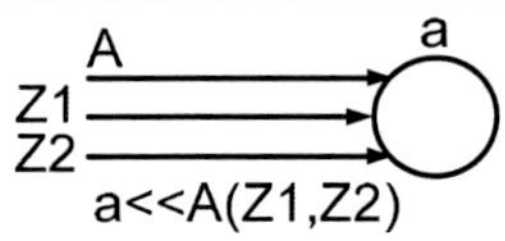

An Instruction of Data WHILE-Read

According to this instruction, a data stream is read from the distributed memory. The length of the stream cannot be defined during translation of the task; it is defined only during execution of the task. The starting addresses are defined by the vector of starting addresses of the statement READ_WHILE. The statement is to refer to the same distributed memory page as the one in which the macromemory is loaded. The increment of the addresses is defined by the vector of address increments of the instruction. The data follows to the output A of the macromemory channel. The data is being read until logical unity enters the input Z of the channel.

An Instruction of Data WHILE-Write

According to this instruction, data streams are written in the distributed memory. The starting address is defined by the vector of starting addresses of the statement WRITE_WHILE. The statement is to refer to the same distributed memory page as the one in which the macromemory is loaded. The increment of the addresses is defined by the vector of address increments of the statement WRITE_WHILE. The data enters the input A of the macromemory channel. Writing cycles are being executed until logical unity enters the input Z of the channel.

An Instruction of Data Read from an External Address

According to this instruction, data streams are read from the distributed memory from the address, consisting of two parts.

The first part of the address enters the input Z1, and the second one enters the input Z2 of the macromemory channel. To read data from the external address, it is necessary to use the statement READ with the keyword EXTERN instead of the name of the vector of starting addresses. The statement is to refer to the same distributed memory page as the one in which the macromemory is loaded. The data follows to the output A of the macromemory channel.

An Instruction of Data Write to an External Address

According to this instruction, data streams are written into the distributed memory to the address, consisting of two parts. The first part of the address enters the input Z1, and the second one enters the input Z2 of the macromemory channel. To write data to the external address, it is necessary to use the statement WRITE with the keyword EXTERN instead of the name of the vector of starting addresses. The statement is to refer to the same distributed memory page as the one in which the macromemory is loaded. The data enters the input A of the macromemory channel.

An Instruction of Data Read with an External Increment

According to this instruction, data is read from the distributed memory from the address, which is the vector of starting addresses of the statement READ. The statement is to refer to the same distributed memory page as the one in which the macromemory is loaded. To read data with the external increment, it is necessary to use the statement READ with the keyword EXTERN instead of the name of the vector of address increments.

The increment enters the inputs Z1 and Z2 of the macromemory channel. The statement is to refer to the same distributed memory page as the one in which the macromemory is loaded. The data follows to the output A of the macromemory channel.

An Instruction of Data Write with an External Increment

According to this instruction, data streams are written into the distributed memory to the starting address which is defined by the vector of starting addresses of the statement WRITE. The statement is to refer to the same distributed memory page as the one in which the macromemory is loaded.

To write data with the external increment, it is necessary to use the statement WRITE with the keyword EXTERN instead of the name of the vector of address increments. The increment enters the inputs Z1 and Z2 of the macromemory channel. The statement is to refer to the same distributed memory page as the one in which the macromemory is loaded. The data enters the input A of the macromemory channel.

Describing Structural Components

A structural component of a cadr is described by the following structure:

Structure *name_of_cadr*
in *list_of_formal_parameters_of_inputs;*
out *list_of_formal_parameters_of_outputs;*

set_of_structure_elements:
Endstructure;

The keyword STRUCTURE marks the beginning of the structural component of the cadr, which may contain macrooperations, macroaccesses or some other structural components. The format of the macrooperation call:

Name_of_macrooperation(list_of_actual_parameters);

As a result, when the actual parameters of the macroaccess are replaced by the actual parameters of the cadr, the translator sets connections between the components of the reconfigurable structure and implements them in the spatial communication system. Hence, the application programmer has no need to program the switch.

Describing Cadrs

A cadr is a set of program sections of the distributed memory controller. All sections are to be executed concurrently. Usually, the program is split into a set of cadrs, which are called successively during execution of the program. In the simplest case, the program contains only one cadr.

The program may pass from one cadr to another not after completing the last instruction of the current cadr, but in the middle of the current cadr. Such a jump will be correct, if all the memory controllers do the same jump, omitting the last instruction of the cadr.

The keyword CADR marks the beginning of the cadr. All internal operators of the memory controller are to be contained solely inside the cadr. All external operators, excluding the statement STOP, are to be placed solely outside the cadr. Directives of assignment and declaration of data types, and general directives may be placed both inside and outside the cadr. The cadr description ends with the keyword ENDCADR.

6.2.3. Programming of the Distributed Memory

The list of programming structures of the distributed memory controller contains directives and operators. The directives of the memory controller are the assembler structures, which form the structure of the controller programs during assembling.

Let us consider the directives of the distributed memory controller. While implementing the task, some modules may be unused. The directive *Module* defines resources, which are needed for implementation of the current task.

The directive *Module* has the following format:

Module BM0[,BM1[,...[,BMi]]];

where *BMi :: = a number of basic module.*

The number is the absolute number of the basic module.

Each basic module contains several (from 8 to 32) distributed memory units or memory pages. To access to the page of the distributed memory controller, it is necessary to point out the number of the basic module, which the page belongs to, and the number of the page.

To load data and programs into the distributed memory, it is necessary to point out the number of the memory page, which they are destined for. The loading statement must contain the number of the only one page of the distributed memory controller.

When a new page of the distributed memory controller is selected, the current page is cancelled automatically. The statements of the distributed memory controller may be internal and external.

The Internal Statements

The internal statements of the distributed memory controller are always to be placed only inside the cadr. The statements are loaded directly to the memory of the page, which they belong to. There are three groups of the statements: statements of data memory access, operators of parameter change, and control statements.

In general case, any internal statement of the distributed memory controller contains the directive, which defines the logical number of the basic module and the logical number of the distributed memory controller, and the internal operator itself:

```
Cadr Cadr1
DMC[№BM , №DMC]#
Read A Step I Repeat N;
Endcadr;
```

where DMC is the directive, which defines belonging of the statement;

№BM is the number of the basic module;

№DMC is the number of the distributed memory controller;

READ is the internal statement of the distributed memory controller.

The access statements may be read statements and write statements. According to the addressing type, the statements may be placed into two groups: the statements with incremental addressing and the statements with threshold addressing. According to the data type, the read/write statements may be placed into two groups: the statements, which read/write numbers, and the statements, which read/write macrooperations and macroswitches.

Table 6.2 contains the formats of these access statements.

The format of *the read statement with incremental addressing*:

Read VSA [Step VAI [Repeat LP]]; [reading of numbers].

A data stream is read concurrently from all channels. The number of read cycles is defined by the loop parameter (LP). The parameters *step* and *repeat* may be omitted, if the stream is the only one. The starting addresses of the array elements are defined by the vector of starting addresses (VSA).

During the first iteration of the statement, the data is being read from the memory cells of the data memory channels according to the addresses, defined by the vector of starting addresses. When the statement is performed, the program passes to the next statement.

A format of *the write statement with incremental addressing:*

Write VSA [Step VAI [Repeat LP]]; [writing of data].

A data stream is written concurrently into all channels. The number of write cycles is defined by the loop parameter (LP). The parameters *step* and *repeat* may be omitted, if the stream is the only one. The starting addresses of the array elements are defined by the vector of starting addresses (VSA).

Table 6.2. The formats of the access statements

Mnemonic	Format	Action
READ	Read VSA [Step VAI [Repeat LP]];	Data read with incremental addressing
WRITE	Write VSA [Step VAI [Repeat LP]];	Data write with incremental addressing
CONTINUE	Continue Step VAI Repeat LP	Additional read/write statement
READ_MOD	Read_Mod VSA [Step VAI [Repeat LP]];	Data read with threshold addressing
WRITE_MOD	Write_Mod VSA [Step VAI [Repeat LP]];	Data write with threshold addressing
READ_WHILE	Read_While VSA [Step VAI];	Data WHILE-read
WRITE_WHILE	Write_While VSA [Step VAI];	Data WHILE-write
READ	Read_Extern [Repeat LP];	Data read with an external address
WRITE	Write_Extern [Repeat LP];	Data write with an external address
READ	Read VSA Step Extern [Repeat LP];	Data read with an external increment
WRITE	Write VSA Step Extern [Repeat LP];	Data write with an external increment
READ_STRING	Read_String VSA [Size N];	Read a string
WRITE_STRIN G	Write_String VSA [Size N];	Write a string
READ	Read_VSA TO *NAME_OF_DEVICE*	Read of a macrooperation (macroswitch) for *NAME_OF_DEVICE* (MAP0-MAP3)
WRITE	Write_VSA TO *NAME_OF_DEVICE*	Write of a macrooperation (macroswitch) for *NAME_OF_DEVICE* (MAP0-MAP3)
MOVE_REG	Move_Reg [@]VSA, name_of_register	Read of a register
MOVE_REG	Move_Reg name_of_register; [@]VSA	Write of a register

During the first iteration of the statement, the data is being written into the memory cells of the data memory channels to the addresses, defined by the vector of starting addresses. When the statement is performed, the program passes to the next statement.

The format of *the read statement with threshold addressing:*

Read_Mod VSA [Step VAI [Repeat LP]]; [reading of data].

The statement performs data reading with threshold addressing. The data is concurrently read from all channels. The number of read cycles is defined by the loop parameter. If during execution of the statement the current data address amounts *A0*, then the next address will possess the value of the BASE register.

The format of *the write statement with threshold addressing:*

Write_Mod VSA [Step VAI [Repeat LP]]; [writing of data].

The statement performs data writing with threshold addressing. The data is concurrently written into all channels. The number of write cycles is defined by the loop parameter. If during execution of the statement the current data address amounts *A0*, then the next address will possess the value of the BASE register.

The format of *the Continue statement:*

Continue [Step VAI] Repeat LP;

The statement is intended for organization of nested cyclic data access, and may be used only after the statements of data read/write.

The format of *the WHILE-read statement:*

Read_While VSA [Step VAI];

The statement performs reading of a data array until a logical unity enters the control input.

The format of *the WHILE-write statement:*

Write_While VSA [Step VAI];

The statement performs writing of a data array until a logical unity enters the control input.

The format of *the statement of reading from an external address:*

Read_Extern [Repeat LP];

The statement performs reading of data from the address, received from the external address input.

The format of *the statement of writing to an external address:*

Write_Extern [Repeat LP];

The statement performs writing of data to the address, received from the external address input.

The format of *the statement of reading with an external increment:*

Read VSA Step Extern [Repeat LP];

The statement performs reading of data from the specified address with the increment, received from the external address input.

The format of *the statement of writing with an external increment:*

Write VSA Step Extern [Repeat LP];

The statement performs writing of data according to the specified address with the increment, received from the external address input.

The format of *the string read statement:*

Read_String VSA [Size N];

The statement reads a block of N values from the starting address. The data block is a continuous sequence. The beginning and the end of the sequence are marked by a start marker (BM) and an end marker (EM) (in contrast to reading of numbers, when each number is marked).

The format of *the string write statement:*

Write_String VSA [Size N],

The statement writes a block of N values to the starting address. The data block is a continuous sequence. The beginning and the end of the sequence are marked by a start marker (BM) and an end marker (EM) (in contrast to writing of numbers, when each number is marked). The elements of the string are transferred as a continuous data stream, i.e. discontinuities between the string elements are inacceptable. The strings are processed by special hardwarily implemented circuit solutions.

In addition to the considered statements, used for programming of the distributed memory controller, there are group statements, that are intended for processing of address arrays.

Owing to the group statements, the directive, that defines the memory page which the statement belongs to, and which the statement will be carried out for, becomes not obligatory.

Table 6.3 contains the formats of the group access statements.

Table 6.3. The formats of the group access statements

Mnemonic	Format	Action
GROUPREAD	GroupRead ARRAY [<Parameters>];	Read of data array with incremental addressing
GROUPWRITE	GroupWrite ARRAY [<Parameters>];	Write of data array with incremental addressing

The field <Parameter> defines the processing range of the statement, and contains the start and the end indices of the address array. The parameter ARRAY of the group statements may be represented by a variable of Vector, Stream or Array type. The variable of any other type, used as ARRAY, will cause a syntax error.

The format of *the group read statement:*

GroupRead ARRAY (StartIndex TO EndIndex);

The statement reads data from the addresses, specified by the field ARRAY, starting from the address StartIndex to the address EndIndex.

The format of *the group write statement:*

GroupWrite ARRAY (StartIndex TO EndIndex);

The statement writes data according to the addresses, specified by the field ARRAY, starting from the address StartIndex to the address EndIndex.

Register Read Statements

The assembler contains 8 registers: a base register (BASE), a return register (RETURN_REG), a status register (STATUS), an operator register (OPER_SEG), a parameter register (PARAM_SEG), a data register (DATA_SEG), a threshold register (MOD_REG) and a segment of macroinstructions (MACRO_SEG). To access these registers, the assembler contains register read statements.

The format of *the register read statement:*

Move_Reg [@]VSA, name_of_register;

The statement reads the specified register and writes the calculated data to the address, defined by the vector of starting addresses. The statement allows indirect addressing.

The format of *the register write statement:*

Move_Reg name_of_register,[@]VSA;

The statement writes data, stored in the memory cells with addresses, defined by the vector of starting addresses, into the corresponding register. The statement allows indirect addressing.

The Operators of Parameter Change

Table 6.4 contains the operators of parameter change.

The format of *the addition operator:*

Add vector1, vector 2.

The operator adds the contents of the source *vector2* to the contents of the target *vector1*, and moves the result to *vector1*.

The format of *the move operator:*

Move target, source;

The operator moves the contents of the *source* to the *target*. The data vector, or the vector of memory addresses of the moved data may be considered as the *source* operand. In the first case, the operator moves the contents of the *source* to the *target*. In the second case, the contents of data memory, stored in the cells with the addresses, specified by the *source*, are moved to the *target*.

The format of *the operator of shift field addition:*

AddoSH label, VAI;

Table 6.4. The operators of parameter change

Format	Action
Addition Add vector1, vector 2;	Recalculation of a vector (vector1). Recalculation of an address vector (increments of addresses).
Move Move vector1, vector2; Move vector1,@ vector2;	 Move of a vector Move of a vector to an indirect address

Addition of operator fields
AddoSH label, VAI;
AddoS label, VAI;
AddoJ label, VAI;
AddoM label, VAI;
	
	Modification of an SH field
	Modification of an S field
	Modification of a J field
	Modification of an M field

Substitution of operator fields
MoveSH label, VAI;
MoveS label, VAI;
MoveJ label, VAI;
MoveM label, VAI;
	
	Substitution of an SH field
	Substitution of an S field
	Substitution of a J field
	Substitution of an M field

Clearing of operator fields ClrSH label; ClrS label; ClrJ label; ClrM label;	 Clearing of an SH field Clearing of an S field Clearing of a J field Clearing of an M field

The operator adds the corresponding shift field of the operand to the null element of the VAI operand.

The format of *the operator of loop field addition:*

AddoS label, VAI;

The operator adds the corresponding loop field of the operand to the null element of the VAI operand.

In this case, it is sufficient to specify only the null element of the vector, because all other elements of the VAI will not be used by the operator.

The format of *the operator of label field addition:*

AddoJ label, VAI;

The operator adds the correspondent label field of the operand to the null element of the VAI operand.

The format of *the operator of modification field addition:*

AddoM label, VAI;

The operator adds the correspondent modification field of the operand to the null element of the VAI operand.

The format of *the operator of shift field substitution:*

MoveSH label, [@]VAI;

The operator substitutes the shift field of the operand by the null element of the VAI operand.

The format of *the operator of loop field substitution:*

MoveS label, [@]VAI;

The operator substitutes the loop field of the operand by the null element of the VAI operand.

In this case, it is sufficient to specify only the null element of the vector, because all other elements of the VAI will not be used by the operator.

The format of *the operator of label field substitution:*

MoveJ label, [@]VAI;

The operator substitutes the label field of the operand by the null element of the VAI operand.

The format of *the operator of modification field substitution:*

MoveM label, [@]VAI;

The operator substitutes the modification field of the operand by the null element of the VAI operand.

The format of *the operator of shift field clearing:*

ClrSH label;

The operator ClrSH clears the shift field of the operator.

6.2.4. The Control Statements

The group of the control statements contains the following statements:

- the statement of an unconditional jump;
- the pause operator;
- the loop statement;
- the statement of a conditional jump.

The control statements are given in Table 6.5.

The format of *the statement of an unconditional jump:*

Goto J;

The statement of an unconditional jump performs a jump to the operator, marked by the label J. The statements *call* and *return* are the special cases of the statement *goto.*

The format of *the call statement:*

Call J;

The statement performs a jump to the label J, saving the address into a stack. The statement is used for organization of subroutines.

Table 6.5. The control statements

Name	Format	Action
Statement of an unconditional jump:	Goto J;	An unconditional jump to the address J
Pause operator:	Pause N;	N-cycle delay of program execution
Loop statement:	Loop S [Goto J];	Loop statement
Statement of a conditional jump:	If A Goto J;	A jump according to the contents of the cell A
	If @A Goto J;	A jump according to the contents of the cell with the address A

The format of *the return statement:*

Return;

The statement extracts an address from a register and jumps to it. The statement is used for return from subroutines.

The format of *the pause operator:*

Pause N;

The operator delays program execution for N cycles.

The format of *the loop statement:*

Loop S [Goto J];

The format of *the statement of a conditional jump:*

If A Goto J;

The statement of a conditional jump performs the following actions. If the memory cell *A* contains a non-nil value, then the statement performs a jump to the label J, else – the next operator is performed.

The second variant of the *statement of a conditional jump:*

If @A Goto J;

The symbol @ means that indirect addressing is used.

The format of *a structure of declaration of address switching*:

Acommutation <Name_of_switching>;
Chan0 (<Number of Bit_0>, ... <Number of Bit_15>);
Chan1 (<Number of Bit_0>, ... <Number of Bit_15>);
Endacom;

In the executable cadr, switchings may be used with the help of the statement of switching loading LOADACOM, and the statement of switching control LOADBCOM. The format of *the LOADACOM statement*:

Loadacom <Name_of_Switching>.

The switching is to be preliminarily declared. The switching, loaded in the specified distributed memory controller, is valid until a new switching is loaded, or until a switching reset instruction is received.

The format of *the LOADBCOM statement:*

Loadbcom <Name_of_Variable>.

The LOADBCOM statement switches external buses, connected to the distributed memory controller.

Address switching may be reset by *the reset statement*:

Resetacom.

As a result, a natural order of address bits is reset.

The statements of loading and reset may be used in the group operations for a group of distributed memory controllers.

6.2.5. External Operators of Distributed Memory Controllers

Table 6.6 contains the formats of the external operators of the memory controller. The distinctive feature of the operators is impossibility of being marked by labels.

Table 6.6. The formats of the external operators of the memory controller

Format	Action
Operators: - input	
Input VSA [Step VAI] [Channels NCH] [Repeat LP] Data {numerical series};	Direct data input
Input VSA [Step VAI] [Channels NCH] [Repeat LP] File "name_of_file";	Data input from an outer file
- output	
Output VSA [Step VAI] [Channels NCH] [Repeat LP] File "name_of_file";	Data output into a file (non-converted data)

Direct data input from the source text of the program is performed by the operator, which has the following format:

Input VSA [Step VAI] [Channels NCH][Repeat LP] Data {numerical_series};

where *numerical_series ::= sequence of numbers*, separated by commas, or a set of expressions.

Here the loop parameter specifies the number of input data vectors. If the values are to be taken from a separate file, then the operator is written as follows:

Input VSA [Step VAI] [Channels NCH] [Repeat LP] File "name_of_file";

where *name_of_file ::= string_of_symbols*.

The OUTPUT operator unloads contents of data memory into a file, and has the following format:

Output VSA [Step VAI] [Channels NCH] [Repeat LP] File "name_of_file";

There are two types of input/output operators in the ARGUS v.3.0: external and internal.

The operators, which are placed outside the cadr, and which control the distributed memory controller from the outside, are called *the external operators*. External control is performed by an application "Loader of interrupt vectors" along with the basic module controller.

The external operators may be placed into two groups: the input/output operators and the task control operators. The input/output operators perform data loading/unloading in/from data memory of the distributed memory controller. Both the data and addresses may be loaded into all channels or into the selected ones. This restriction is defined by the vector of the selected channels. The data may be loaded from the program or an outer binary file. If the data is defined in the program text, then it may be represented in any numerical system, supported by the ARGUS v.2.0. The input/output operators may be placed solely outside the cadr, in a so-called "intercadr space". If the input/output operator is placed before the first CADR directive, then the operator is called the "Operator of initial loading (unloading)". If the input/output operator is placed between the cadrs, i.e. it is to be performed when a certain cadr is already executed, then the operator is called the "Operator of data loading (unloading)". If the ARGUS V.3.0 finds such operator, it sets an attribute of "goto with a halt" to the last operator of the previous cadr. If the program is being executed, then after carrying out the operator with such type of jump, the STOP statement will be performed in the current cadr in all memory controllers. The handler of interrupt vectors will carry out input/output in/from the memory controller, that was set in the program. When the input/output operations are carried out, the loader will yield the general START to continue functioning of the basic module. The number of the controller is specified by analogy with cadr programming.

Data may be loaded/unloaded in/from all distributed memory channels or the selected ones. The input operators do not allow external parameters, when data is loaded directly from the text of the program. If data is loaded from some outer file, then the input/output operator may contain the external parameters.

Since the loop parameter of the input/output operator is non-modifiable, then it may be an identifier, a number or an expression.

The operators, that are placed inside the cadr and control functioning of the distributed memory controller "from the inside", are called *the internal operators*. The group of the internal control operators contains the statement of data read from a memory channel, and the statement of data write into a memory channel.

6.3. THE HIGH-LEVEL PARALLEL PROGRAMMING LANGUAGE COLAMO

The high-level programming language COLAMO is used for development of applications on the base of structural-procedural computations in the MPCS. One of the problems, that the application programmer faces with during development of any parallel program, is a semantic gap between the parallel algorithm and its implementation in the computer system. This problem can be solved with the help of structural-procedural computations.

Effective implementations of computational algorithms on multiprocessor systems require a high-level programming language (HLPL), which must be effectively implemented on various computer systems, and which must provide description of any algorithm without any changes of its initial parallel structure [130-132]. Using the language, the application programmer must be able to describe different kinds of parallelism in the most simple and concise form. In addition, the language must contain standard structures, that are usually contained in traditional high-level programming languages [133, 131].

New languages, considered in well-known papers [121,134-137], are the traditional procedural high-level programming languages with extensions, intended for programming of parallel computations. Other languages are oriented on specified computer systems. In any case, the number of users of these languages is limited. Analysis of these languages showed that existing methods of parallelization are too complicated and laborious, and cannot provide adequate mapping on the given computer system, therefore, it is necessary to develop the languages with capabilities of explicit description of independent parts of the program [138].

Traditional methods of algorithm parallelization require the multilevel structure of the task [102]. The example of such programming language is OCCAM, which is widely used in transputer systems. The application programmer may describe any multilevel structure, using the instructions *par*, *seq*, which mark the fragments of parallel and sequential processes [137].

However, the programs, that describe the multilevel structure of the task, have several shortcomings, especially low readability of the program text and complexity of debugging. This is caused by concurrent execution of several threads, i.e. there are executed several programs, that must exchange data between each other and synchronize their data streams. It is particularly difficult to do, when processing elements of the computer system are functioning asynchronously.

Usually, the problems of process synchronization are solved with the help of structures like *fork-join* [41], Dijkstra's semaphores [139], rendezvous [121], preemptive process determination [137]. Each method has its own disadvantages, considered in [140].

In practice, it is impossible to obtain the rectangular multilevel structure of any task [102]; hence, it leads to degradation of system effectiveness. In most cases, well-known multiprocessor computer systems with fixed communication channels between the system components are not similar to the structure of a solving problem. This circumstance considerably complicates parallelization of the problem, and significantly reduces system effectiveness. Possibility of programming of communication channels between the components of a multi-object MPCS allows the application programmer to program the system, and hence, to achieve similarity between the structure of the solving problem and the architecture of the system. Besides, it provides high granularity of parallelism of data processing.

A qualitatively new method may be used for the reconfigurable multipipeline computing structure with structural-procedural organization of computations. A task is split into hardwarily implemented independent, functionally self-contained fragments. Selection and execution of the computing structures is controlled by the parallel application. Hence, only one computational fragment (structure) may be executed.

The problem, referring to programming of connections between system components [141] is one of the most complicated in high-level programming languages of multiprocessor systems. For instance, in the OCCAM language this problem is solved by means of description operators of channels between the processing elements and the elements of distributed memory [142]. The system SWARD contains switching means, such as two machine instructions *send* and *receive*, and the objects "port" [133]. In medias res, such methods lower the high-level programming languages to the level of machine languages.

The considered language COLAMO contains no explicit forms of parallelism description. Tasks may be parallelized by declaration of variable types, and by indexing of array elements. The single-assignment rule [143] is a natural method of implicit description of parallelism in the parallel application, and is widely used in data flow languages. According to the rule, a variable may receive a value in the program only once [144]. The rule leads to a contradiction with traditional programming principles [145]. As a result, the rule is restricted: the single-assignment rule is valid for the special structures of the suggested language, that describe computing structures, defined by application programmers.

The programming language COLAMO contains three groups of operators:

- operators of computing structure description;
- operators of computational process control;
- operators of description of data and firmware resources.

The fundamental type of a computation structure in the language is the structure "*cadr*". The *cadr* corresponds to a set of arithmetic-logical instructions, performed by various processing elements and distributed memory controllers, that are connected according to the informational structure of the algorithm, and provide the maximum possible parallel and asynchronous computations [144].

In general case, the cadr is described as follows:

```
Cadr name;
P
Endcadr;
```

where *Cadr*, *Endcadr* are the keywords of the language;
name is the name of the cadr;
P is the list of the cadr operators.

Assignment operators, represented as *Name:=f;*, where *Name* is the name of the variable, which receives the value of the arithmetic-logical expression *f*, can be only internal. Statements of unconditional jump such as *Goto lab*, where *lab* is the label of a target operator, can be only external. The computational structures are considered as large program-indivisible units, and it is forbidden to jump from one structure component to another inside the structure.

The cadr body may contain several assignment operators. The application programmer does not mark out parallel fragments of computations. As a result, after translation of the program, parallelism, which is contained in the description of computations, is naturally derived from it on the base of connections between data.

According to the method of storage, the variables in the COLAMO may be declared as memory variables (MEMory), switching variables (COMmutation) and register variables (REGister). The *memory variable* is stored in a memory cell and, hence, it keeps its value until the next assignment (see Fig. 6.3).

```
Var a,b,c,d,e,f,g : Array Integer [10 : Stream] Mem;
Var I: Number;
Cadr F;
  For I := 1 To 5 Do
   Begin
     A[I] := B[I]·C[I] + F[I];
     D[I] := B[I]·C[I] + E[I]·G[I] + G[I-1];
   End;
Endcadr;
```

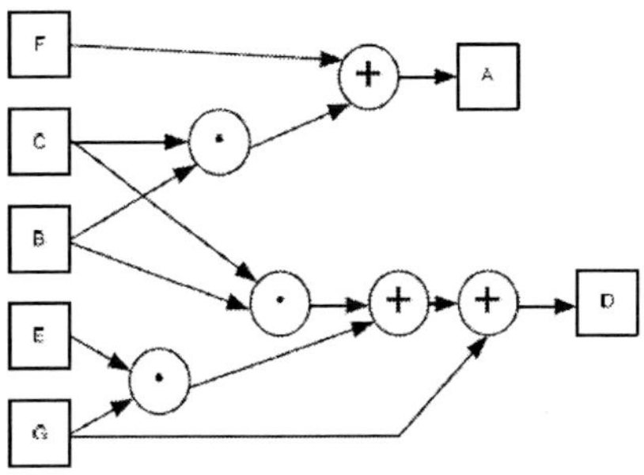

Figure 6.3. Use of a memory variable.

The only restriction of use of the memory variables is the rule of single substitution. In this case, it is forbidden to use the variables, stored in the same memory channels, in the left and right parts of computational expressions.

The *switching variable* is used for description of the channels between the computational elements. The values of switching variables are inaccessible for the application programmer (see Fig. 6.4).

```
Var a,b,c,d,e,f,g : Array Integer [10 : Stream] Mem;
Var Z : Integer Com;
Var I : Number;
Cadr F
  For I := 1 To 5 Do
   Begin
     Z := B[I]·C[I];
     A[I] := Z+F[I];
     D[I] := Z + (E[I]·G[I]) + G[I-1];
    End;
Endcadr;
```

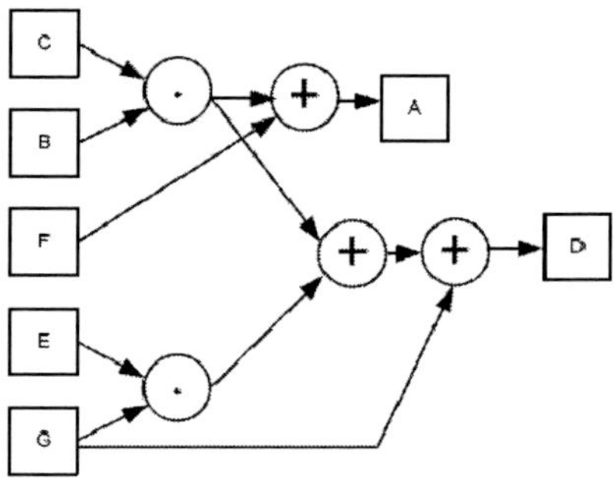

Figure 6.4. Use of a switching variable.

The *register variable* is used for storage of data, available for the application programmer at any time (see Fig. 6.5).

```
Var a,b,c,d,e,f,g : Array Integer [10 : Stream]
Mem;
Var Z : Integer Reg;
Var I : Number;
Cadr Cadr1;
  For I := 1 To 5 Do
   Begin
     Z := B[I]·C[I];
     A[I] := Z+F[I];
     D[I] := Z + (E[I]·G[I]) + G[I-1];
    End;
Endcadr;
```

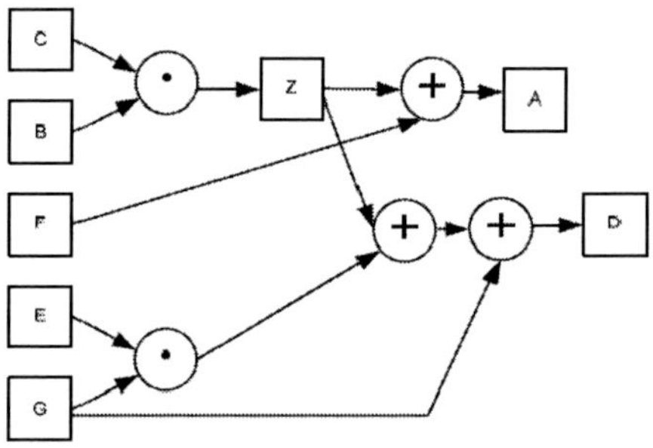

Figure 6.5. Use of a register variable.

Use of the register variables may lead to ambiguity of the application execution. All cadr components are performed asynchronously and independently, therefore, it is impossible to precisely say, which cadr component will be performed first – reading or writing of the register variable, and hence, it is not known whether the operations are correct in this case.

The conflict may be solved by splitting the incorrect cadr into several cadrs. In this case, the order of cadr execution regiments the values of the obtained variables:

```
Var M, B, C, A, D, E, G : Array Integer [100 : Stream] Mem;
Var I : Number;
Cadr S1;
For I := 1 To 5 Do
```

```
    Begin
        M[I] := B[I] · C[I];
    End;
Endcadr;
Cadr S2;
For I := 1 To 5 Do

    Begin
        A[I] := M[I]+F[I];
        D[I] := M[I] + (E[I]·G[I]) + G[I-1];
    End;
Endcadr;
```

Such structure provides a deterministic yield of results, but reduces the number of arithmetic-logical operations of the cadr and, consequently, the number of processing elements (PE), that implement the cadr. This leads to degradation of the real performance during the task execution, and to reduction of effectiveness of the whole MPCS.

Owing to the switching or register variables, the graphs of the computational structures, described by the application programmer in the COLAMO language, may be simplified.

It should be noted, that use of the computational structure, that requires reading and writing of one and the same variable in one cadr, is incorrect. This principle is rather easily implemented for scalar variables, but use of array elements leads to certain difficulties.

During translation it is impossible to define, whether the expression

```
X[i]:=f(X[j]);
```

is correct, because it is impossible to define beforehand, whether the situation, when *i* is equal to *j*, will occur during execution of the cadr, and will break the single-assignment rule. Besides, several readings and writings are not allowed in one cadr for both memory and register variables.

One of the problems, that were set during development of the high-level programming language COLAMO, is one-to-one correspondence between computational structures and the COLAMO records. This implies a paradoxical conclusion, according to which non-deterministic language structures may be acceptable. The translator marks such language structures by warnings. The conflict may be solved by splitting the incorrect cadr into several cadrs. It provides deterministic results but, as it was stated above, it leads to degradation of the real performance during execution of the task. Owing to the switching variables, the considered task may be implemented with no unwanted side effects. Types of variables are declared with the help of the VAR statement, which, in general case, has the following format:

```
Var p1 a1, p2 a2,..., pn : an;
```

where *pi* is the list of variables of the group *i* with the correspondent list *an* of properties,

Var is the keyword.

The items of the variable list are separated by commas. The items of the property list are separated by spaces. The property list contains a storage type of each variable (memory, register, switching) and a type of each variable (real (REAL), complex (COMPL), character (CHAR), bit (BIT), address (ADDR), logical (LOGIC), control (CONTR), integer (INTEGER)).

In addition to scalar variables, the language allows use of arrays, that are declared according to the following format:

VAR a1, a2, …, an : Array TypeVariable [p1, p2, ... , pn] Access,

where *ai* is a variable from the list of the declared variables;

pi is the list of properties of the *i*-th dimension of the array;

TypeVariable is the array type;

Access is the type of the array storage.

In monoprocessor computers multidimensional arrays are represented as one-dimensional arrays. In the MPCS with structural-procedural computations all arrays are represented as two-dimensional arrays. In this case, the first coordinate corresponds to the numbers of the distributed memory channels, and the second one – to the memory cells in the channel.

The COLAMO works with data vectors (VECTOR) and data streams (STREAM). The similar data types were implemented in the programming language IVTRAN [146]. On default, arrays are considered as vectors. Streams are the arrays of variables, whose items may be processed only sequentially. Vectors are the arrays of variables, whose items may be processed concurrently.

In practice, the application programmers usually use parallel-serial processing of data arrays. The granularity of parallelism depends on the specified algorithm and hardware resources, available to the application programmer. The granularity of parallelism is the number of concurrently performed elementary operations.

The loop statement is represented according to the following format:

For[j:=i1] [To i2] [Step i3] [While i4] Begin p End;

where *FOR*, *TO*, *STEP*, *WHILE* are the keywords;

j is the name of the loop variable;

i1 is the list of initial values of the loop variable;

i2 is the expression, which defines the increment limit of *j*;

i3 is the increment of the loop variable;

i4 is a logical expression; if the expression is false, then the loop terminates;

BEGIN, *END* are the keywords, which bound the loop body;

P is the list of the loop operators.

All instructions, written for vectors and placed in an inner loop, are multiplied and mapped on the computing structure according to the loop parameters. If the instructions are written for streams, then the stream elements are sequentially processed according to the instructions, included into the loop. Hence, semantics of syntactically similar cadr components is defined not by the text of the program, but by the data structure, that uniquely specify the rule of access to the array items.

The loop statement, which is an outer one to the computational structure description, cyclically performs the operators of the cadrs, including those, placed inside the loop body. The outer loop statement is similar to the loop statement of standard programming languages.

Figure 6.6 shows the program, that performs parallel-serial addition of arrays with the help of the inner and outer loop statements, and the graph of the computational structure, that will be obtained after translation of the program.

```
Var A,B,C : Array Integer [10: Vector] Mem;
Var I,J : Number;
For I:=1 To 9 Step 2 Do
  Cadr summa;
    For J:=1 To I+1 Do
       C[J] :=A[J]+B[J];
  Endcadr;
```

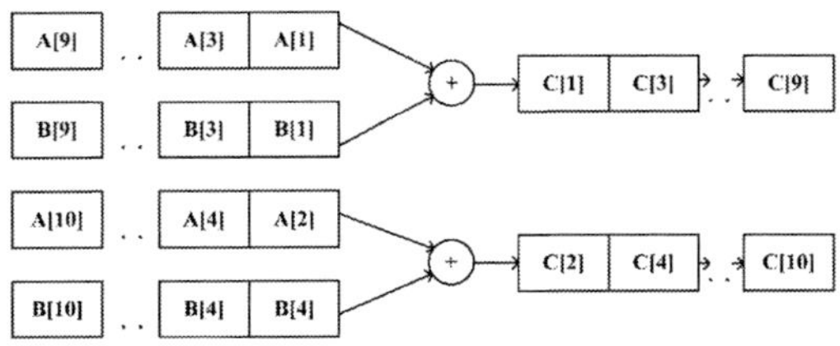

Figure 6.6. The program and the equivalent graph of the computational structure.

In practice, it is quite often necessary to access and to sample array items. In general case, a sample has the following format:

A[l:k:s],

where *A* is the name of the array;

l is the index, which defines the sample starting item;

k is the index which defines the sample final item;

s is the step of the sample index.

All these parameters are positional. If any parameter, except the last one, is missed, it is necessary to put a certain number of colons.

The default value of the sample start index is equal to the initial value of array declaration; *k* is equal to the end value of the array declaration. For example, the declaration A[i, 4:8] means that the sample is made according to the second coordinate of the array, from the item 4 to the item 8 with the step 1.

Structural mapping of the sample is specified by the type of array item processing, according to which the sample is made. The sample of the vector component of the array corresponds to the specified numbers of distributed memory channels, from which the data is concurrently read and transferred into the computing structure, or into which the data is concurrently written. For a profile of the stream component, the data is read (written) from (into) all distributed memory channels, that correspond to the arrays.

The sample of the stream component specifies a cycle of access to the memory cells of the distributed memory channels; the profile specifies sequential access to all elements of the distributed memory.

Arrays of variables may be declared as memory, switching and register ones. The vectors of the switching type are the group of data channels with the same name. By means of a switching stream, a delay of operands may be organized.

Figure 6.7 shows the programs, written in the high-level programming language, and the graphs of the computational structures, that will be obtained as a result of translation of the programs.

When calculation of some variable in one cadr requires access to several items of an array, it means that if the variable is declared as a vector, then the distributed memory channels will be connected according to specified rules. If the variable is declared as a stream, then the array items will be delayed according to the structure of the task.

```
Var A : Array Integer [10 : Vector] Mem;
Var C : Integer Reg;
Var I : Number;
Cadr a;
  For I:=2 To 9 Do
     C:=A[I]+A[I-1]+A[I+1];
Endcadr;
Var A : Array Integer [10 : Stream] Mem;
Var C : Integer Reg;
Var I : Number;
Cadr b;
  For I:=2 To 9 Do
     C:=A[I]+A[I-1]+A[I+1];
Endcadr;
```

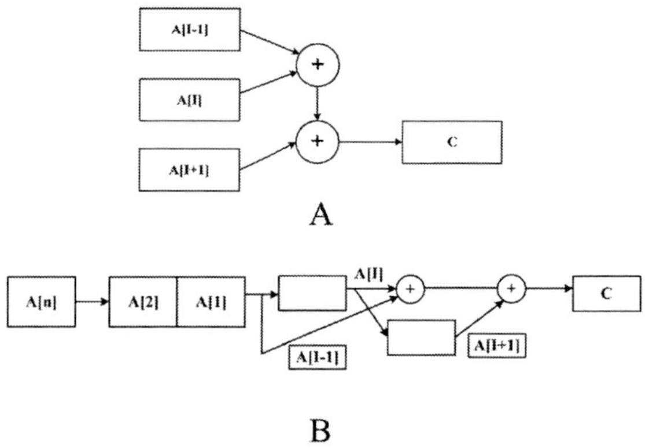

Figure 6.7. Access to the vector items (a) and to the stream items (b).

It should be noted, that if the program contains, for example, the following fragment:

```
Var X : Array Integer [100 : Stream] Mem;
Var I : Number;
Cadr ST;
For I := 1 To 50 Do
    C:= X[i] + X[j];
Endcadr;
```

then the translator is to declare such structure as an incorrect one, because it is impossible to define the buffer size for the stream X. The difference between the indices of the stream items, that are accessed in the cadr, is to be either a constant or, at least, a value, that remains constant during execution of the cadr.

Assume that it is necessary to calculate the expression $A_i = (B_i + B_{i-1}) * C$. The program that performs the required calculations, using implicit buffering, may be represented as follows:

```
Var A,B,C : Array Integer [4 : Stream] Mem;
Var I : Number;
Cadr A;
    For i:=2 To 4 Do;
            A[i] := (B[i]+ B[i-1])*C[i];
Endcadr;
```

The program, given below, implements the same example, using explicit buffering:

```
Var A,B,C : Array Integer [4 : Stream] Mem;
Var R : Integer Reg;
Var I : Number;
Cadr BO;
    R := C[1];
 Endcadr;
 Cadr B;
    For i:=2 To 4 Do
            Begin
                A[i] := (B[i]+R)*C[i];
                R := B[i];
             End;
 Endcadr;                                              (6.1)
```

At first sight, data access is incorrect: one and the same variable is read and written inside the cadr. However, the identifier *R* declares the register variable, which is specially used for implementation of delays. The value of the register variable will be available if and only if some value is written in it, or at the beginning of the cadr execution.

So, the assignment operator *R:=B[i]*, contained in the fragment of the program (6.1), is carried out concurrently with the assignment operator *A:=(B[i]+R)*C[i]*; here the values *B[i-1]* correspond to *R*.

For the example, given in Fig. 6.8b, buffering may be similarly represented in an explicit form.

```
Var A : Array Integer [10 : Stream] Mem;
Var R1, R2 : Integer Reg,
Var I : Number;
    Cadr Cadr1;
          R1:=A[1];
          R2:=A[2];
     Endcadr;
     Cadr Cadr2;
     For i=3 To 10 Do
     Begin
          C:=A[i]+R1+R2;
          R1:=A[i];
          R2:=R1;
     End;
     Endcadr;
```

Since a one cycle delay corresponds to a register variable, then *A[i–1]* corresponds to *R1*, and *A[i–2]* corresponds to *R2*, which is a delayed *R1*. The example shows the special case of use of register variables.

Use of the rule of single substitution in the cadr leads to difficulties with organization of recursive computations. Since recursive computations in the cadr are available only to register variables, then it is necessary to have special programming skills in the COLAMO language programming. For instance, the fragment

```
Var A, B : Integer Mem;
    Cadr Cadr1
        A:=B;
        B:=A;
    Endcadr;
```

is semantically incorrect, because it does not correspond to the principles of data stream programming. The variables *A* and *B* are to be concurrently read and written, but memory variables allow only one memory access in the cadr. An apparent sequence of instructions is not fulfilled here, because all instructions of the cadr are performed concurrently. The fragment

```
Var A, B : Integer Reg;
Cadr Cadr2
    A:=B;
    B:=A;
Endcadr;
```

is semantically correct. After execution of the program, *A* and *B* will be modified. This unusual action may be explained by concurrent execution of all instructions of the cadr.

Only one access to the memory variable may be performed in the cadr, therefore, the program examples, given below, are incorrect:

```
Var A, C, D: Array Integer [100 : Stream] Mem;
Var I : Number;
    Cadr B;
        For i:=1 To 100 Do
        A[i]:=f1(C[i]);
        For j:=10 To 100 Do
        A[i]:=f2(D[j]);
    Endcadr;
```

In this example the application programmer tried to write data into one and the same memory cell twice. It is violation of the single-assignment rule, and in this case, the translator will yield a syntax error message. Since there is no information dependence between the i-counter loop and the j-counter loop, the both structures may be executed concurrently. The text of the fragment must be corrected.

In the next example the application programmer made a semantic mistake of another type:

```
Var A: Array Integer [100 : Stream] Mem;
```

```
    C: Array Integer [100 : Stream] Mem;
    D: Array Integer [100 : Stream] Mem;
Cadr B;
    For i:=1 To 100 Do
    A[i]:=f1(C[i]);
    For j:=1 To 100 Do
    D[j]:=f2(A[j]);
Endcadr;
```

Here the application programmer tries to perform two processes in one cadr: in the i-counter loop the data array *A* is being written into memory, and in the j-counter loop *A* is being read from memory. Both of these loops are performed independently: it does not matter what the values of *A* are to be read – the initial or the one, modified by the function f_2. The translator will yield a syntax error message.

It is more difficult to understand the fragment, that implements two loops of reading in one cadr:

```
Var A: Array Integer [100 : Stream] Mem;
    C: Array Integer [100 : Stream] Mem;
    D: Array Integer [100 : Stream] Mem;
Cadr B;
    For i:=1 To 100 Do
    C[i]:=f1(A[i]);
    For j:=1 To 100 Do
    D[j]:=f2(A[j]);
Endcadr;
```

The distributed memory controller cannot perform two independent processes with the memory variables; the translator will yield a warning message. The result of the program execution is non-predictable. The given fragment ought to be transformed into the fragment with a single process:

```
Cadr B;
    For i:=1 To 100 Do
    Begin;
    C[i]:=f1(A[i]);
    D[j]:=f2(A[j]);
    End;
Endcadr;
```

The fragment may also be represented as two cadrs:

```
Var A: Array Integer [100 : Stream] Mem;
    C: Array Integer [100 : Stream] Mem;
    D: Array Integer [100 : Stream] Mem;
Cadr B1;
```

```
    For i:=1 To 100 Do
    C[i]:=f1(A[i]);
Endcadr;

Cadr B2;
    For j:=1 To 100 Do
    D[j]:=f2(A[j]);
Endcadr;
```

Different cadrs, i.e. independent computational structures, may perform various processes.

It should be noted, that computational restriction is valid only for the arrays of Stream type. If the array is a vector, considered as a set of independent data channels, then the restriction is not valid, because each vector item is an independent distributed memory controller.

Programming will be simpler, if one cadr contains several reading processes with different constant values, as it is written in the following example for the items of the stream array *A*:

```
For i:=1 To n Do
        f(A[i], A[i+n], A[i-n]);
```

Here *n* and *m* are constants or variables, not contained in the cadr. In addition, it is possible to use indirect addressing f(A[i], C[A[i]], C[A[i-n]], C[A[i+n]]).

Branching in programs is performed by the statement of a conditional jump IF. Regarding the cadr, the IF-statement may be inner or outer. The format of the statement of a conditional jump:

If *q* Then *a* [Else *b*];

where *If, Then, Else* are the keywords;

q is some logical expression;

a, b are the lists of instructions, bounded by the keywords *BEGIN*, *END*, and performed according to *q*.

If *a* or *b* consists of one instruction, the keywords *BEGIN*, *END* may be omitted. Nested statements of a conditional jump are also available:

If *q*1 Then If *q*2 Then If *q*3 .... Else *d*;

The semantics of the outer statement of a conditional jump is similar to the semantics of that that is used in procedural high-level programming languages.

The inner statement of a conditional jump is performed as follows: the lists *a* and *b*, that represent the subgraphs of the cadr graph, are performed concurrently. The results of execution of the both lists enter the switch, that operates according to the logical expression *q*. In other words, the inner statement of a conditional jump is a structural analog of a conditional jump. Owing to the statement of a conditional jump, used in the cadr, it is possible to naturally implement parallelism of the algorithm, and to obtain the result as fast as possible.

Figure 6.8 shows an example of the inner statement of a conditional jump.

```
Var A,B: Array Integer [100 : Stream] Mem;
Var I : Number;
Cadr m;
  For I:=0 To 99 Do
    If A[I]>0 Then
      B[I]:=Sin(A[I]);
    Else
      B[I]:=Exp(A[I]**2);
 Endcadr;
```

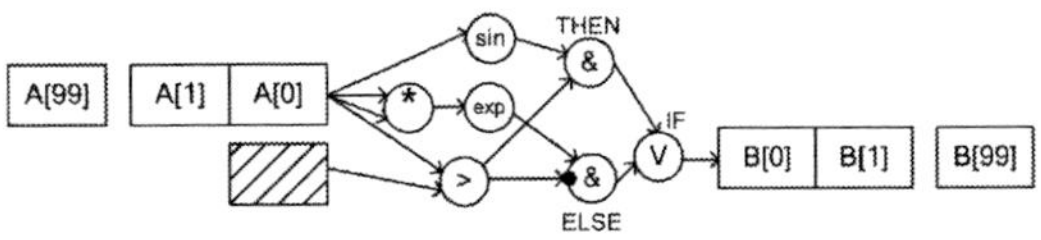

Figure 6.8. Use of the inner statement of a conditional jump.

The considered example may be written in another way:

```
Var A,B : Array Integer [100 : Stream] Mem;
Var I : Number;
    For I:=0 To 99 Do
    If A[I]>0 Then
            Cadr b1;
            B[I] := Sin(A[I]);
            Endcadr;
    Else
            Cadr b2;
            B[I] := Exp(A[I]**2);
    Endcadr;                                          (6.2)
```

Here the statement of a conditional jump is placed outside the cadr, and the task is implemented with the help of two computational structures. During execution of the program one of the structures is being initialized according to the logical expression *A[I]>0*. In comparison with the previous one, the current implementation is slower, but it requires less number of processing elements for the task. The abovementioned reasoning only refers to the cases, when the inner statement of a conditional jump contains both alternative branches *THEN* and *ELSE*. If the inner statement *IF* contains only *THEN*-branch, then translation will be performed in another way. Depending on the types of the variables, the statement *If q Then a=b* will be translated into different variants of computational structures. The previous value of the switching variable does not exist, hence, the fragment

```
Var A : Integer com;
Cadr L1;
    A := If q Then b;
Endcadr;
```

is similar to A=q&b.

If *q* is not fulfilled, then the switching variable receives a logical zero. While implementing parallel computations by means of cadrs, the self-contained subgraphs of the cadrs may be isomorphic. In this case, use of the subcadrs, that are structural analogs of subroutines, is more rational. The statement of the subcadr declaration has the following format:

Subcadr Name (F) <List_of_statements> Endsubcadr;

where *Subcadr*, *Endsubcadr* are the keywords;

Name is the name of the subcadr;

f is the list of formal parameters, bounded by round parenthesis.

The subcadr is called by the name which is followed by the list of actual parameters, bounded by round parenthesis. If the subcadr is some function, then the statement *Var* of the data declaration must contain the subcadr name and the variable type. Recursion, including hidden recursion, is not allowed in subcadrs, i.e. it is forbidden to call the subcadr from itself, or to call the sequence of subcadrs, that ends with the call of the current subcadr, from the current subcadr. To prevent subcadr side effects, only switching variables are to be used in the subcadr. When the subcadr is being called, the data exchange channels, that correspond to the actual parameters, are combined with the formal parameters of the subcadr. In general, subcadrs are declared and described as the cadrs. It is forbidden to use the transfer-of-control statements and the statements of computational structure description in subcadrs.

Figure 6.9 shows the program with subcadrs, and the graphs of the computational structures, that will be obtained as a result of translation of the program.

Mismatch of the types of the formal and actual parameter names may lead to errors of the program execution.

```
Subcadr Pirsum (In: A,K) : Integer;
Var A: Array Integer [N: Vector] Com;
Var C: Array Integer [N/2: Vector, N/2: Vector] Com;
Var T,L: Integer Reg;
Var M: Integer Com;
Var I,J: Number;
M:=log2(N);
For I:=1 To K/2 Do
  Begin;
    C[1,I]:=A[T]+A[T+1];
    T:=T+2;
  End;
L:=2;
For J:=2 To M Do;
  Begin;
    For I:=l To K/2 Step l Do
      C[J,I]=C[J-1, I]+C[J-1, I-L/2];
      L=L*2;
  End;
Result := C[M, K/2];
Endsubcadr;
```

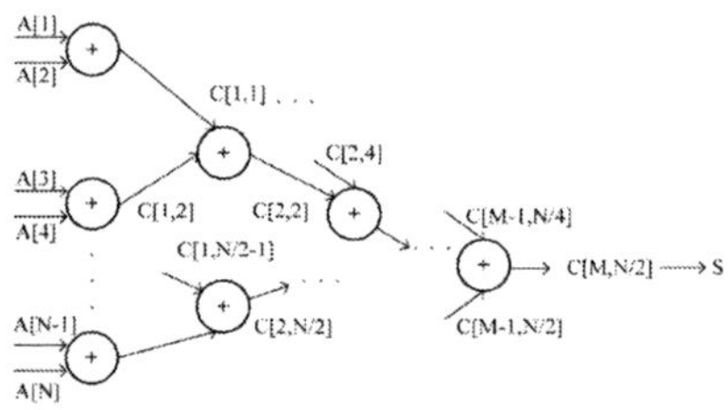

Figure 6.9. The program and the flow-graph of the subcadr of addition of array items.

If the subcadr has several outputs, then it is forbidden to substitute one and the same name of the actual parameter for the formal parameters.

It leads to a semantic error, i.e. it breaks the rule of single substitution in the cadr. An example of use of the subcadrs is given below.

```
Var A,B,C,D,E,F,G : Real Mem;
Var K: Integer Com;
Subcadr BETA (In: A1,B1; Out : C1,D1);
Var A1,B1,C1,D1: Integer Com;
    C:= A1+B1/2; D1:=A1-B1*2;
Endsubcadr;
Subcadr ALFA (In : A2,B2,C2);
Var A2,B2,C2: Integer Com;
    C2=A2*3.14+B2*B2;
Endsubcadr;
Cadr A1;
    Beta (A*2,4-B,D,C);
    Alfa(D*C,E*E,F);
Endcadr;
Cadr A2;
    Beta (F,F,G,K);
    Alfa (K,E,A);
Endcadr;
```

The example demonstrates two particular features of subcadr use, that should be analyzed in details: dynamic organization of an array and substitution of a parallel parameter. According to dynamic organization of an array, the array length in the subcadr *pirsum* is declared as a variable: *VAR A[N]*, where *N* is the parameter, specified by the application programmer. It is evident, that the array may be both VECTOR and STREAM. Substitution of the stream array is similar to passing the array name as a subroutine parameter in traditional programming languages. The list of parameters includes the array starting address and the number of array items, specified by the application programmer. Besides, the address of the distributed memory channel, that stores the array, is also required. So, calling the subcadr, the application programmer must specify the list of actual parameters, that consists of the stream address in the form of an ordered pair *<Ak, Ag>*, where *Ak* is the address of the distributed memory channel, *Ag* is the data starting address. During processing of the data stream, *Ak* remains constant and *Ag* is varying during access to the array items.

If the actual parameter of the called cadr is a vector array, then *Ag* remains constant, and *Ak* is varying during access to the array items. The address *Ak=<Akb, Akk>* of the distributed memory channel consists of two components, such as the address *Akb* of the distributed memory channel itself, and the set of addresses *Akk* for the multicascade switch, that provides access of the required distributed memory channel to the computation field of processing elements of the computational structure. In addition to subcadrs, the language contains traditional subroutines. In general case, a subroutine is described as follows

Subroutine Name (F) P Endsubroutine;

where *Subroutine*, *Endsubroutine* are the keywords;

F is the list of the subroutine formal parameters;

P is the list of the subroutine instructions.

The subroutine may contain several cadrs and control operators. The subroutine is called by the name that is followed by the list of actual parameters.

Implementation of parallel computations by means of cadrs may require use of a fixed computing structure, that is not re-loaded during program execution and cadr changing. Such computational structure is to be a self-contained isomorphic graph. The fixed computational structure may be implemented with the help of a structure LET, that is the structural analog of the subcadr, supported by the COLAMO language. The structure LET is declared by the following statement:

LET <Name> (<List_of_parameters>) <List_of_instructions> ENDLET;

In contrast to subcadrs, the structure LET cannot be called from cadrs or subcadrs, it is an independent local computational structure. When the structure LET is declared, the data exchange channels are combined with its formal parameters.

All the variables of the structure LET except those, declared as the formal parameters, are local and have no relation to the variables, declared in cadrs.

Recursion, including hidden recursion, is not allowed in the LET structures, i.e. it is forbidden to call the LET structure from itself, or to call a sequence of subcadrs, cadrs or subroutines from the LET structure. It is recommended to use switching variables in the LET structure. It is forbidden to use the transfer-of-control statements and the statements of computational structure description in the LET structure. During program execution the LET structure may be cancelled by the STOPLET statement, which has the following format:

STOPLET <Name_of_LET_structure>;

The COLAMO language contains the statement MCADR, intended for effective control of cadr execution. Using the statement MCADR, the application programmer can describe all declared cadrs as the single array, whose items may be accessed by the array index. The statement MCADR has the following format:

MCADR : <Name> [<List_of_cadrs>];

Owing to the MCADR statement, the application programmer may change cadrs in any sequence. The cadrs, used in MCADR, are to be declared before declaration of the MACDR statement. The MCADR statement may be used both inside and outside the cadr. If it is used outside the cadr, then the called cadrs are processed successively. If the MCADR statement is used inside the cadr, all called cadrs are united into the single cadr, and are executed concurrently. The example of the MCADR statement use is given below.

```
Cadr Cadr1
............
Endcadr;

Cadr Cadr2
............
Endcadr;
.............
Cadr CadrN
............
Endcadr;

Mcadr : MasCadr [Cadr1, Cadr2,....., CadrN];
// Successive processing of the called cadrs
    For I := 0 To 5 Step 2 Do
    MasCadr[I];
// Parallel processing of the called cadrs
Cadr Cadr0
    For J := 0 To 5 Step 1 Do
    MasCadr[J];
Endcadr;
```

The high-level programming language of reconfigurable multipipeline computing structures with structural-procedural organization of computations provides effective implementation of any type of parallel processing. Use of implicit description of parallelism, virtual communication systems and distributed memory simplifies programming. The language reflects all features of structural-procedural organization of computations and provides effective development of MPCS application software.

COLAMO Translation

Translation of COLAMO-programs into ARGUS-programs consists in generation of two components: a structural-procedural ARGUS-program, executed on a reconfigurable structure, and a control program, executed on a host-computer.

During translation, the translator analyses the source code of the parallel program and identifies syntax and semantic errors. After analysis of the program source code, the translator generates four components: structural, stream, procedural and control.

When all these components are successively generated, they are united into the parallel structural-procedural ARGUS-program. The structural and stream components represent the cadr form of the task. The procedural component adjusts the reconfigurable structure for execution of various cadrs. The control component is an all-sufficient executable program (exe-file), that loads and unloads program and numerical information into the reconfigurable structure from the host-computer and controls the task execution. Figure 6.10 shows the flow-chart of the translator algorithm.

The structural component of the parallel-pipeline program describes the hardwarily implemented information subgraphs of the task. While generating the structural component, each operation or macrooperation of the cadr is mapped on separate computational resources.

Hardware resources, allocated for the operation (macrooperation), cannot be used for implementation of other operations during execution of the cadr. If the program structures of the cadr, written in the COLAMO language, contain loop statements with loop parameters, that are used as indices of arrays with parallel type of access (VECTOR arrays), then the loop bodies are being multiplied in the reconfigurable structure according to the loop parameters. It should be noted, that during implementation of the conditional jump statement, all alternative branches of calculations and choice units will be hardwarily implemented within the cadr. Figure 6.11 shows the algorithm of generation of the structural component.

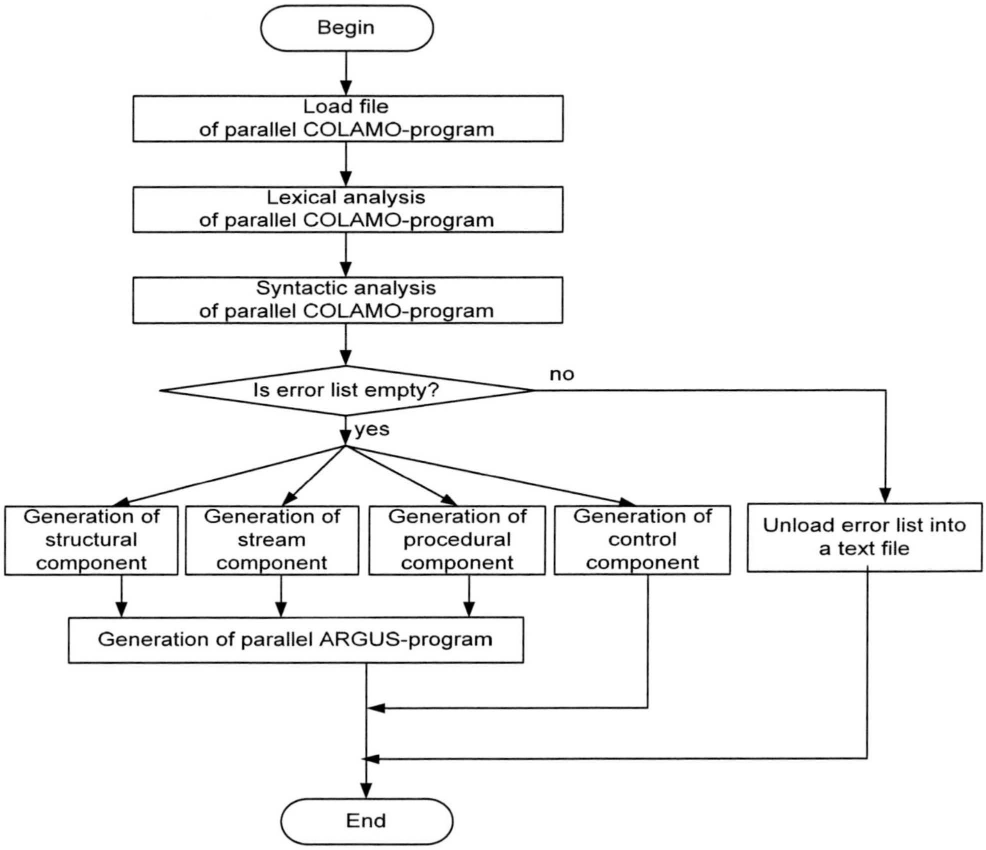

Figure 6.10. The outline algorithm of the COLAMO translator.

Figure 6.12 shows an example of transformation of some parallel program into the equivalent information graph.

A vertex, designated as a circle, corresponds to an input/output vertex of the graph, which, in its turn, correlates with distributed memory. The graph vertices, designated as light circles, are arithmetic-logical operations, implemented in processing elements. The graph vertices, designated as rectangles, signify delays of information of the specified numbers of cycles. The stream component of the structural-procedural program provides passing of data streams through the computational structure, specified by the structural component, and conflict-free parallel reading and writing from/into distributed memory channels.

Figure 6.13 shows the algorithm of generation of the stream component.

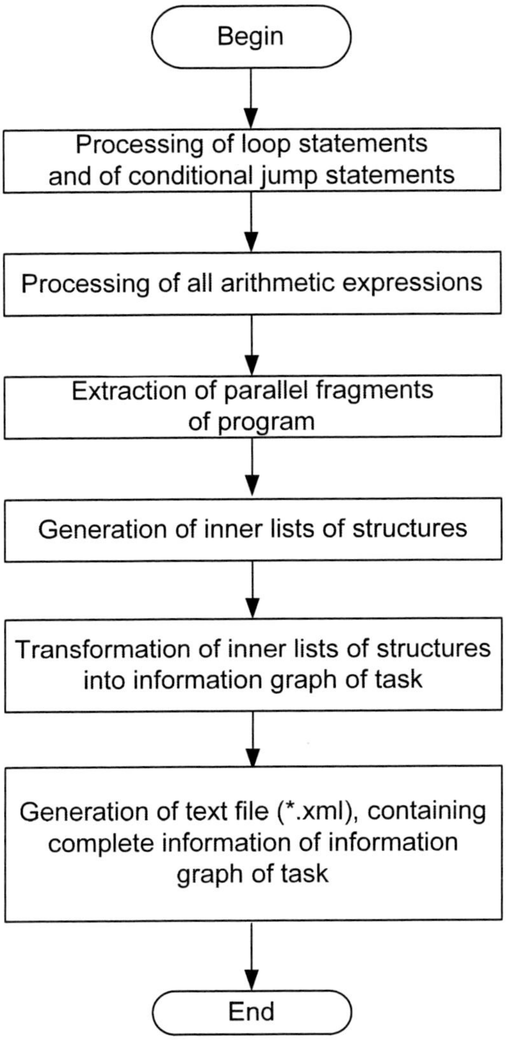

Figure 6.11. The algorithm of generation of the structural component.

```
Var a,b,c,d,e,g : Array Real [2 : Vector 10 : Stream] Mem;
Var I, J : Number;
        Cadr Cadr1
          For I := 0 To 1 Do
            For J := 0 To 9 Do
             Begin
               A[I,J] := B[I,J] + B[I,J + 3];
               D[I,J] := C[I,J] + (B[I,J] * G[I,J]);
             End;
        Endcadr;
```

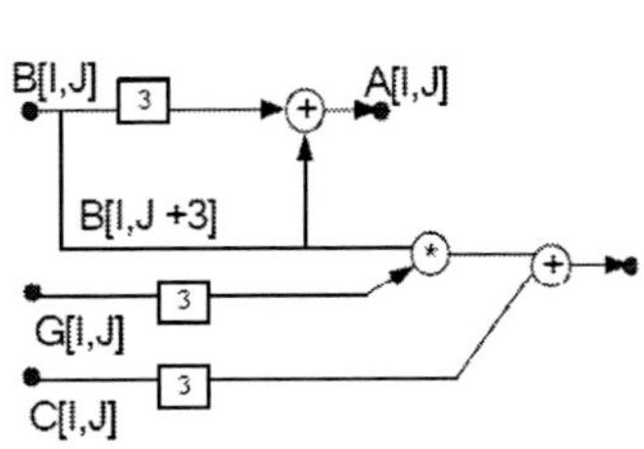

Figure 6.12. A program and its equivalent flow-graph.

The procedure of data stream extraction consists in sorting of the information vertices of the graph according to their connection with memory channels, and according to the processing sequence.

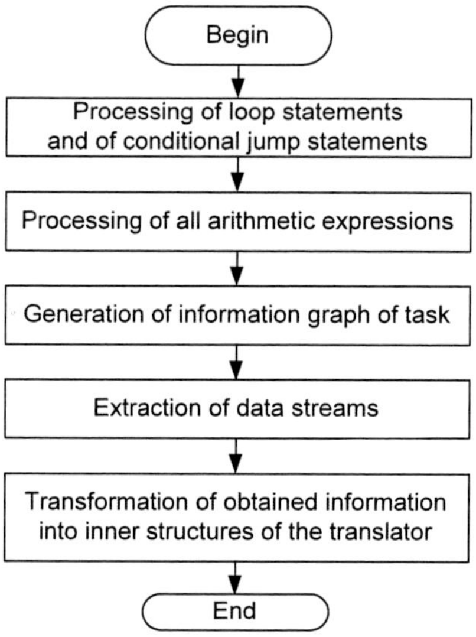

Figure 6.13. The algorithm of generation of stream component.

The following example shows a fragment of the stream component, generated as a result of translation of some parallel program.

VAR a,b,c,d,e,g:Array Integer	Var A,B,C,D,G : Param Adress Mem;
[2: Vector, 10: Stream] Mem;	Cadr Cadr1
Var I, J : Number;	DMC[0.B1] Read B[0] Step 1 Repeat 12;
Cadr Cadr1;	DMC[0.C1] Read C[0] Step 1 Repeat 12;
For I := 0 To 1 Do	DMC[0.G1] Read G[0] Step 1 Repeat 12;
For J := 0 To 6 Do	DMC[0.A1] Write A[0] Step 1 Repeat 12;
Begin	DMC[0.D1] Write D[0] Step 1 Repeat 12;
A[I,J] := B[I,J] + B[I,J + 3];	
D[I,J] := C[I,J] + (B[I,J] * G[I,J]);	DMC[0.B2] Read B[1] Step 1 Repeat 12;
End;	DMC[0.C2] Read C[1] Step 1 Repeat 12;
Endcadr;	DMC[0.G2] Read G[1] Step 1 Repeat 12;
	DMC[0.A2] Write A[1] Step 1 Repeat 12;
	DMC[0.D2] Write D[1] Step 1 Repeat 12;
	Endcadr;

The statement *Read* reads a data stream from a memory controller into a special-purpose computing structure. The statement *Write* writes the data stream.

The procedural component of the parallel program is intended for description of the program execution order. According to the rules of the ARGUS, the program execution order is defined by the cadr sequence of the program.

Translation of the parallel program into the procedural component consists in transformation of the COLAMO intercadr statements.

Figure 6.14 shows the algorithm of generation of the procedural component.

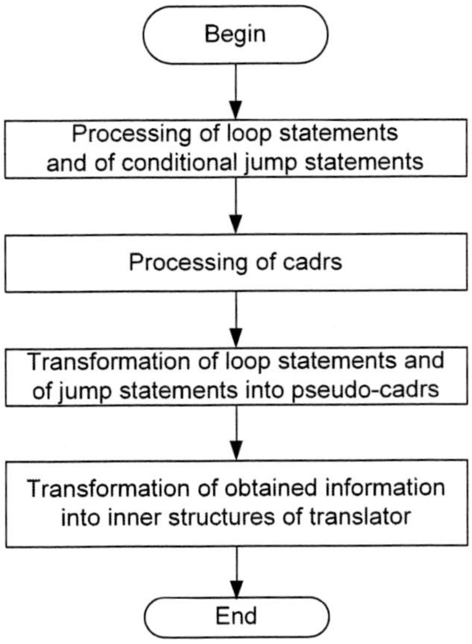

Figure 6.14. The algorithm of generation of the procedural component.

The following example shows some fragment of the procedural component, generated as a result of translation of the parallel program.

VAR a,b,c,d,e,g : Array Integer [2 : Vector 10 :	Program Promer;
Stream] Mem;	Var A,B,C, D : ParamAddress MEM;
Var I : Number;	Var N, K, M : Number;
For I:= 0 To 5 Do	Define N = 5; Define K = 6; Define M = 1;
Begin	Label M0, M1;
Cadr Cadr1;	M0 : Cadr Cadr1
A[0,0] := B[0,0] + B[0,3];	<Stream statements>
D[0,0] := C[0,0] + (B[0,0] * G[0,0]);	Endcadr;
Endcadr;	Cadr CadrIF
If A > 0 Then	If Mem Goto M1;
Cadr Cadr2;	Goto M2;
A[0,0] := B[0,0] + B[0,3];	Endcadr;
D[0,0] := C[0,0] + (B[0,0] * G[0,0]);	M1 : Cadr Cadr2
Endcadr;	<Stream statements>
End;	Endcadr;
	Cadr CadrFOR
	M2: Loop 4 Goto M0;
	Endcadr;

The pseudocadr *CadrIf* implements the statement of a conditional jump (IF).

The pseudocadr *CadrFor* implements the outer loop statement of the program.

The control component of the parallel program is a program, written in a programming language of a monoprocessor computer, for instance, in the Pascal. The result of the parallel program translation into the control component is an executable program.

The algorithm of translation of the COLAMO-program into the control component consists in translation of the data load/unload statements and cadr structures into the

PASCAL-program. The result file is compiled by traditional language compilers, and the executable program is generated.

The following example shows a fragment of the control component, generated as a result of the parallel program translation.

Program Primer;
Var A,B,C,D:Array Integer [N: Stream, K: Stream]:Mem;
Var I, J : Number;
ReadFile (A, FF_STR, "C:\Input.txt");
 	
Cadr Cadr1;
For I := 0 To N-1 Do
For J := 1 To K-1 Do
D[I, J] := (A[I,J] + A[I,J-1]) + 5;
Endcadr;
 	
WriteFile (D, FF_STR, "C:\Output.txt");
End_Program.
	Program 1;
	Uses SysUtils;
	TempLongword : PLongWord;
	Data64: PInt64; InputFile: File; OutputFile: File; Str: String;
	Begin
	LoadFile(D:\Result\OutFile.out);
	{Loading data from file into distributed memory}
	// Start of all DMC
	RunAllDMC(0, 16);
	// Processing STOP
	Repeat
	ReadCOReg(0, 2, TempLongword);
	bit5 := Integer(TempLongword^) and 32;
	Until Bit5 = 32;
	{Data unloading from distributed memory into file}
	End;

6.4. Design Environments of Parallel-Pipeline Programs

The Development Environment of Parallel Programs

The development environment Argus IDE is intended for creation of parallel programs, using the language of structural-procedural parallel computations [17]. One of particular features of the environment is a multiwindow user interface, based on two types of windows (see Fig. 6.15).

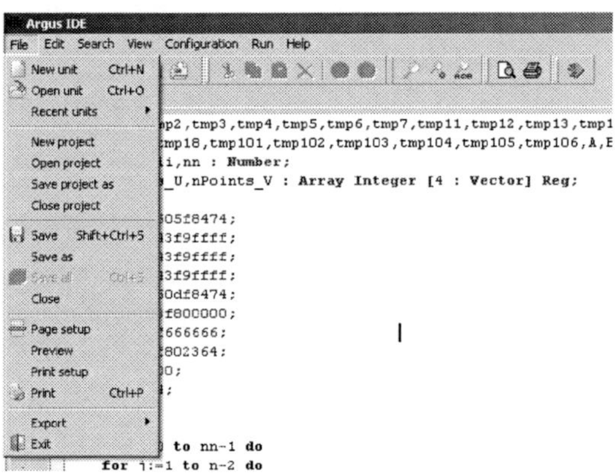

Figure 6.15. The menu “File”.

The first type windows are windows of a special-purpose text editor, intended for input of parallel programs. The second type windows are auxiliary windows, that provide access to various items, that control development of parallel programs.

The control items of program development are the following: a message list, an object model of a program, a property table of program objects, and a log of events.

The help system contains information about the principles of the development environment functioning, and about the structural-procedural parallel programming language Argus.

The main aim of the Argus IDE is to hasten development of parallel programs for computing structures of various configuration, owing to integration of all necessary tools into the single environment, that provides interactive creation of source codes and executable files of parallel solutions. The user interface contains:

1. *Menus and toolbars*. The main menu and toolbars provide access to the main functions of the development environment (see Fig. 6.17).

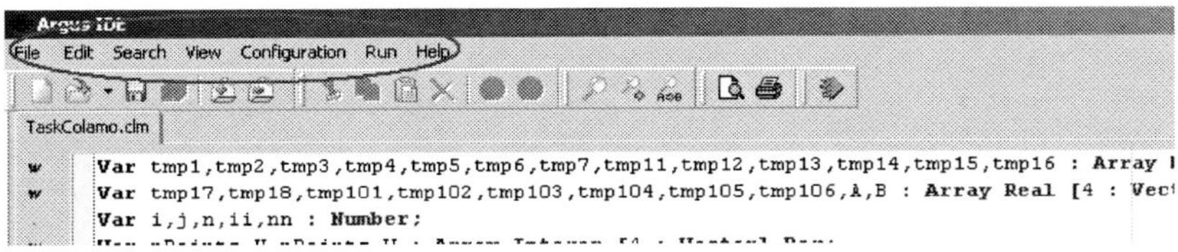

Figure 6.17. The main menu of the environment.

2. *The special-purpose text editor*. The text editor is used for development of programs in the structural-procedural computations language Argus (see Fig. 6.18).

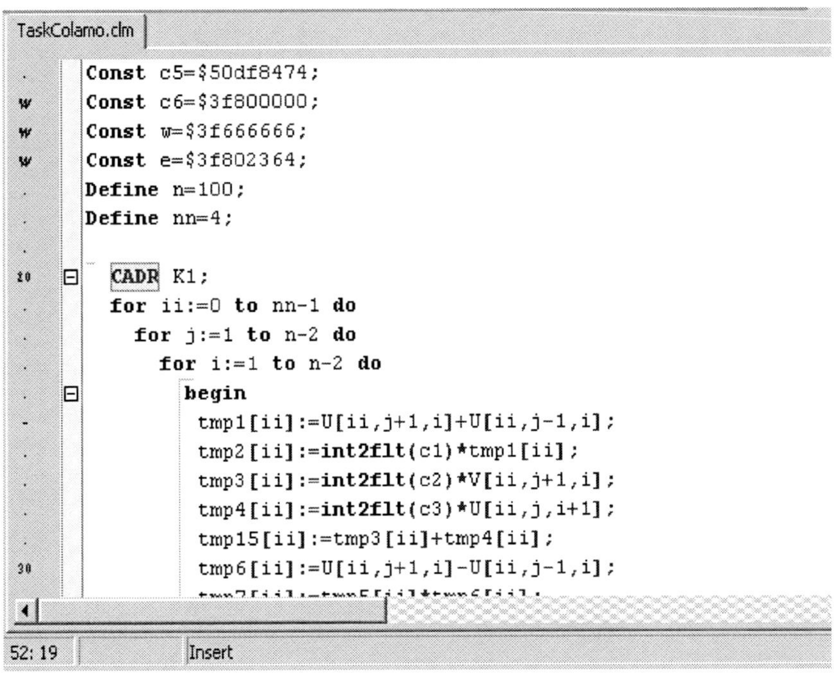

Figure 6.18. The special-purpose text editor.

The distinctive features of the text editor:

- source code attributes (such as highlighted keywords and errors, code folding, etc.), corresponding to the ARGUS syntax;
- interactive use of the control items for quick context navigation in the source code;
- printing the source code with all the source code attributes;
- search and replacement of lines.

3. *The fields of the editor*. The fields indicate various attributes of the source code lines, such as breakpoints, warnings or errors, the operator that is being executed.

4. *The object tree of the parallel program*. The object tree of the parallel program is generated during translation, and gives a compact hierarchical representation of all objects of the parallel program (see Fig. 6.19).

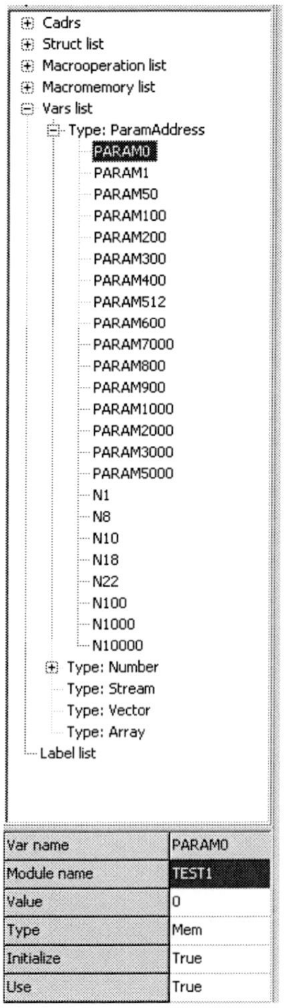

Figure 6.19. The object tree of the parallel program.

In addition, the object tree provides navigation in the text of the parallel program (see Fig. 6.20).

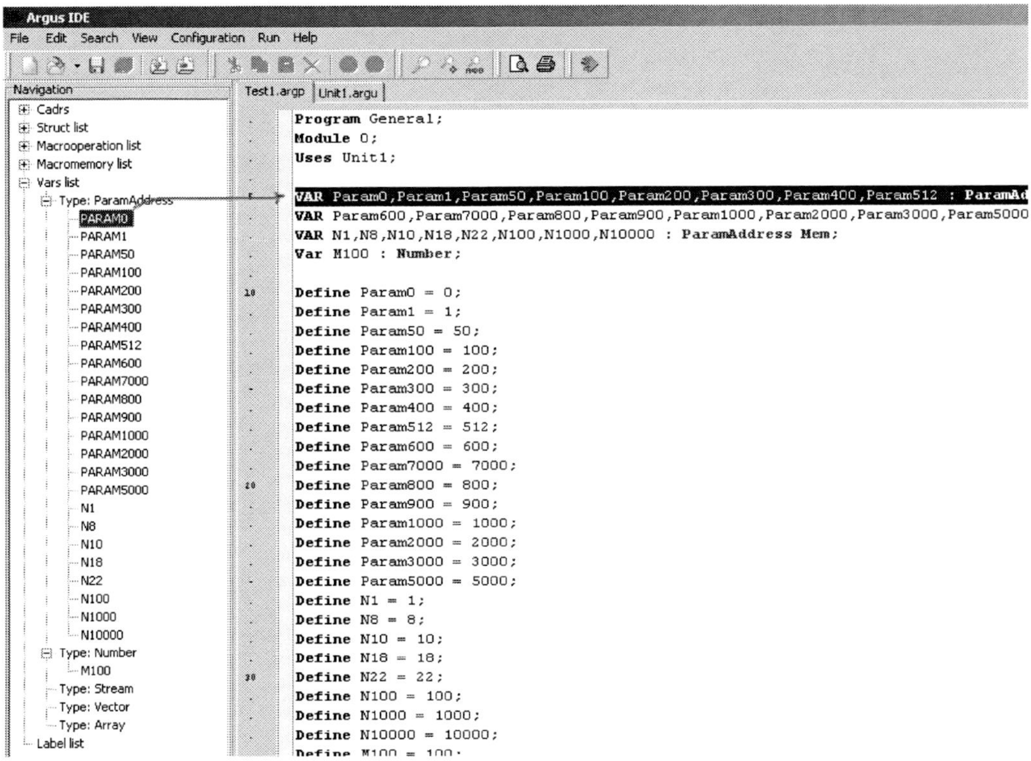

Figure 6.20. Navigation in the text of the parallel program by means of the object tree.

5. *The property table of program objects*. The property table contains information about program objects, and corresponds to the selected item of the object tree of the parallel program (see Fig. 6.21).

Var name	PARAM0
Module name	TEST1
Value	0
Type	Mem
Initialize	True
Use	True

Figure 6.21. The property table of program objects.

6. *The information panel of the program text editor*. The panel shows information about the cursor position, the insert mode indicator, the text modification indicator (see Fig. 6.22).

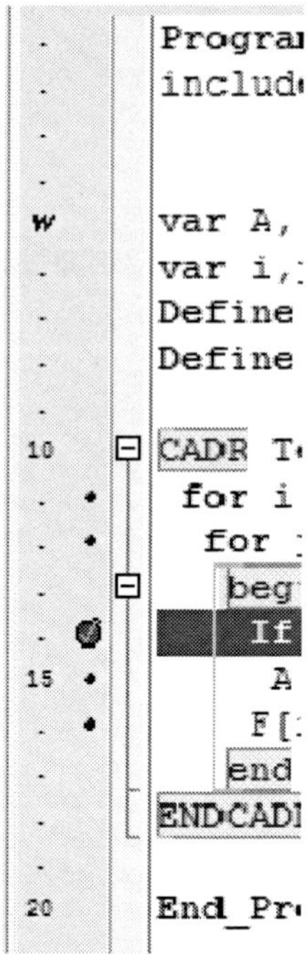

Figure 6.22. The information panel of the program text editor.

7. *The list of messages*. The list of lines, generated during translation, that contains information about errors and warnings. The list of messages may be used for the source code navigation (see Fig. 6.23).

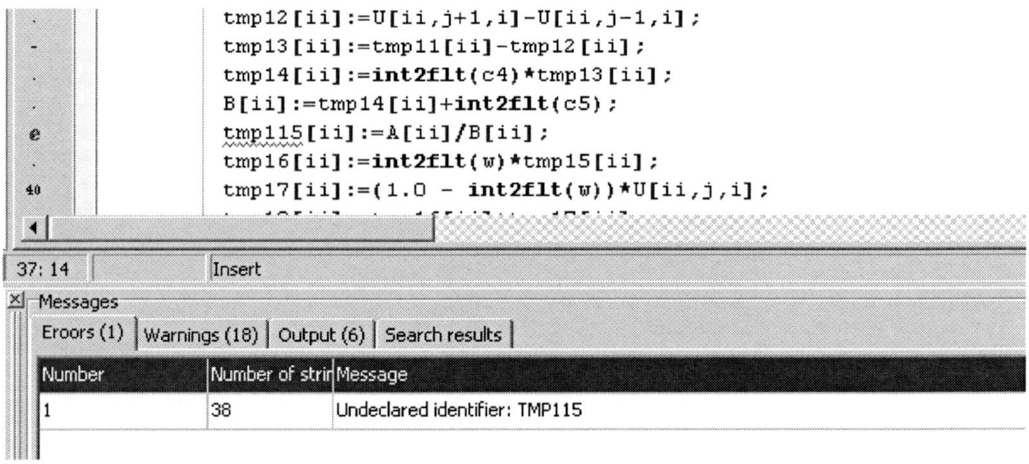

Figure 6.23. The panel "List of messages".

Using the window of messages, the application programmer can navigate in the source code and move to the correspondent lines of the program. The fields of the text editor reflect the current state of each line. According to content, the lines may contain errors and warnings. Auxiliary windows, such as the window of messages, and the object model tree, may be placed at the discretion of the application programmer. The interface of the development environment is standard and similar to that of Microsoft Visual Studio or Borland Delphi. The parallel program source code may be printed with keeping all attributes of description of structural-procedural computations.

The Argus IDE for development of programs and program components in the language of structural-procedural computations provides simultaneous work with several source files by page switching.

The application programmer can control program development by the main menu and toolbars of the environment. The menus “File”, “Edit”, “Search”, “View”, “Help” contain items that are standard for such environments. Compilation of source files and run of executable files is performed by the menu “Run” (see Fig. 6.24).

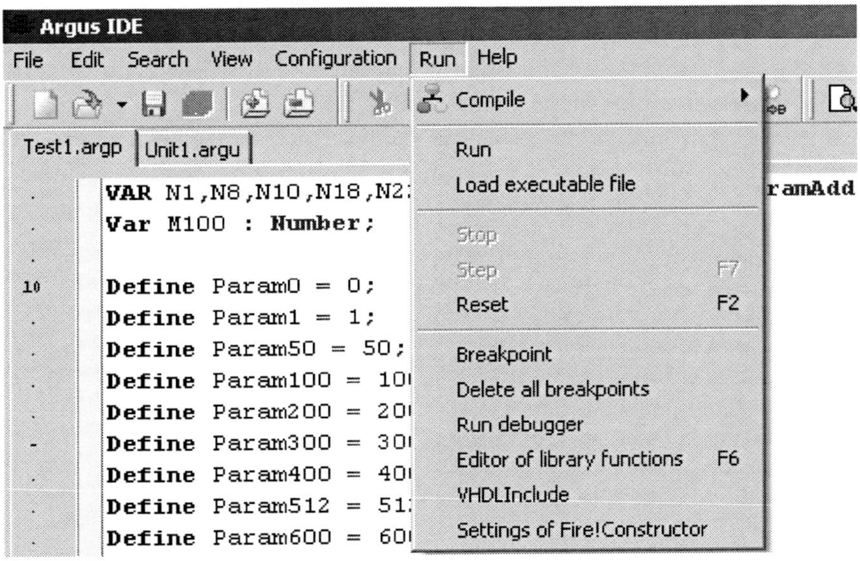

Figure 6.24. The menu “Run”.

The menu contains the following items:

1. *"Compile"*. The compile procedure consists of the following steps:

- saving source files;
- the current file is considered as the main file;
- translation of the current file;
- generation of the executable file, if translation is a success, i.e. if the program contains no errors;
- activation of the message window, if the message list is not empty after translation;
- generation of the object model of the parallel program.

2) *"Run"*. The procedure of execution of the current parallel program on the reconfigurable structure consists of the following steps:

- compilation of the current program, if the executable file does not exist;

- compilation of the current program, if the data of the executable file last modification is older than the data of the source file;
- if the executable file is generated, then it is loaded and executed on the reconfigurable structure.

3) *"Load"*. The executable file of the current parallel program is being loaded into the reconfigurable structure.

During translation, after successful syntax analysis and generation of the program object model, the translator calls a subsystem, that maps the graph of the task on the architecture of the reconfigurable structure and specifies loading of hardware resources of its basic modules according to the structure of the task (see Fig. 6.25).

The subsystem exempts the application programmer from sufficiently complicated and laborious explicit programming of the communication structure.

The structural-procedural programming language Argus contains structures, that may be used both for direct programming of connections between basic module elements, and for circuit specification of virtual components of parallel structural-procedural programs, that will be configured automatically.

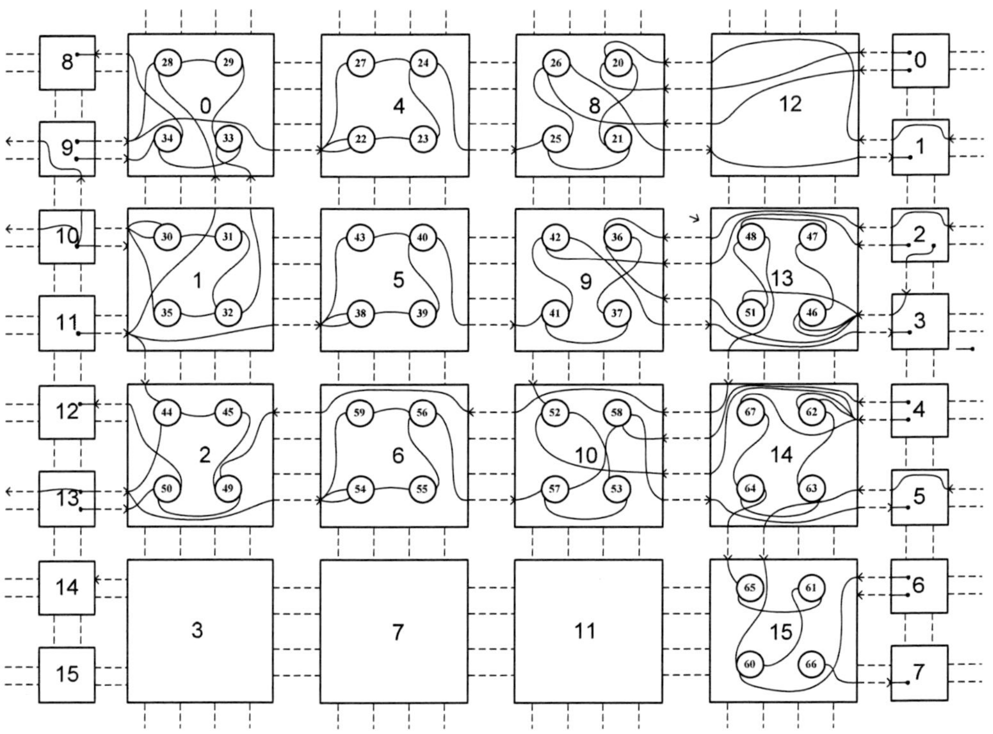

Figure 6.25. Mapping of the information graph of the task on the basic module of the MPCS with the orthogonal communication structure.

The Development Environment of Computing Structures

The development environment of reconfigurable structures Fire!Constructor was created for visual image and mapping of information graphs of parallel-pipeline programs on hardware resources of reconfigurable structures.

The Fire!Constructor IDE represents information graphs of parallel programs as flow-graphs. The language of flow-graphs is used for equivalent description of the parallel program components, whose text, written in the structural-procedural parallel programming language, is not so clear as graphic representation.

A flow-graph is an alternative auxiliary form, that is used for description of the structural component of the task, as opposed to the languages Argus and COLAMO, which are used for description of the procedural component. With the help of the text editor, the application programmer sequentially prints text of parallel programs. With the help of the graphics editor, the application programmer sequentially inputs graphic components of parallel programs, i.e. objects of flow-graphs (see Fig. 6.26). The work file of the text editor contains lines of program code, written in the Argus or COLAMO. The work file of the graphics editor or the file of flow-graphs consists of pages with graphic components of parallel programs.

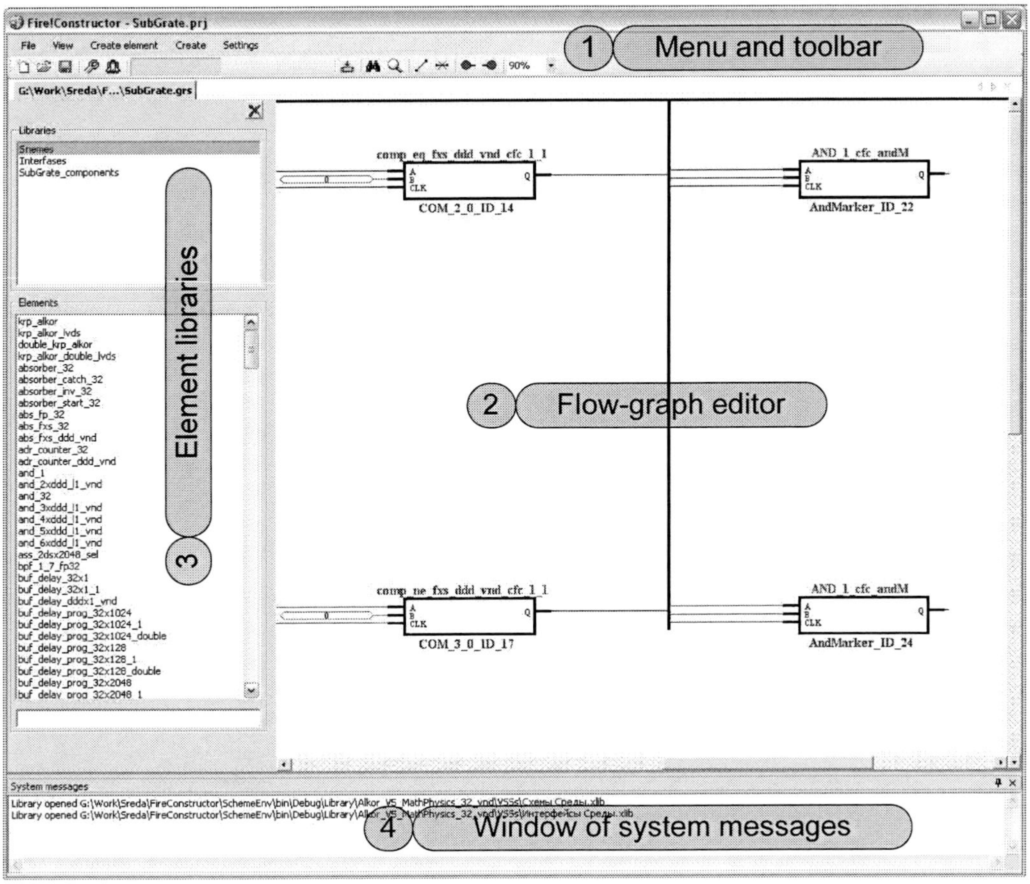

Figure 6.26. The graphics editor of source units of parallel programs.

The flow-graph elements are computational units and interfaces. The outputs of the computational units may only be connected with the interface inputs, when the interface outputs may only be connected with the inputs of the computational units.

Pictorial symbols of the computational units and interfaces, such as rectangles, ellipses and other geometrical objects, simplify readability of flow-graphs. Figures 6.27 *a, b* show the pictorial symbols, that represent some computational unit and interface in the Fire!Constructor.

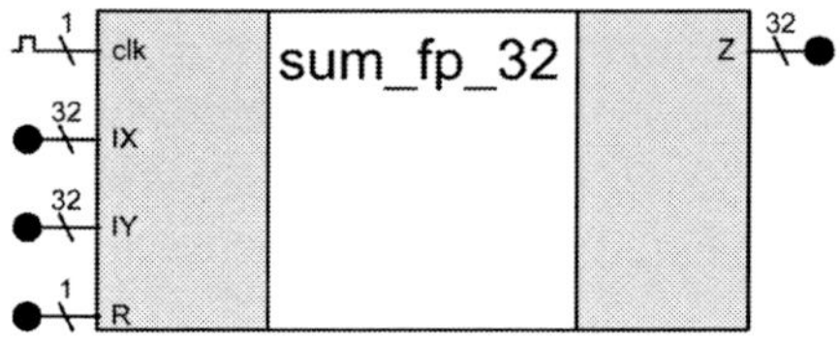

a

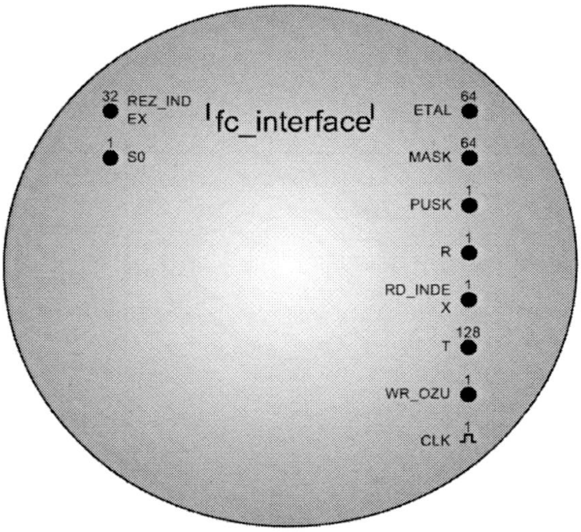

b

Figure 6.27. Pictorial symbols of some computational unit and interface in the Fire!Constructor.

Graphic objects of flow-graphs are assembled from various graphic primitives, such as lines, text fields, geometric figures, etc. A special class of open-ended graphic primitives, called *connections*, is used for definition of structural connections between the flow-graph objects. The connections between the objects are created by setting connecting points with coordinates that depend on the coordinates of the objects themselves.

The developed library of basic elements contains all elements, that may be required for effective creation of the parallel program flow-graphs. The library of basic elements contains both separate functional units (counters, adders, multipliers), and hardwarily implemented large and self-contained macrooperations, such as calculators of elementary functions, etc.

The environment Fire!Constructor combines capabilities of a general-purpose graphics editor with capabilities of a synthesizer, intended for mapping of parallel program information graphs on hardware resources of reconfigurable structures. As a result of mapping, the information graph of the parallel program is partitioned into disjoint subgraphs, each of which is structurally implemented in a separate FPGA chip of the selected reconfigurable structure.

The algorithm of partition of the information graph of the application problem is based on a sequential layout algorithm. According to the algorithm (see Fig. 6.28), the vertices of the information graph and the FPGA chips, used for placement of the vertices, are sequentially selected in pairs.

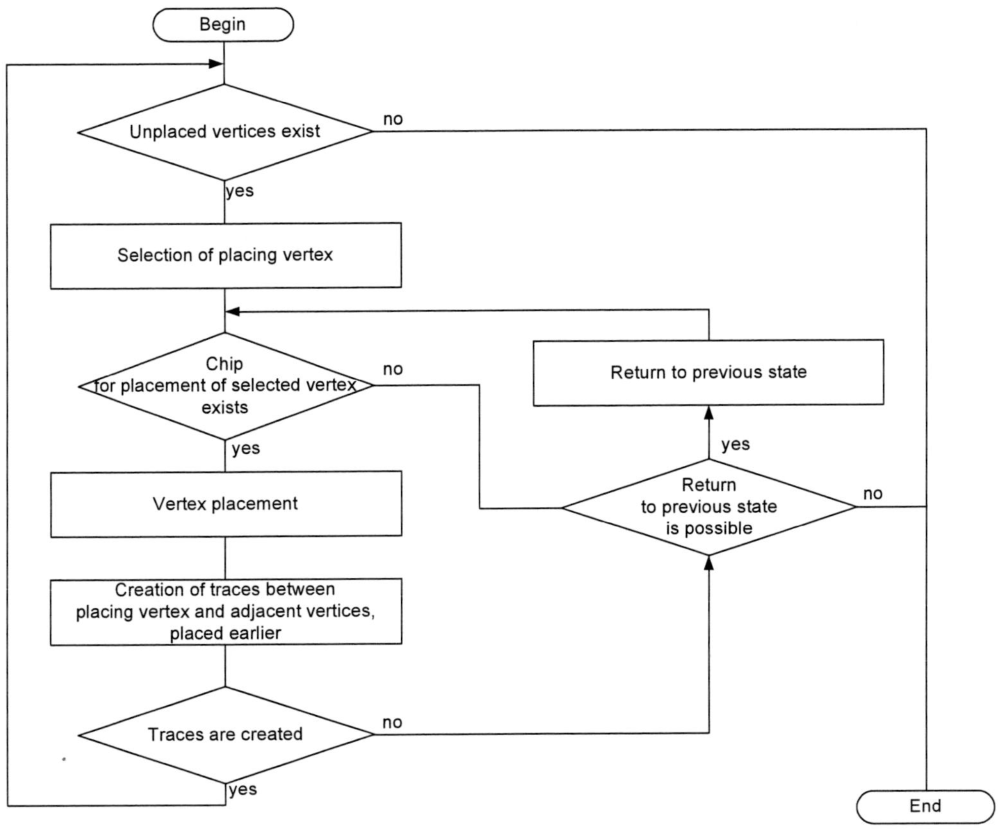

Figure 6.28. The layout algorithm of the application information graph.

Tracing of connections is performed during each iteration of the algorithm after selection of a vertex and an FPGA chip for placement. In case, when the pair "vertex/chip" is not selected, or tracing is failed, the algorithm returns to the previous step: one or several vertices, placed previously, become unplaced, and the chips, that contained the vertices, are marked as forbidden. The vertices, that were marked as unplaced during return, are attempted to be placed again.

This is the exhaustive search algorithm with the running time, considerably depending on the number of the graph vertices and on the number of the FPGA chips, selected for implementation of the application problem. Hence, before starting execution of the described layout algorithm, it is necessary to perform procedures, that, firstly, limit the FPGA fields, intended for placement, according to the task dimension, and secondly, limit selection of the vertex groups of the information graph, that are to be placed in one and the same FPGA chip.

When the information graph of the application problem is partitioned into the disjoint subgraphs, each of which is to be structurally implemented in a separate chip, the Fire!Constructor generates VHDL-files and files of time and topological restrictions for each FPGA chip. These files are added to the project, that is designed in the integrated development environment Xilinx ISE, and that contains all circuit solutions of the computational and interface units, used for creation of the task information graph. Then the Fire!Constructor generates configuration files for each FPGA chip.

The Debugger of Parallel Programs

Using the debugger of parallel programs Argus Debugger, the application programmer can interactively debug the application on various basic modules (see Fig. 6.29).

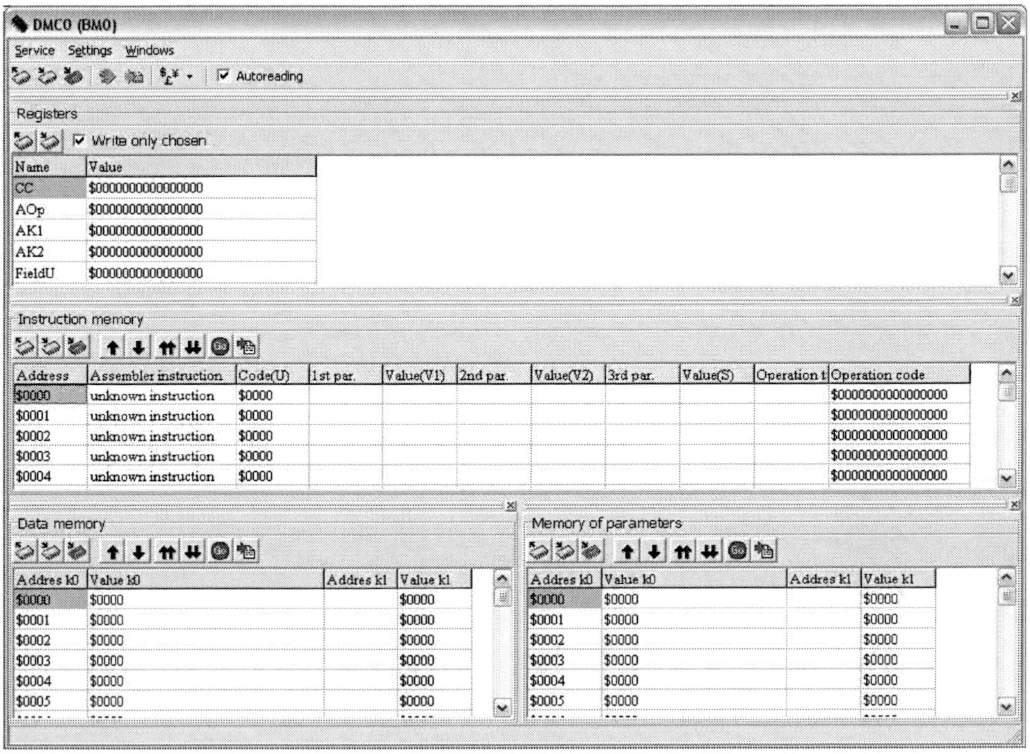

Figure 6.29. The debugger of parallel programs Argus Debugger (a window of a distributed memory controller).

Distinctive features of the Argus Debugger are capability of interactive access to the distributed memory controller of basic modules, to its registers and memory segments, controlled by the registers. Each memory controller has its own specialized window for view and input of values into the registers and memory segments.

Control functions of basic module calculations are available through the menu of the Debugger, and may also be called by pressing hot-keys. Using the Argus Debugger, the application programmer can load executable modules of parallel programs and unload them into the specified file.

The development environments of parallel-pipeline programs of reconfigurable multipipeline computing structures provide wide support and considerably simplify work of the application programmer.

6.5. USE OF THE MPCS SYSTEM SOFTWARE FOR DEVELOPMENT OF APPLICATION PROGRAMS

As an example, let us consider the problem of fluid filtering in a porous medium. The mathematical definition of the problem is described in subsection 5.4.1.

The parallel program in the high-level programming language COLAMO with implicit description of parallelism is developed on the base of the equation (5.17).

The memory variables $p1$ and $p2$ contain the values of the n-th and (n-1)-th approximation of the grid function $p_{i,j,k}^{(n)}$. For the first iteration (and for all the following odd iterations) the initial values of the grid function are stored in $p1$, and the results will be kept in $p2$. For the second iteration (and for all the following even iterations) the initial values of the grid function are stored in $p2$, and the results will be kept in $p1$.

The memory variables *DD* and *LL* keep the values of oil mobility $\lambda(p)$ and the items of the vector $D(x, y, z)$, respectively. The equation (5.17) is implemented in the *Let*-structure, called *LetZond.* The calculations, carried out according to the equation (2.2), correspond to the switching variables *c1*, *c2*, *c3*, *c4*.

The text of the parallel program is given below:

```
Program Zond;
Include MyLibNewElement;
VAR p1,p2,DD,LL : Array Real [1:Vector, 20:Stream, 20:Stream, 20:Stream] Mem;
VAR it : Number;
VAR x, y, z : Number;
VAR bCont : Boolean Mem;
Const Nit=10000;
Const vect=1;
Const XX=20;
Const YY=20;
Const ZZ=20;
Const dx=0.000002;
Const dy=0.000002;
Const dz=0.000002;
Const w=0.9;
Const e=0.001;
Let LetZond(In: p, px_1, px1, py_1, py1, pz_1, pz1, d, dx_1, dx1, dy_1, dy1, dz_1, dz1, l,
lx_1, lx1, ly_1, ly1, lz_1, lz1; Out: pout, lbCont);
```

```
VAR p, px_1, px1, py_1, py1, pz_1, pz1, d, dx_1, dx1, dy_1, dy1, dz_1, dz1, l, lx_1, lx1,
ly_1, ly1, lz_1, lz1, pout : Array Real [1:Vector] Com;
VAR c1, c2, c3, c4 : Array Real [1:Vector] Com;
VAR v, i : Number;
VAR lbContRg : Boolean Reg;
VAR lbCont : Boolean Com;
For i:=0 To vect-1 Do
Begin
   If i=0 Then
        lbContRg:=false;
   Else If i=1 Then
   Begin
      For v:=0 To vect-1 Do
      Begin
         c1[v]:=w/((lx1[v]-lx_1[v])/dx+(ly1[v]-ly_1[v])/dy+(lz1[v]-lz_1[v])/dz);
         c2[v]:=(lx1[v]*(px1[v]-(dx1[v]-d[v]))+lx_1[v]*(px_1[v]+(d[v]-dx_1[v])))/dx;
         c3[v]:=(ly1[v]*(py1[v]-(dy1[v]-d[v]))+ly_1[v]*(py_1[v]+(d[v]-dy_1[v])))/dy;
         c4[v]:=(lz1[v]*(pz1[v]-(dz1[v]-d[v]))+lz_1[v]*(pz_1[v]+(d[v]-dz_1[v])))/dz;
         pout[v]:=c1[v]*(c2[v]+c3[v]+c4[v])+(1.0-w)*p[v];
         If (p[v]-pout[v])>e Then
            lbContRg:=true;
         End;
       End;
   Else
     lbCont:=lbContRg;
   End;
   EndLet;
   For It:=0 To Nit-1 Do
   Begin
      Cadr Cadr1;
           For z:=0 To ZZ-1 Do
                 For y:=0 To YY-1 Do
                      For x:=0 To XX-1 Do
                            LetZond(p1[*,x,y,z],p1[*,x-1,y,z],p1[*,x+1,y,z],
                                    p1[*,x,y-1,z],p1[*,x,y+1,z],p1[*,x,y,z-1],p1[*,x,y,z+1],
                                    DD[*,x,y,z],DD[*,x-1,y,z],DD[*,x+1,y,z],
                                    DD[*,x,y-1,z],DD[*,x,y+1,z],DD[*,x,y,z-1],DD[*,x,y,z+1],
                                    LL[*,x,y,z],LL[*,x-1,y,z],LL[*,x+1,y,z],
                                    LL[*,x,y-1,z],LL[*,x,y+1,z],LL[*,x,y,z-1],LL[*,x,y,z+1],
                                    p2[*,x,y,z],bCont);
   Endcadr;
     Cadr Cadr2;
        For z:=0 To ZZ-1 Do
              For y:=0 To YY-1 Do
                  For x:=0 To XX-1 Do
                      LetZond(p2[*,x,y,z],p2[*,x-1,y,z],p2[*,x+1,y,z],
```

```
                p2[*,x,y-1,z],p2[*,x,y+1,z],p2[*,x,y,z-1],p2[*,x,y,z+1],
                DD[*,x,y,z],DD[*,x-1,y,z],DD[*,x+1,y,z],
                DD[*,x,y-1,z],DD[*,x,y+1,z],DD[*,x,y,z-1],DD[*,x,y,z+1],
                LL[*,x,y,z],LL[*,x-1,y,z],LL[*,x+1,y,z],
                LL[*,x,y-1,z],LL[*,x,y+1,z],LL[*,x,y,z-1],LL[*,x,y,z+1],
                p1[*,x,y,z],bCont);
    Endcadr;
End;
End_Program.
```

Figure 6.30 shows a screenshot of the Argus IDE v.3.0 and parallel program fragment, written in the COLAMO.

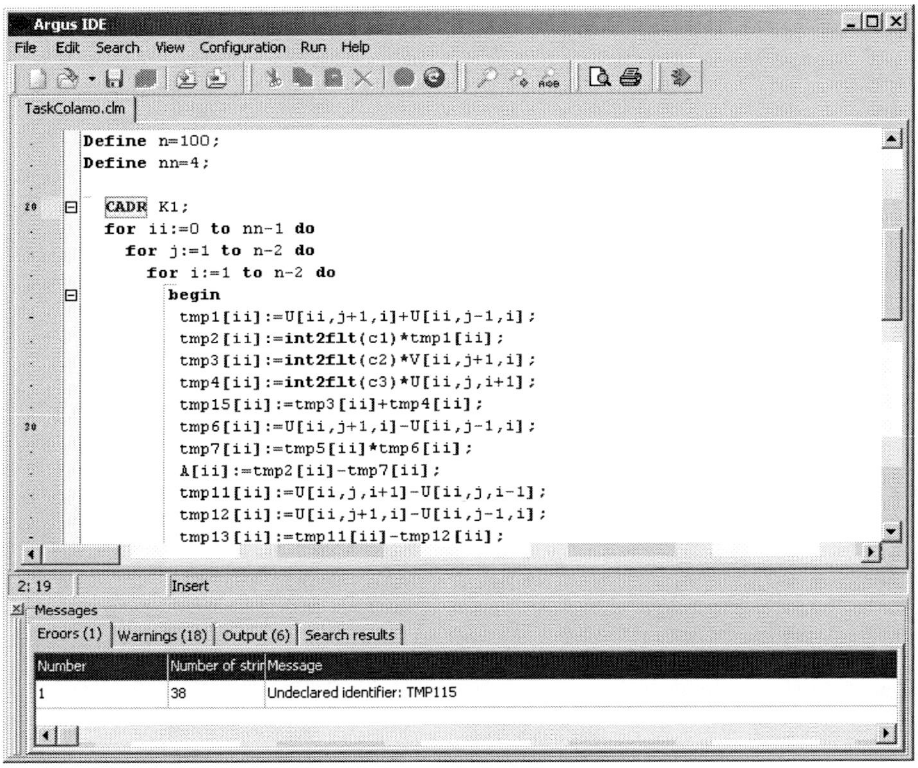

Figure 6.30. A screenshot of the Argus IDE v.3.0 and parallel program fragment, written in the COLAMO.

To implement the operators and functions of the languages (adders, multipliers, functions of comparison, trigonometric functions, etc.), used in the parallel program, circuit engineers create circuit solutions, called library elements. The library elements are designed in the integrated development environment of digital devices Xilinx ISE, or in any other IDE, compatible with the Xilinx ISE, and are plugged in the library of the COLAMO-translator.

Using the text of the parallel program, the COLAMO-translator generates the description of the information graph of the application problem, and pass it into the development environment Fire!Constructor, which, taking into account the library elements, creates the flow-graph (see Fig. 6.31), which corresponds to the information graph.

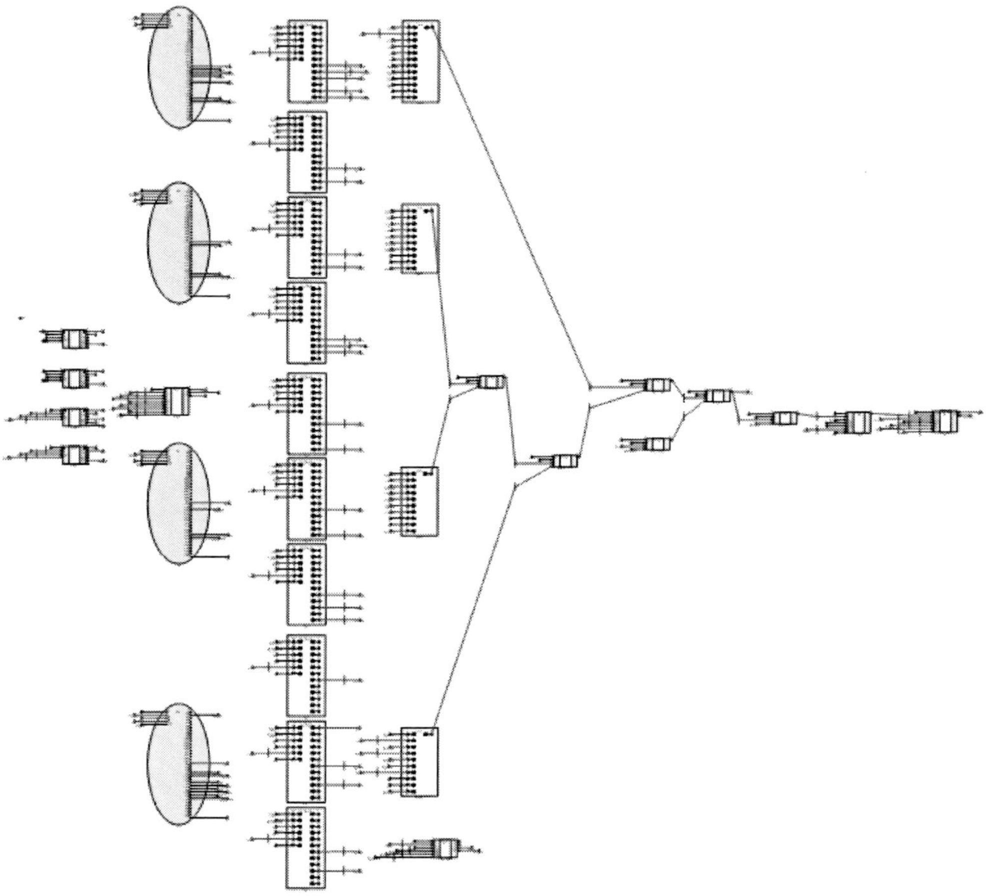

Figure 6.31. The flow-graph of computations in the node point of the problem of fluid filtering in a porous medium.

Figure 6.32 shows the same flow-graph in the window of the Fire!Constructor.

In addition to creation of the flow-graph, the Fire!Constructor partitions it into disjoint subgraphs, each of which will be structurally implemented in separate FPGA chips of the reconfigurable structure.

Figure 6.33 shows mapping of the information graph, which performs calculations in the node point of the problem of fluid filtering, on the basic module 16V4-50.

The layout results (see Fig. 6.33) show that the information graph of the task takes only 4 FPGAs of the basic module 16V4-50, and 4 distributed memory units. Proceeding from this, let us assume, that 4 independent structures, performing concurrent computations in 4 node points of the problem of fluid filtering in a porous medium, may be implemented on the basic module. The application programmer can use two alternative methods: to modify the COLAMO-program, or to use capabilities of the Fire!Constructor. Let us consider both of these variants.

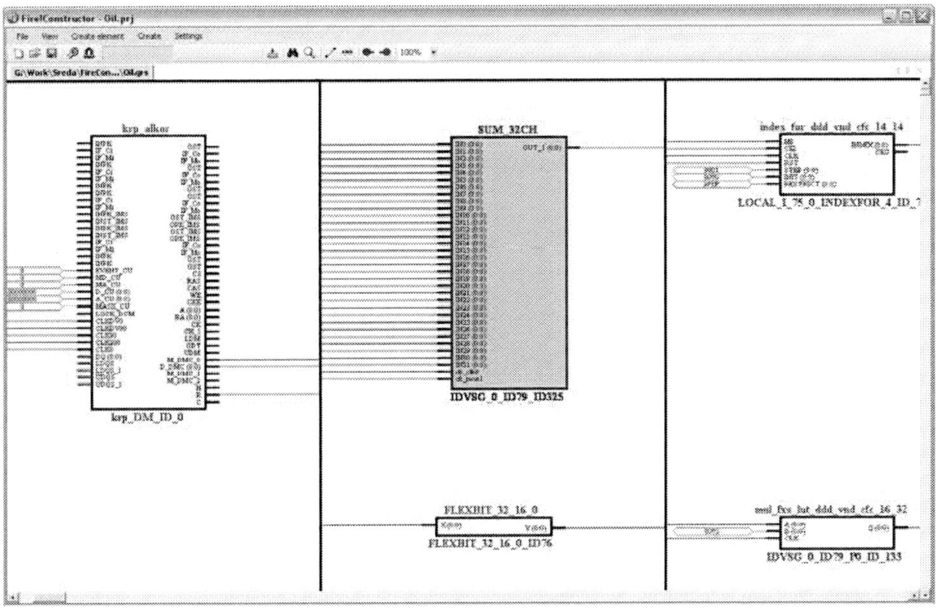

Figure 6.32. The flow-graph of the problem of fluid filtering in a porous medium, generated by the Fire!Constructor.

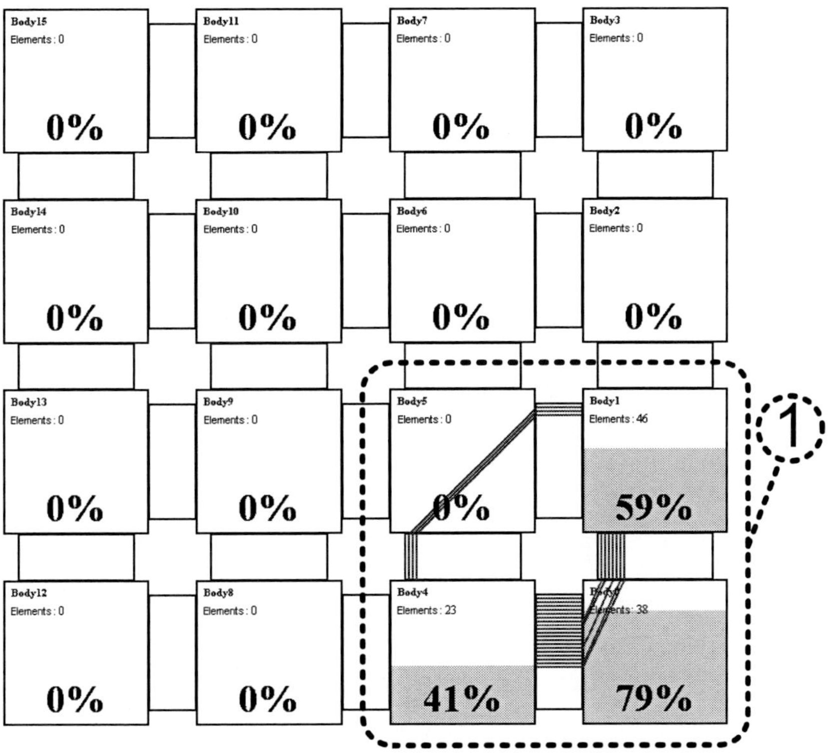

Figure 6.33. Mapping of the information graph on the basic module 16V4-50.

Modification of the COLAMO-program. It is sufficient to modify vector (parallelizing) components of arrays (to replace 1 by 4, as it is shown in Fig. 6.34) in the text of the parallel COLAMO-program, which implements the problem of fluid filtering.

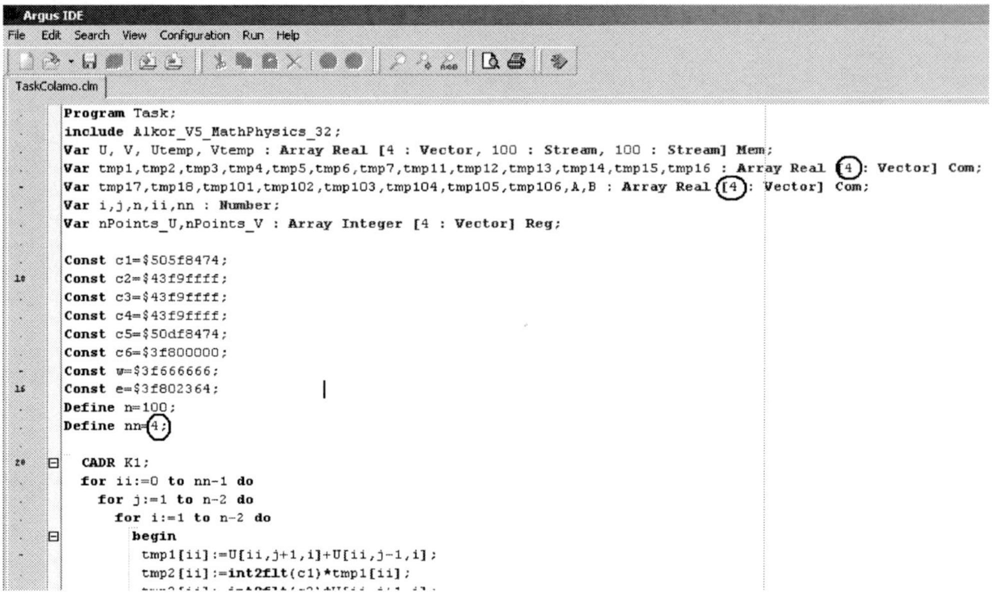

Figure 6.34. The result of modification of vector components of arrays (1 is replaced by 4).

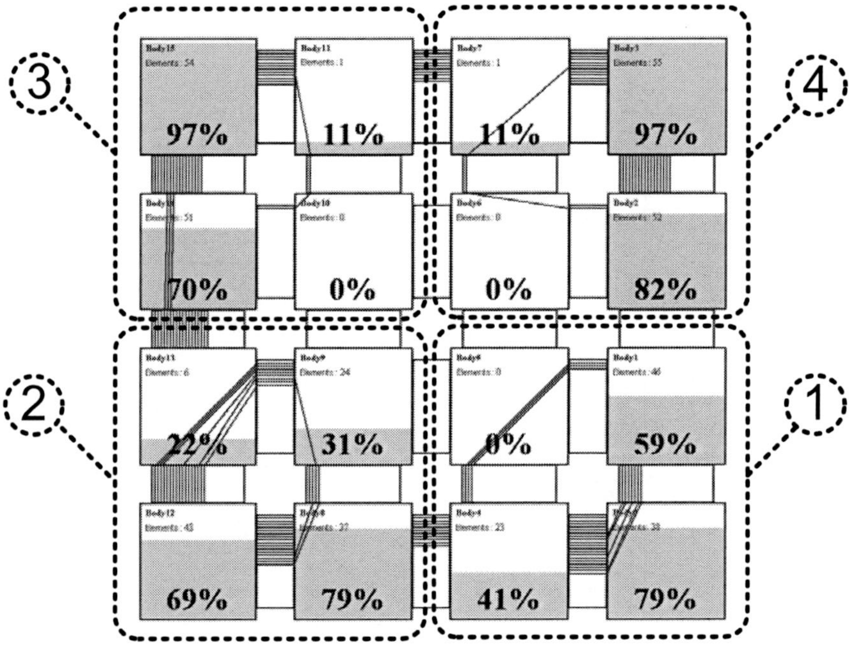

Figure 6.35. Mapping of the problem of fluid filtering with the granularity of parallelism 4 on the basic module 16V4-50.

During translation of the program, the translator generates the information graph, containing 4 subgraphs for implementation of computations in 4 node points of the problem of fluid filtering.

Figure 6.35 shows mapping of 4 independent computing structures of the problem of fluid filtering in a porous medium with the granularity of parallelism 4 on the basic module 16V4-50. Such solution corresponds to the first variant of the task parallelization, described in subsection 5.4.2.

Use of capabilities of the Fire!Constructor for the task parallelization. For this variant, it is necessary to select a replicated part in the flow-graph (the component "Top" of the project designed in the Fire!Constructor), and then to press mouse right button, and to select the item "Parameters" in the pop-up menu.

Then it is necessary to find the parameter "Granularity of parallelism" in the window of object parameters, to set it equal to 4, as it is shown in Fig. 6.36, and to press the button "OK".

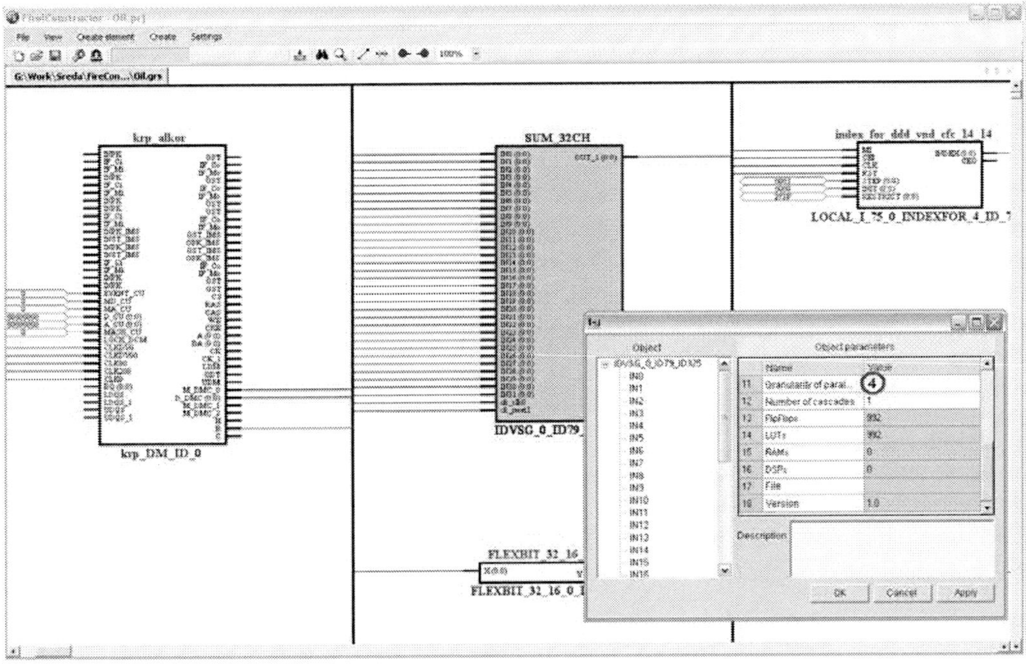

Figure 6.36. Parallelization of the task by the Fire!Constructor.

Solving the problem of fluid filtering on basic modules of other structure. The parallel program, written in the COLAMO language, is resource independent, and may be translated and mapped on resources of any reconfigurable structure.

Selection of a file of hardware structure description in the Fire!Constructor specifies the hardware platform of a reconfigurable structure for mapping of the task information graph. The file of structure description may be set in the window of project parameters (see Fig. 6.37), which is activated by selection of the item "Settings" – "Project parameters" of the main menu of the Fire!Constructor.

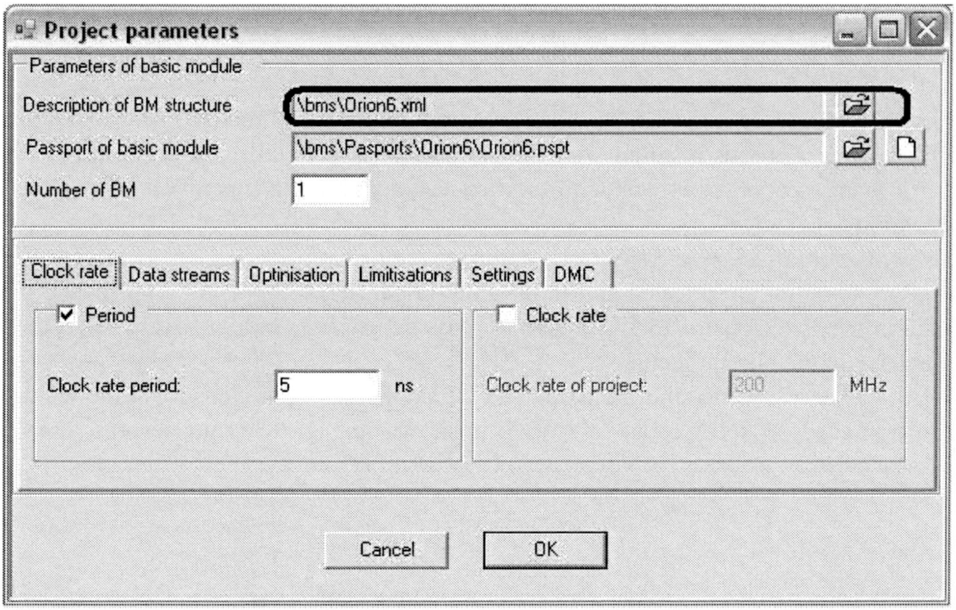

Figure 6.37. The window of project parameters.

In the window of project parameters let us select the files of structure description of the basic module 4V4-25, which is described in subsection 4.4.3.3. The basic module 4V4-25 contains 4 Virtex-4 Xilinx FPGAs and 8 distributed memory units.

Circuit resources of 4 FPGAs of the basic module 4V4-25 are sufficient for implementation of 4 independent structures of the problem of fluid filtering in a porous medium.

However, 8 distributed memory units are sufficient for implementation of only 2 computing structures for 2 node points of the problem. To map the information graph of the solving problem on the basic module 4V4-25, it is necessary to modify the values of vector components to 2 in the parallel program.

Figure 6.38 shows mapping of the information graph of the problem of fluid filtering with the granulation of parallelism 2 on the basic module 4V4-25.

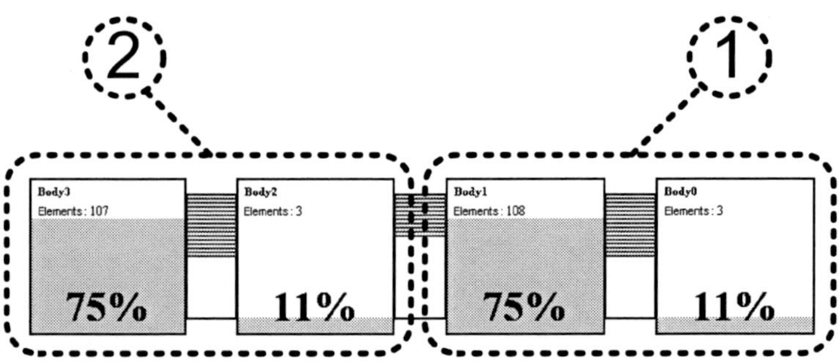

Figure 6.38. Mapping of the information graph of the problem of fluid filtering in a porous medium (the granularity of parallelism 2) on the basic module 4V4-25.

In the same way the problem of fluid filtering in a porous medium may be mapped on the basic module Alkor 16V5-75, containing 16 Virtex-5 Xilinx FPGAs and 24 distributed memory units. The description of the Alkor 16V5-75 is given in subsection 4.4.4.1.

Circuit resources of the basic module are sufficient for implementation of 8 independent structures of the problem of fluid filtering in a porous medium. However, the number of distributed memory units is sufficient for implementation of only 6 independent structures of the problem of fluid filtering.

Therefore, in the parallel program let us set the number of vector components of arrays equal to 6.

Figure 6.39 shows mapping of the information graph of the problem of fluid filtering in a porous medium with the granularity of parallelism 6.

So, owing to the wide capabilities of the suggested firmware tools, the application programmer can develop effective applications for different architectures and configurations of reconfigurable computing structures, containing various information-dependent FPGA chips.

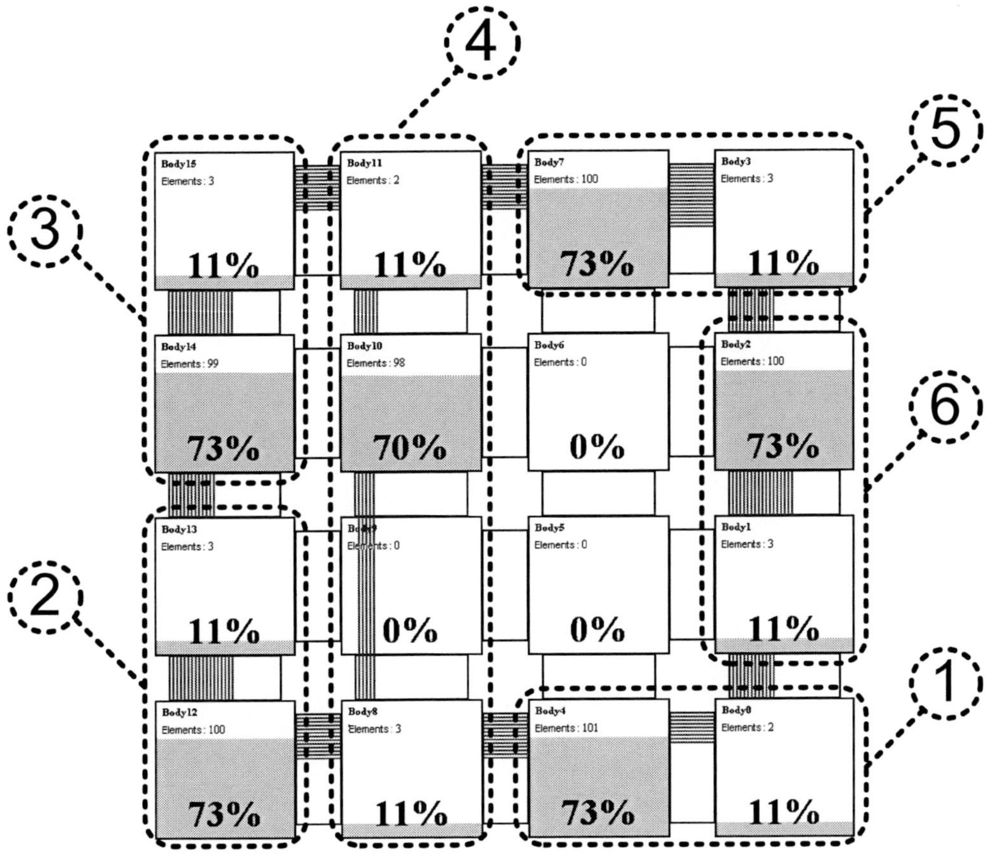

Figure 6.39. Layout of the information graph of the problem of fluid filtering (the granularity of parallelism 6).

Owing to the developed system software, applications may be ported to other hardware platforms, and the cost of applications may be considerably reduced. Hence, the field of use of reconfigurable computing structures may be considerably widened.

CONCLUSION

Highly-performance multiprocessor computer systems are intended for solving complicated and computationally laborious problems that hardly ever can be solved on traditional personal computers, because their solution will take unacceptably huge period of time, or even may be absolutely impossible. However, in contrast to personal computers, whose real performance is practically proportional to the clock rate of their processor, the real performance of a multiprocessor system depends on many characteristics, including its architecture, organization of its processor nodes, organization of data exchange between its processors, its clock rate, etc. Hence, the real performance of various multiprocessor systems may differ in ten times for execution of one and the same task.

Until recently the main direction of development of supercomputer technologies was a large-size assembly, when performance indices were achieved by means of technological innovations in design of microprocessors and communication systems, but not owing to architectural and circuit solutions. As a result, multiprocessor systems of cluster type, designed on the base of widely-available computer nodes and communication networks, have become the most widespread during the last years. This is caused by their relatively low price and simple programming, because the software, developed and debugged on personal computers, can be easily adapted to cluster systems. It may seem, that opportunity of use of cluster MCSs by various specialists, such as research workers, solving problems of modeling, design, computing experiment, etc., are quite perfect. But attention should be paid on such a nuance: cluster MCSs provide high real performance only for those tasks, that can be partitioned into separate loosely-coupled fragments, each of which can be assigned to the separate cluster node. While executing tasks of other types, the real performance of cluster MCSs drops down and, what is more, may even decrease, when the number of processors of the system grows.

As a result, in some ways things have reached a revolutionary situation: cluster systems cannot provide high real performance for various problems, and vendors of cluster MCSs do not want to change anything. This "revolutionary situation" may be solved only by extending user opportunities of not only procedural programming of supercomputers, but of architectural (structural) programming of system hardware as well. These are the opportunities, that can be provided by reconfigurable multipipeline computing structures. If, using a cluster MCS, the developer tries to adapt the structure of the solving task to the structure of the MCS, then the situation is quite opposite for reconfigurable systems, when the system architecture is adapted to the structure of the solving task. Besides, the hardware (structural) analogue of the information graph of the solving task is generated within the system, that provides the theoretical maximum of real performance for the task solution.

But everything must be paid for. In this case, the high real performance is achieved by rather complicated programming of the reconfigurable MCS, because it is necessary to program the system on both levels – procedural and structural (hardware). Therefore, reconfigurable multipipeline computing structures are perfectly suit for solving data flow problems, i.e. the problems of processing of huge data arrays (streams) according to a single algorithm. In this case, it is possible to provide the highest performance of the huge data array processing, spending time only once for the architecture programming of the reconfigurable multiprocessor computing structure.

The concept of design of reconfigurable computing structures appeared 30 years ago and was accepted with enthusiasm. However, after some time, technical difficulties of practical implementation of the concept cooled out the enthusiasm, and the idea was forgotten. More than 20 years passed, when appearance of FPGAs of large-scale integration brought the idea of design of reconfigurable multipipeline structures back to life.

At present, developers of supercomputers show greater and greater interest in the methods of design of FPGA-based reconfigurable multipipeline computing structures. Theoretical and practical aspects of creation of such computer systems are a matter of active discussions during international conferences, and FPGA-based computational nodes are widely used in various highly-performance systems.

It can be claimed, that the “ice age” of architectural programming of supercomputers on the level of hardware parallelism is going to its end. And if western developers have only recently begun to use the technology of reconfigurable computing, more than 30 years’ experience in domestic research and development in this area gives us a good starting point and we can reasonably expect to gain the leading position in the global market of reconfigurable computing multipipeline structures. And we do hope that this monograph will serve this important goal of strategic development of Russia.

REFERENCES

[1] Kalyaev, A.V. (1984). Multiprocessor systems with programmable architecture. *Radio i svyaz,* Moscow.

[2] Yevreinov, E.V. (1981). Homogeneous computer systems, structures and environments. *Radio i svyaz*, Moscow.

[3] Pranghishvili, I.V. and others. (1967). Microelectronics and homogeneous structures for design of logical and computational devices. *Nauka,* Moscow.

[4] Yevreinov, E.V., Khoroshevskiy, V.G. (1978). Homogeneous computer systems. *Nauka,* Novosibirsk.

[5] Bogachev, M.P. (1981). Architecture of computer system with homogeneous structure, *Homogeneous computing environments,* p.15-21. Lvov.

[6] Korneev, V.V. (1983). Architecture of computer systems with programmable architecture, p. 116. *Nauka,* Novosibirsk.

[7] Shmoilov, V.I. (1993). Architecture of homogeneous computing environments, p. 289. *Integral,* Lvov.

[8] Burtsev, V.S. (2000). Data-processing systems with automatic parallelization of computational processes. *Proceedings of international conference "Parallel calculations and control problems".*

[9] Aladyshev, O.S. and other. (2004). Supercomputers: fields of application and performance requirements. *Proceedings of Institutes of Higher Education. Electronics,* vol. 1, p. 13-17.

[10] Yevreinov, E.V., Pranghishvili, I.V. (1974). Digital automatons with adjustable structure. *Energiya,* Moscow.

[11] Kalyaev, A.V., Levin, I.I. (2003). Modular-scalable multiprocessor systems with structural-procedural organization of computations. Yanus-K, Moscow.

[12] Yevreinov, E.V., Kosarev, Y.G. (1966). High-performance homogeneous general-purpose computer systems. *Nauka,* Novosibirsk.

[13] Kalyaev, A.V. (1978). Homogeneous communication register structures. *Sovietskoye Radio,* Moscow.

[14] Kalyaev, A.V., Levin, I.I. (1999). Multiprocessor systems with reconfigurable architecture: concepts of development and application. *Science for industry,* vol. 11, p. 11-19.

[15] Kalyaev, A.V., Stanishevskiy, O.B. (1990). Principles of development of firmware tools for supermacrocomputers. Information science. *Design automation,* no. 2, p. 13-21.

[16] Kalyaev, A.V., Levin, I.I., Fomin, S.Y. (1987). About effectiveness of tsunami propagation modeling on multiprocessor system. *Multiprocessor computing structures,* no. 9, p. 14-18.

[17] Kalyaev, I.A., Levin, I.I. (2007). High-performance modular-scalable multiprocessor systems based on reconfigurable components. *Computational approaches and programming,* vol. 8, no. 1, p. 181-190. Moscow University Publishing, Moscow.

[18] Kalyaev, A.V., Kalyaev, I.A., Levin, I.I. (2004). FPGA-based multiprocessor computer systems with programmable architecture. *Herald of the SSC RAS,* p. 24-33.

[19] Levin, I.I. (2007). Structural-procedural organization of parallel computations. Proceedings of the 4th International Youth Scientific School "High-performance computer systems", p. 49-68. *TIT SFU Publishing,* Taganrog.

[20] Palagin, A.V., Opanasenko, V.N. (2006). Reconfigurable computer systems. *Prosvita,* Kiev.

[21] Bolotski, M. (1996). Abacus: A Reconfigurable Bit Parallel Architectures. *Ph.D.Thesis.* Massachusetts Institute of Technologies.

[22] Levin, I.I. (2004). Methods and firmware tools of parallel structural-procedural computations. *D.Sc.Thesis.* Taganrog.

[23] Valiev, K.A. (2005). Quantum computers and quantum computations, vol. 175(1), p. 3.

[24] Varshavskiy, V.I. and other. (1973). Homogeneous structures. *Energiya,* Moscow.

[25] Voevodin, V.V., Voevodin, Vl.V. (2002). Parallel computations. *BHV-Petersburg,* St.-Petersburg.

[26] Dimitriev, Y.K., Khoroshevskiy, V.G. (1982). Computer systems based on minicomputers. *Radio i svyaz,* Moscow.

[27] Gritsik, (ed.). (1985). *Parallelization of data processing,* vol. 2. Lvov.

[28] Kuch, D.J. (1968). ILLIAC IV Software and Application Programming. *IEEE Jrans. Comput.,* vol. 6-17, no. 8, p. 758-770.

[29] Babb, R.G., (ed.). (1988). Programming Parallel Processors. *Addison-Wesley.*

[30] Kalyaev, A.V. (1970). Theory of digital integrating devices and structures. *Sovietskoye radio,* Moscow.

[31] Pranghishvili, I.V., Todua, D.A., Abramova, N.A. and other. (1978). Computer PS-300. *Control systems and devices,* no. 10.

[32] Rusin, B.P., Kuzio, M.N., Shmoilov, V.I. (2002). Pulsing information lattices – a new generation of homogeneous computing environments. *Automation and computer engineering,* no. 1, p. 60-71. Riga.

[33] Sedov, V.S., Sobolev, V.N., Samchinskiy, A.A. and other. (1989). *The HCE components.* Lvov.

[34] Shmoilov, V.I. (1991). Organization of computational process in multipipeline computing structure. *Integral,* Lvov.

[35] Shmoilov, V.I., Rusyn, B.P., Kuzio, M.N. (2001). Homogeneous computing environments and pulsirs. *Mercator,* Lvov.

[36] Kung, S.Y. (1988). VLSI array processors. *Prentice-Hall, Englewood Cliffs,* New Jersey.

[37] Kung, H.T. and Leiserson, C.E. (1978). Systolic arrays (for VLSI). In *Sparse Matriz Symposium, SIAM,* p. 256-282.

[38] Bacherikov, G.I. (ed.) and other. (1985). Multipipeline computing structures based on homogeneous environments. Lvov.

[39] Kalyaev, A.V., Stanishevskiy, O.B., Fradkin, B.G. (1990). Macroprocessor VLSI set. Proceedings of the 1st All-Union conference "Homogeneous computing environments and systolic structures", vol. 2, p. 33-46. Lvov.

[40] Vectorization of programs: theory, approaches, implementation. (1991). *Collected articles,* p. 246-267. Mir, Moscow.

[41] Voevodin, V.V. (1986). Mathematic models and methods of parallel processes. *Nauka,* Moscow.

[42] Voevodin, V.V. (1991). Mathematical basics of parallel computations. *MSU Publishing,* Moscow.

[43] Kalyaev, A.V., Kalyaev, I.A., Levin, I.I. (1997). The Basic Module of Multiprocessor System with Structural- Procedural Organization of Computing. Parallel Computing Technologies. *Proceedings of the 4th International Conference, PaCT-97,* p. 394-396. Yaroslavl, Russia.

[44] Yevreinov, E.V. (1962). About microstructure of elementary devices of a computer system. *Computer systems,* no. 4, p. 3-16. Novosibirsk.

[45] Yevreinov, E.V., Kosarev, Y.G. (1963). About high-performance computer systems. *Proceedings of the Russian Academy of Sciences "Technical cybernetics",* no. 4, p. 10-12.

[46] Yevreinov, E.V. (1965). Theoretical basics of general-purpose computing environment design. *Computer systems,* no. 16, p. 3-72. Novosibirsk.

[47] Pranghishvili, I.V. and other. (1973). Homogeneous microelectronic associative processors. *Sovietskoye radio,* Moscow.

[48] Kukharev, G.A., Shmerko, V.P., Zaitseva, E.N. (1990). Algorithms and systolic processors for multidigit data processing. *Nauka i tekhnika,* Minsk.

[49] Frumkin, M.A. (1990). Systolic computations. *Nauka,* Moscow.

[50] Bankovskiy, A.G., Karpa, N.M., Shmoilov, V.I. and other. (1990). Library of microprogram modules of arithmetic and logical operations for fixed-point numbers. Part 1. *Integral,* Lvov.

[51] Bankovskiy, A.G. and other. (1990). Library of microprogram modules of arithmetic and logical operations for fixed-point numbers. Part 2. *Integral,* Lvov.

[52] Vasilkevitch, A.V. and other. (1991). Library of microprogram modules for solving image processing problems on HCE. *Integral,* Lvov.

[53] Golichenko, E.P. and other. (1990). Microprogram modules of arithmetic and logical operations for floating-point numbers. *Integral,* Lvov.

[54] Golichenko, E.P. and other. (1991). Library of microprogram modules of arithmetic and logical operations for integer numbers. *Integral,* Lvov.

[55] Timchenko, O.V. (1999). Difference methods of digital filtration. *Feniks,* Lvov.

[56] Aleksin, Y.P., Bartish, M.Y., Shmoilov, V.I. and other. (1986). Multipipeline computing structures based on homogeneous environments. Lvov.

[57] Norenkov, I.P., Manichev, V.B. (1983). Computer-aided design systems for electronic and computing devices. *High School,* Moscow.

[58] Almashi, A.G., Reznikov, D.Y., Shimkin, A.P. (1991). HCE programmer manual. *Integral,* Lvov.
[59] Antonenko, A.I. and other. (1991). Environment of HCE microprogram module development. *Integral,* Lvov.
[60] Bacherikov, G.I. and other. (1990). *HCE software.* Lvov.
[61] Golichenko, E.P. and other. (1989). Programming of homogeneous computing environments. *Integral,* Lvov.
[62] Malanina, N.M., Shmoilov, V.I. (1989). Microprogramming in multipipeline computing structures. *Integral,* Lvov.
[63] Shmoilov, V.I., Sloboda, M.Z. (1999). Divergent continued fractions. *Merkator,* Lvov.
[64] Shmoilov, V.I. (1986). Microprogramming in multipipeline computing structure. Computer system, structures and environments, p. 116-164. *Naukova dumka,* Kiev.
[65] Bankovskiy, A.G., Ghilburd, M.M., Shmoilov, V.I. (1989). Homogeneous computing environments with instruction stream. *Integral,* Lvov.
[66] Kuzio, M.N., Samtchinkiy, A.A., Fedortsiv, V.I., Shmoilov, V.I. (1991). Homogeneous computing environment for image processing. *Integral,* Lvov.
[67] Glazkov, M.A., Malakhov, V.V., Savostin, M.F. (1990). HCE-based videoprocessor. *Integral,* Lvov.
[68] Kuzio, M.N., Samtchinskiy, A.A., Shmoilov, V.I. and other. (1990). HCE-based image processing suit for VLSI control. *Integral,* Lvov.
[69] Bogachev, M.P. and other. (1987). Homogeneous computing structures in communications. *Military Academy of communication,* Leningrad.
[70] Bogner, R.E., Constantinides, A.G. (1975). Introduction to digital filtering. *Wiley.*
[71] Isaenko, V.A., Kuzio, M.N., Shmoilov, V.I. and other. (1991). High-performance image analysis suit with processor based on homogeneous computing environment. *Integral,* Lvov.
[72] Kuzio, M.N., Rusin, B.P., Shmoilov, V.I. (2000). Homogeneous environments – components of high-performance computer systems. *Control systems and devices,* no. 4, p. 52-62. Kiev.
[73] Levin, I.I., Ponomarev, I.M., Fradkin, B.G. (1995). Analysis of effectiveness of structural-procedural organization of computational process at solving applications on MCS. *Collection of abstracts and scientific articles,* p. 23-28. Moscow.
[74] Litviniuk, A.A., Pogribnaya, I.M., Rozhankovskiy, I.V. and other. (1991). Algorithms, structures and library of microprogram modules of digital filters with delta-modulation on HCE. *Integral,* Lvov.
[75] Pogribnoi, V.A., Bobalo, S.I., Kalmuk, Y.S. and other. (1990). HCE-oriented methods of correlation analysis with sign delta-modulation. *Integral,* Lvov.
[76] Pogribnoi, V.A., Timchenko, A.V., Rozhankovskiy, I.V. and other. (1991). Digital signal filtration with delta-modulation on HCE. *Integral,* Lvov.
[77] Pogribnoi, V.A., Litviniuk, A.A., Rozhankovskiy, I.V. and other. (1990). HCE-oriented methods of non-recursive filtration with delta-modulation. *Integral,* Lvov.
[78] Bondarev, A.N., Yeliseyev, V.G., Shmoilov, V.I. and other. (1991). Library of microprogram modules of arithmetic operations for complex numbers. *Integral,* Lvov.
[79] Litviniuk, A.A. (1988). HCE-based calculator. *Integral,* Lvov.
[80] Sedov, V.S. (1991). Matrix of homogeneous processors. *Integral,* Lvov.

[81] Yeleiko, V.I., Yeleiko, Y.I., Shmoilov, V.I. and other. (1991). HCE without standby processing elements. *Integral,* Lvov.

[82] Yeleiko, Y.I., Samtchinskiy, A.A., Shmoilov, V.I. (1991). Design of fault-tolerant homogeneous computing environment. *Integral,* Lvov.

[83] Kamsha, V.P., Kuzio, M.N., Shmoilov, V.I. and other. (1990). HCE-on-chip. *Integral,* Lvov.

[84] Kokurin, S.L., Kuzio, M.N., Shmoilov, V.I. (1991). Design of homogeneous computing environment on chip. *Integral,* Lvov.

[85] Kuzio, M.N., Rusin, B.P., Shmoilov, V.I. (2001). Pulsing information lattices. Optoelectornic information-power technologies, no. 1, p. 51-78. *Vinnytsia,* Ukraine.

[86] Kuzio, M.N., Rusin, B.P., Shmoilov, V.I. (2001). Homogeneous computing environments. *Optoelectornic information-power technologies*, no. 2, p. 19-37. Vinnytsia, Ukraine.

[87] Podrubniy, O.V., Samtchinskiy, A.A., Shmoilov, V.I. (1990). Development of fault-tolerant HCE components. *Integral,* Lvov.

[88] Rusin, B.P., Kuzio, M.N., Shmoilov, V.I. (2000). Reconfigurable high-performance HCE-based systems. *Automation and computer engineering,* no. 3, p. 72-81. Riga.

[89] Samtchinskiy, A.A., Sedov, V.S., Shmoilov, V.I. (1991). Architecture of homogeneous computing environment. *Integral,* Lvov.

[90] Shmoilov, V.I., Rusin, B.P., Kuzio, M.N., Zayats, I.A. (1999). Pulsing information lattices. *Merkator,* Lvov.

[91] Shmoilov, V.I., Rusin, B.P., Kuzio, M.N., Kapshiy, O.V. (2000). Design of pulsing information lattices. *Merkator,* Lvov.

[92] Shmoilov, V.I., Rusin, B.P., Kuzio, M.N. (2001). A cell of pulsing information lattices. *Merkator,* Lvov.

[93] Shmoilov, V.I., Adamatskiy, A.I., Kuzio, M.N. and other. (2004). Pulsing information lattices. *Merkator,* Lvov.

[94] Shmoilov, V.I., Adamatskiy, A.I. and other. (2003). Information lattices. *Merkator,* Lvov.

[95] Shmoilov, V.I., Rusin, B.P., Kuzio, M.N. (2004). Pulsing information lattices – a new generation of homogeneous computing environments. *Control systems and devices,* no. 2, p. 23-38, Kiev.

[96] Dijkstra, E.W. (1968). Cooperating Sequential Processes. *Programming Languages, Genuys* (ed.), p.43-112. Academic Press.

[97] Kalyaev, I.A., Levin, I.I., Semernikov, E.A., Shmoilov, V.I. (2008). Reconfigurable multipipeline computing structures. *SSC RAS Publishing*, Rostov-on-Don.

[98] Levin, I.I. (1992). Research and development of a problem-oriented multiprocessor system for structural-procedural realization of problems of mathematical physics. *Ph.D. thesis.* Taganrog.

[99] Antipov, V.N., Goryainov, V.T. (ed.), Kulin, A.N. and other. (1988). Radar stations with digital synthesizing of antenna aperture. *Radio i svyaz,* Moscow.

[100] Dordopulo, A.I., Kalyaev, I.A., Levin, I.I., Semernikov, E.A. (2007). The family of multiprocessor computer systems with dynamically reconfigurable architecture. *Proceedings of the 4th International Youth Science School "High-performance computer systems",* p. 68-74. TIT SFU Publishing, Taganrog.

[101] Pospelov, D.A. (1970). Arithmetic basis of digital computers. *High school,* Moscow.

[102] Pospelov, D.A. (1972). Introduction to theory of computer systems. *Sovietskoye radio,* Moscow.

[103] Dijkstra, E.W. (1976). A Discipline of Programming. *Prentice-Hall Series in Automatic Computation.*

[104] Samarskiy, A.A. (1987). Introduction to numerical methods. *Nauka,* Moscow.

[105] Kalyaev, A.V., Levin, I.I., Fomin S.Y. (1988). Multiprocessor systems for effective simulation of hydrophysical processes. *Collection "Firmware tools of computer simulation",* p. 70-77. Moscow.

[106] Kalyaev, A.V., Levin, I.I., Fomin, S.Y. (1989). About estimation of effectiveness of solving problems of mathematical physics on multiprocessor systems. *Electronic modelling,* no. 6, p. 11-15.

[107] Levin, I.I. (1990). Organization of computational process for physical-topological modeling of VLSI transistor structures on multiprocessor system. Collection "Mathematical modeling of elements and fragments of LSI". Riga.

[108] Levin, I.I. (1991). Co-processor for structural-procedural realization of problems of mathematical physics by means of structural-procedural method of parallelization. Dubrov Y.A. (ed.). Analysis of effectiveness of computer systems, p. 14-21. *Integral,* Lvov.

[109] Levin, I.I., Ponomarev, I.M., Fradkin B.G. (1995). Analysis of effectiveness of structural-procedural organization of computational process at solving applied problems on MCS. *Collection of abstracts and scientific articles,* p. 23-28. Moscow.

[110] Levin, I.I., Ponomarev, I.M. (1995). FFT realization on multiprocessor system with structural-procedural organization of computations. *Electromechanics,* no. 4, p. 72-74.

[111] Rabiner, L.R., Gold B. (1975). Theory and application of digital signal processing. *Englewood Cliffs, N.J., Prentice-Hall,* Inc.

[112] Levin, I.I., Ponomarev, I.M. (2000). Methodology of organization of high-performance parallel computations in multiprocessor systems. *Proceedings of International conference "Artificial intelligence-2000",* p.142-144. *TSURE Publishing,* Taganrog.

[113] Levin, I.I., Ponomarev, I.M., Shmatok, A.V. (2003). Methods of parallel program transformation according to structure of computer system. Artificial intelligence, no. 3, p. 227-231. *Nauka i osvita,* Donetsk, Ukraine.

[114] Korneyev, V.V. (2004). Computer systems. Gemos ARV, Moscow.

[115] Rivkin, M.N. (1991). Vector operations for modeling procedures of data transformation. *Programming,* no. 3.

[116] Shteinberg, B.Y. (1995). Parallelization of recurrent loops with conditional statements. *Automatics and telematics,* no. 6, p. 176-184.

[117] Levin, I.I., Ponomarev, I.M. (1995). Realization of wave routing algorithm on multiprocessor system with structural-procedural organization of computations. *Scientific Research Institute of Multiprocessor Computer Systems at Taganrog State University of Radio Engineering,* Taganrog.

[118] Levin, I.I., Ponomarev, I.M. (2003). Structural-procedural realization of the routing problem. Artificial intelligence, no. 3, p. 121-129. *Nauka i osvita,* Donetsk, Ukraine.

[119] Ercal and Lee H.C. (1997). Time-efficient maze routing algorithms on reconfigurable mesh architectures. *Journal of Parallel and Distributed Computing,* 44(2), p. 133-140.

[120] Tahsin M. Kurç, Cevdet Aykanat, Fikret Erçal. (1991). Parallelization of Lee's Routing Algorithm on a Hypercube Multicomputer, p.244-253. EDMCC.

[121] Kotov, V.Y. (ed.). (1983). Elements of parallel programming. Radio I svyaz, Moscow.

[122] Oppenheim, A.V. (ed.). (1978). Applications of digital signal processing. *Prentice-Hall,* Inc., Englewood Cliffs, New Jersey.

[123] Kung, S.Y. (ed.), Whitehouse, H.J. (ed.), Kailath, T. (ed.). (1985). VLSI and modern signal processing. *Prentice-Hall, Inc.,* Englewood Cliffs, New Jersey.

[124] Semernikov, E.A., Doronchenko, Y.I., Trunov, I.L. (2004). Pipeline DSP macroprocessor for multiprocessor computer systems with structural-procedural organization of computations. Artificial intelligence, vol. 3, p. 157-163. *Nauka i osvita,* Donetsk, Ukraine.

[125] Semernikov, E.A., Doronchenko, Y.I. (2005). Pipeline macroprocessor of digital signal processing with structural-procedural organization of computations. Herald of computer and information technologies, no. 6, p. 49-55. *Mashinostroyeniye,* Moscow.

[126] Levin, I.I., Semernikov, E.A. (2005). Stability of pipeline recursive filters. *Herald of the Southern Scientific Centre of the Russian Academy of Sciences,* vol. 1, no. 2, p. 28-40.

[127] Semernikov, E.A. (2005). The problem of stability of pipeline recursive filters. Artificial intelligence, vol. 4, p. 713-723. *Nauka i osvita,* Donetsk, Ukraine.

[128] Semernikov, E.A., Doronchenko, Y.I., Kovalenko, V.B. and other. (2006). Generation of address sequences for pipeline FFT calculators. *Artificial intelligence,* vol. 3, p. 68-78. *Nauka i osvita,* Donetsk, Ukraine.

[129] Semernikov, E.A., Semernikova E.E. (2006). Replacement of sequents in fast algorithms of digital signal processing. *Herald of the Southern Scientific Centre of the Russian Academy of Sciences,* no. 4, vol. 2, p. 25-30. SSC RAS Publishing.

[130] Myers, G.J. (1982). Advances in computer architecture, part 1. John Wiley & Sons, Inc.

[131] Hwang, K. Advanced parallel processing with supercomputer architectures. *Proceedings of the IEEE,* vol.75, no. 10, October 1987.

[132] Jones, G. (1987). Programming in Occam. *Prentice-Hall.*

[133] Hockney, R.W., Jesshope, C.R. (1981). Parallel computers. Architecture, programming and algorithms. *Adam Hilger* Ltd., Bristol.

[134] Gehani, N. (1984). Ada. An advanced introduction including reference manual for the Ada programming language. Prentice-Hall, Inc., *Englewood Cliffs.*

[135] Kutepov, V.P., Korablin, Y.P. (1978). Language of parallel algorithm flowgraphs. *Programming,* no. 1, p. 1-16.

[136] Futi, K., Suzuki, N. (1988). Programming Language and VLSI scheme. *Mir,* Moscow.

[137] Hoare, C.A.R. (1985). Communicating sequential processes. *Prentice-Hall International,* UK.

[138] Dijkstra, E.W. (1968). Cooperating Sequential Processes. *Programming Languages, Genuys* (ed.). Academic Press.

[139] Dijkstra, E.W. (1976). A Discipline of Programming. *Prentice-Hall Series in Automatic Computation.*

[140] Babenko, L.K., Makarevitch, O.B., Tchefranov, A,G. (1988). Principles of description and organization of asynchronous large-block computations in multiprocessor system. – Electronic modelling, 1988, №3, p. 13-17.

[141] Sterling, L., Shapiro, E. (1986). The art of PROLOG. *The MIT Press Cambridge,* Massachusetts.

[142] Jones, G. (1987). Programming in Occam. *Prentice-Hall.*

[143] Syre, J.-C. (1982). The data flow approach for MIMD multiprocessor systems. Evans, D.J. (ed.). *Parallel processing systems.* Cambridge University Press.

[144] Kotov, V.E. (1974). Theory of parallel programming. *Applied aspects,* no. 1, p.1-16. Kibernetika.

[145] Faronov, V.V. (2004). Development system Delphi. BHV-Petersburg, Saint-Petersburg.

[146] Levin, I.I. (2000). Language of high-level parallel programming for structural-procedural organization of computations. *Proceedings of All-Russian scientific conference,* p.108-112. MSU Publishing, Moscow.

INDEX

I

J

L

M

N

O

P

R

S

T

U

V

W

Y